Springer

Berlin
Heidelberg
New York
Hong Kong
London
Milan
Paris
Tokyo

Coupling of Fluids, Structures and Waves in Aeronautics

Proceedings of a French–Australian Workshop in Melbourne, Australia 3 – 6 December 2001

Noel G. Barton, Jaques Periaux (Editors)

Springer

Dr. Noel G. Barton
CSIRO Mathematical and Information Science
Locked Bag 17
North Ryde NSW 1670
Australia

Professor Jacques Periaux
Pôle Scientifique
Dassault Aviation/UPMC
78 Quai Marcel Dassault
92214 Saint-Cloud
France

ISBN 3-540-40222-5 Springer-Verlag Berlin Heidelberg New York

Library of Congress Cataloging-in-Publication-Data

Coupling of fluids, structures, and waves in aeronautics : proceedings of a French-Australian Workshop in Melbourne, Australia, 3-6 december 2001 / Noel G. Barton, Jaques Periaux (editors).
p. cm. -- (Notes on numerical fluid mechanics and multidisciplinary design ; v. 85)
Includes bibliographical references.
ISBN 3-540-40222-5 (alk. paper)
1. Air flow--Mathematics--Congresses. 2. Coupled problems (Complex systems)--Congresses. 3. Airplanes--Design and construction--Congresses. I. Barton, Noel G. II. Periaux, Jacques. III. Notes on numerical fluid mechanics and multidisciplinary design ; v. 85)
TL574.F5C68 2003
629.132'32--dc21

Springer-Verlag Berlin Heidelberg New York
a member of BertelsmannSpringer Science+Business Media GmbH
http://www.springer.de

Printed in Germany

Typesetting: camera-ready by editors
Cover design: deblik Berlin
Printed on acid free paper 62/3020/M - 5 4 3 2 1 0

NNFM Editor Addresses

Prof. Dr. Ernst Heinrich Hirschel
(General editor)
Herzog-Heinrich-Weg 6
D-85604 Zorneding
Germany
E-mail: e.h.hirschel@t-online.de

Prof. Dr. Kozo Fujii
Space Transportation Research Division
The Institute of Space and Astronautical Science
3-1-1, Yoshinodai, Sagamihara, Kanagawa, 229-8510
Japan
E-mail: fujii@flab.eng.isas.ac.jp

Dr. Werner Haase
Höhenkirchener Str. 19d
D-85662 Hohenbrunn
Germany
E-mail: wug.haase@t-online.de

Prof. Dr. Bram van Leer
Department of Aerospace Engineering
The University of Michigan
Ann Arbor, MI 48109-2140
USA
E-mail: bram@engin.umich.edu

Prof. Dr. Michael A. Leschziner
Imperial College of Science, Technology and Medicine
Aeronautics Department
Prince Consort Road
London SW7 2BY
U. K.
E-mail: mike.leschziner@ic.ac.uk

Prof. Dr. Maurizio Pandolfi
Politecnico di Torino
Dipartimento di Ingegneria Aeronautica e Spaziale
Corso Duca degli Abruzzi, 24
I - 10129 Torino
Italy
E-mail: pandolfi@polito.it

Prof. Dr. Jacques Periaux
Dassault Aviation
78, Quai Marcel Dassault
F-92552 St. Cloud Cedex
France
E-mail: periaux@rascasse.inria.fr

Prof. Dr. Arthur Rizzi
Department of Aeronautics
KTH Royal Institute of Technology
Teknikringen 8
S-10044 Stockholm
Sweden
E-mail: rizzi@aero.kth.se

Dr. Bernard Roux
L3M – IMT La Jetée
Technopole de Chateau-Gombert
F-13451 Marseille Cedex 20
France
E-mail: broux@l3m.univ-mrs.fr

Preface

This Volume contains the Proceedings of a French – Australian workshop held in Melbourne, Australia from 3-6 December 2001.

Entitled "Coupling of Fluids, Structures and Waves in Aeronautics (CFSWA)", the workshop was principally organised by CSIRO Mathematical and Information Sciences, Dassault Aviation and CNRS. The main purpose was to explore new computational methods and tools for efficient multi-disciplinary design in aeronautics. This enterprise strongly depends on modelling of coupled disciplines and development of associated multi-disciplinary simulation tools. As an example of the requirements, the original logo for the workshop illustrated coupled aeroelasticity, scattered radar waves and high lift problems. Optimisation is required with respect to multiple objectives under conflict.

Among the many challenges to be addressed are mathematical, numerical and engineering problems involving coupled flows, structures and waves. Future progress on these topics strongly depends on the physical and mathematical modelling of coupled disciplines and the development of associated multi-disciplinary simulation tools. Strong couplings require appropriate exchanges of different kinds of information: physical and geometrical description of models, coupling of Partial Differential Equations (PDEs) at interfaces, and specification of boundary conditions or multi-objective functions in optimisation or control problems. Validation of multi-physics software also requires ever more efficient and accurate graphic visualisation tools representing experimental and computational data stored in databases.

In particular, the scope of the workshop included the following topics:

- mathematical modeling of fluids, structures and waves
- optimal control
- flight dynamics and control
- robotics and autonomous systems
- advanced computational aerodynamics
- advanced computational electromagnetics
- advanced aircraft structure and materials
- multi-disciplinary design optimisation
- drag, noise and signature reduction
- flow induced noise and turbulence
- aero-acoustics and aero-elasticity
- aircraft and UAV design
- intelligent fluid and material systems
- database tools for graphic visualisation

The workshop was attended by 62 delegates originating from Australian and French universities, Australian and French government laboratories (CNRS, CSIRO, DSTO, INRIA, ONERA), and private industry.

The lectures were thematically grouped into sessions on fluids, structures, waves and numerical methods. Many speakers spoke on the coupling of these themes in the setting of multi-physics. The lectures covered a wide spectrum from fluid dynamics, structures, electromagnetics, optimisation, mathematical and computational aspects, and aircraft design. Within the timetable, there was good opportunity for active discussion by delegates after each presentation.

The Round Table discussions entitled "Future Tools for Design and Manufacture of Innovative Products in the Aeronautics Industry" was a highlight of the workshop and was chaired and moderated by Dr Ron Sandland, Deputy Chief Executive of CSIRO. The outcomes are presented in this volume.

We wish to acknowledge the support of the three principal organising bodies:

- CSIRO Mathematical and Information Sciences
- Dassault Aviation
- CNRS

and also the generous sponsorship and guidance received from

- Academy of Technological Sciences and Engineering
- ANZIAM (Australian and New Zealand Industrial and Applied Mathematics)
- DSTO Aeronautical and Maritime Research Laboratory
- Embassy of the Republic of France in Australia

Dr Nick Stokes (CSIRO Mathematical and Information Sciences) is especially thanked for his organisational assistance. Dr Mourad Sefrioui is thanked for valuable assistance in the Round Table discussions.

Finally, we are grateful to the French - Australian Chamber of Commerce and Industry for arranging a very pleasant banquet held at The Australian Club, which was hosted in a convivial and stylish fashion by Mr Daniel Fremont (*Direction Generale Internationale* of Dassault Aviation).

Noel Barton
CSIRO Mathematical and Information Sciences

Jacques Periaux
Dassault Aviation

April 2003

Forewords

Mathematical models have been used for years with the help of powerful computers and numerical analysis to give new possibilities for simulation of complex systems in many scientific domains and industrial application fields, including aeronautics in particular.

But in most cases, modelling tools remain enscribed in their peculiar scientific disciplines. A new step is now necessary to satisfy the demand for more efficient models of real complex systems. It consists in integrating - within the same numerical simulation - the different physical phenomena, which are not simply superposed but which are coupled by the nonlinear characteristics of the system or the subsystems. Simplified models with separated physical effects are not adequate for these real complex systems.

The complexity of such systems is not only in their structure, but also in their behavior. As a consequence, multiscale representation must be introduced in the models. Aeronautics requires a scientific effort with both multidisciplinary and multiscale approaches.

The Australian and French scientific and technical communities are strongly involved in solving these problems and both have the capabilities for such a challenge. The CFSWA workshop gave, with the help of CSIRO, CNRS and Dassault Aviation, a sharp evaluation of the state of the art of mathematical tools for aeronautics and established the basis for fruitful future cooperation between the two countries. These Proceedings present this evaluation for the wider scientific community.

J.-J. Gagnepain
CNRS

As a discipline, Mathematics possesses its own rules of elegance, order, consistency and applicability. These have served the discipline well since the time of Newton. And they will continue to do so.

However the emergence of challenging new industrial paradigms creates an opportunity to rethink both the importance and relevance of Mathematics in this context. In parallel, the inexorable march of Moore's Law has created a computational environment in which massive simulations are quite feasible and the most complex visualisation requirements can be met.

The drivers of the new industrial paradigms include quality, cost and integration of design and manufacture. Quality means meeting customers' needs reliably and efficiently. In aeronautics this means designing aeroplanes that are "more affordable, safer, quieter and cleaner" (the European vision of future aeronautics). Mathematical modelling offers a new path to achieving this commendable goal since realistic simulation of all aspects of the manufacturing process and operational performance is now quite achievable. This is absolutely critical in aeronautics. It also offers the ability to dramatically reduce cycle times in both design and manufacture, to design for manufacturability and to innovate around design parameters in a manner that is unachievable in the previous paradigm - constructing and testing prototypes and experimental manufacturing processes.

New computational paradigms, involving exponentially increasing speed and memory, permit greater degrees of realism through the incorporation of higher levels of physics and chemistry in models.

In all of the above a strong value proposition is emerging based on improved quality and reduced cycle times. It can only be realised however in a multidisciplinary environment in which this high level of mathematical modelling and computational science operates in a manner which is informed by the business context (what are the key problems being faced?) and engineering/manufacturing considerations. Once this environment has been established the sky (pun intended) is the limit!

The CFSWA workshop was refreshing in that those present were clearly aware of the potential of the field but also highly cognisant of the challenges of the business environment that will enable this potential to be realised. CSIRO is proud to have partnered with CNRS and Dassault Aviation as principal organisers of this workshop and grateful to an excellent organising team for bringing it to fruition.

Ron Sandland
CSIRO

CONTENTS

Chapter 1 Fluids

Chapter 2 Structures

Chapter 3 Electromagnetics

Chapter 4 Optimisation

Chapter 5 Mathematical Methods and Tools

Chapter 6 Aircraft design

Round Table Discussion

The Effect of Noise and Vibration on Separated Flow Regions in Hypersonic Flow

G.A. BIRD
GAB Consulting Pty. Ltd. 144/110 Sussex Street Sydney, NSW 2000, Australia

Summary

The direct simulation Monte Carlo (DSMC) method has been applied to the flow past a 25°/50° double wedge at a range of Mach and Knudsen numbers. The wedge segments are of approximately equal length and the region of separated flow at the junction of the wedges can extend over a significant fraction of the surface in uniform steady flow. The extent of the separation was found to be strongly dependent on the Mach and Knudsen numbers. A noise field was imposed through the imposition of a series of strong compression and expansion waves on the freestream flow. The locations of the separation and attachment points were affected, but the extent of the separation did not change appreciably. Calculations were also made with an oscillation in the streamwise direction imposed on the body. The extent of the separation increased when the peak oscillation speed ceased to be small in comparison with the stream speed.

1 Introduction

The upstream propagation of pressure disturbances in the subsonic region of a boundary layer in an otherwise supersonic flow frequently leads to a region of separated flow when there is a sharp change in the slope of the surface. A number of experimental studies have been made of the separated flows regions that appear in hypersonic flows about biconics, flared cylinders and double wedges. These have been proposed [1] as critical test cases for CFD code validation studies and it is necessary to determine whether the results are dependent on the noise level in the wind tunnel. The motivation for the study was to determine whether the noise and vibration in a hypersonic shock tunnel could, in some cases, affect the extent of separated flow regions and cause the flow to be unrepresentative of that in free flight.

The time that is required for the freestream flow to move a distance equal to the linear size of the model, or transit time, is generally much less than a millisecond and model vibrations with frequencies up to a kilohertz will generally lead to quasi-steady flows. The sensitivity of the separated region to changes in the Mach and Knudsen numbers is investigated in Section 2. This gives an indication of possible effects due to low frequency vibrations as well as those due to systematic changes in the flow conditions in the working section during the run.

The two-dimensional flow past a 25°/50° double wedge has been chosen as the test case and the calculations have employed the direct simulation Monte Carlo (or DSMC) method [2] to compute the flow at the molecular level. The calculations were made with the interactive visual DSWT program [3] that automatically sets the computational parameters. The user does have to specify the number of megabytes that are available for the calculation and the program does not allow the computation to proceed unless the available memory permits a sufficient number of simulated molecules for a valid calculation to be made.

Streamlines and contours of constant pressure for air at a Mach number of five are shown in Figure 1. The Knudsen number based on the semi-height of the body is 0.0056 and the flow falls within the continuum regime. With one exception, the results can be expected to be equivalent to a solution of the Navier-Stokes equations. The exception is that the DSMC method provides a physically accurate solution for the internal structure of the shock wave while the Navier-Stokes solution would predict an excessively thin wave. This calculation employed approximately 1.5 million simulated molecules and required a run time of several days on a 2 GHz Pc.

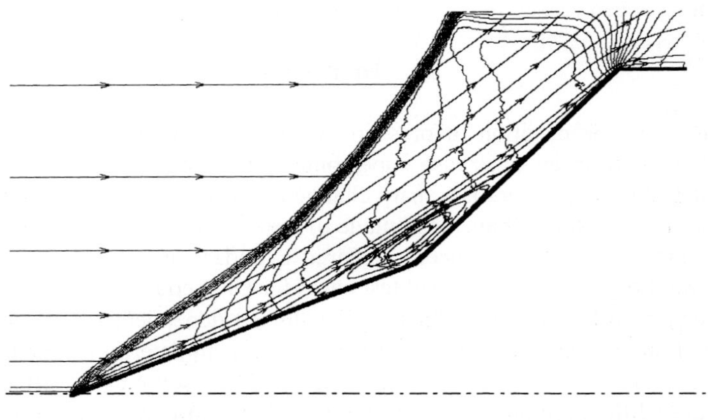

Figure 1 Separated flow region in the flow past a double wedge at M=5 and K_n=0.0056.

The separated flow region leads to an effectively smooth concave surface and there is a single attached shock wave that is progressively strengthened by the compression waves from the concave surface. Note that the 50° angle of the second wedge is well above the limiting angle of 41° for an attached shock wave in air at a Mach number of 5. The flow without a boundary layer may be obtained with the DSMC method by specifying specular reflection for the molecules striking the surface. The resulting flow past this geometry is shown in Figure 2. While the specular reflection allows complete velocity slip at the surface, the transport properties are unaffected and the internal structure of the shockwaves remains physically correct.

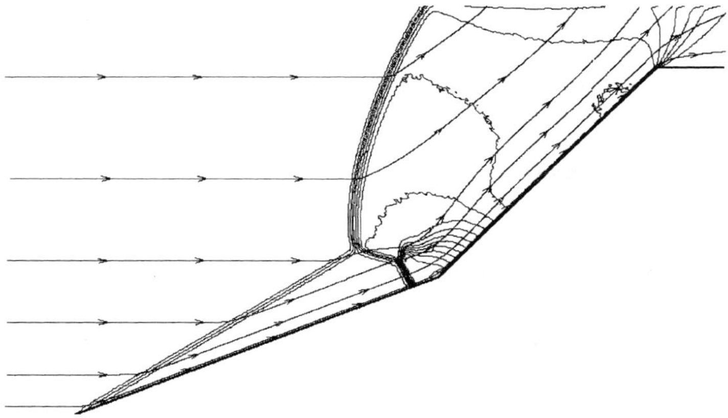

Figure 2 Double wedge flow at M=5 and Kn=0.0056 with specularly reflecting surface.

The extent of the separated flow region may be determined either directly from the streamlines or from the sign of the shear stress on the surface. The shear stress for the case shown in Figure 1 is plotted in Figure 3.

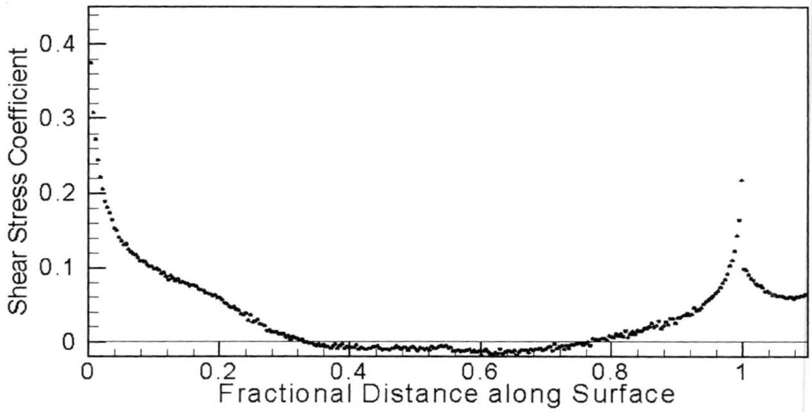

Figure 3 The shear stress distribution on the surface of the flow in Figure 1.

The effect of steady flow disturbances on the extent of the separation region is investigated in Section 3 and the flows over double wedges with an imposed vibration in the stream direction are studied in Section 4.

2 The Effect of Variations in Mach and Knudsen Number

The Mach and Knudsen number dependence of the extent of the separation also establishes the Reynolds number dependence because the Reynolds number in a diatomic gas is equal to 1.21 times the Mach number divided by the Knudsen number. The Reynolds number of the flow shown in Figures 1 and 3 is therefore equal to 1,080. The separated flow region as a fraction of the surface length of the wedges (*i.e.* excluding the shoulder) is 0.44. This case was repeated with the density decreased by a factor of three to give a Knudsen number of 0.0168 and the fractional separation distance decreased to 0.24. A further reduction in the density by a factor of three to a Knudsen number of 0.0504 caused the separated region to disappear completely. A factor of three change from the K_n=0.0168 case therefore causes either a near doubling or the disappearance of the separated region. This has been chosen as the reference case and calculations were also made for cases with a factor of two variation from the reference case. In addition a much larger calculation with five million simulated molecules was made at a Knudsen number of 0.0028. The results are summarized in Figure 4. The region of separated flow is absent in rarefied flows but increases rapidly in the continuum regime. It is possible that the flow becomes completely separated at a sufficiently low Knudsen number and the correspondingly high Reynolds number.

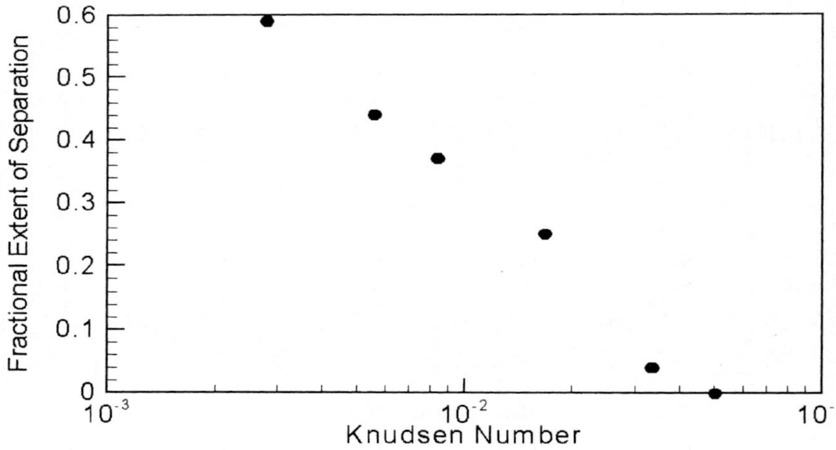

Figure 4 The effect of Knudsen number on the separation region at M=5.

The calculation at the reference Knudsen number of 0.0168 was repeated for a range of supersonic and hypersonic Mach numbers and the results are shown in Figure 5. The extent of the separation for the reference case at M=5 is near the maximum. There is a systematic reduction in the size of the separation region as the Mach number increases. The much sharper reduction as the Mach number is reduced is accompanied by a change in the flow pattern from that of Figure 1 to one in which the shock wave is completely detached.

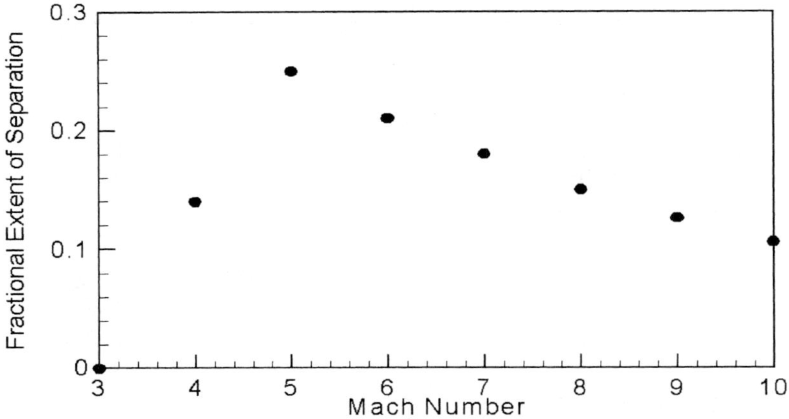

Figure 5 The effect of Mach number on the separation at K_n=0.0168.

The case at a Knudsen number of 0.0168 and a Mach number of 5 was also the reference case that was chosen for the calculations that include disturbances and vibrations. The shear stress distribution for this case is shown in Figure 6. As noted earlier the extent of the separation is only just over half that in the higher density case shown in Figure 3. Also, there is no longer a barely perceptible discontinuity in the shear stress coefficient at the junction of the wedges. A comparison of these figures shows other important consequences of the lower density. The sharp peak in the shear stress at the apex of the body occurs within the attached oblique shock wave and this is affected by the factor of three increase in the thickness of the wave. In addition, the gradients adjacent to the sharp shoulder are more gradual.

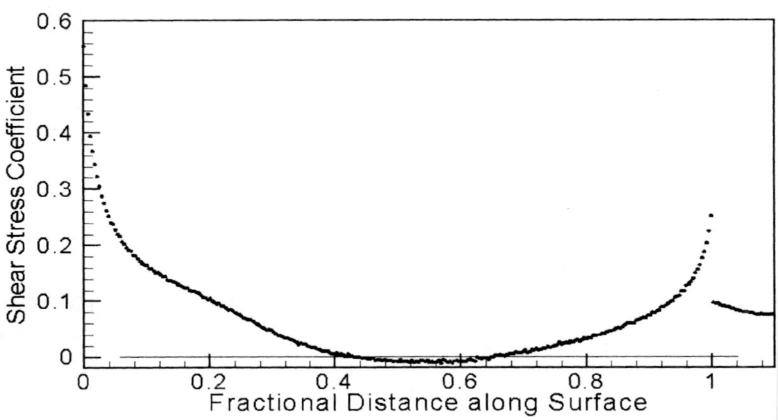

Figure 6 The shear stress distribution for the reference case with K_n=0.0168 and M=5.

3 Flows with Steady Disturbances in the Stream

Irregularities on the walls of supersonic and hypersonic wind tunnels produce disturbances that interact with the model. These were simulated in the reference flow by extending the simulated flow further upstream and placing a series of short flat plates at incidence. Three plates with a chord width equal to one twentieth of the combined wedge surface length were placed in line at a set height above the centerline. The spacing between the plates was equal to the chord width. The plates were positioned such that the waves impacted in the general vicinity of the separated region and calculations were made for the plates in three separate streamwise locations.

The incidence of the plates was 10° and the disturbances were much stronger than those that can be expected in a wind tunnel. As shown in Figure 7, the disturbances were sufficient to produce a 10% increase in the pressure coefficient over most of the surface. On the other hand, Figure 8 shows that there was very little change in the shear stress distribution and the extent of the separated flow region was almost unchanged.

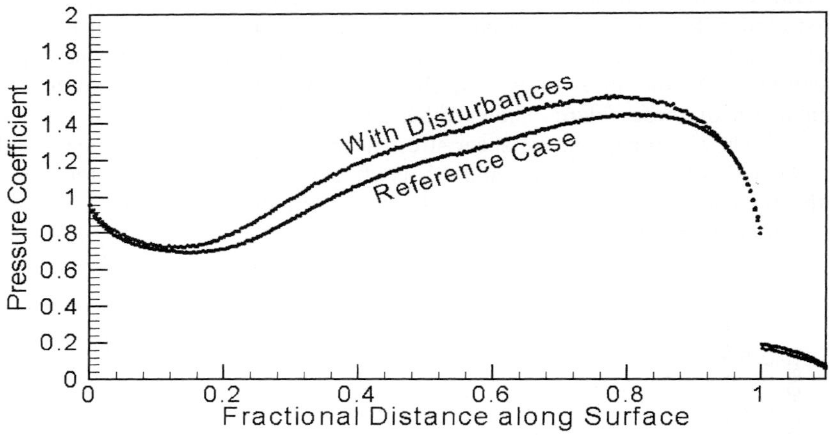

Figure 7 The effect of the disturbances on the pressure distribution.

A problem with these calculations is that, while the overall Knudsen number of 0.0168 based on the vertical distance between the shoulder and the centerline is in the continuum regime, the Knudsen number of the flow past each plate based on the chord width of the plate is 0.2. The chord width of the plate is less than half the thickness of the shock wave and the pressure modulation from the successive compression and expansion waves decays rapidly. This explains the lack of modulation in the pressure distribution on the surface. A qualitatively different result could well be obtained for very small Knudsen numbers.

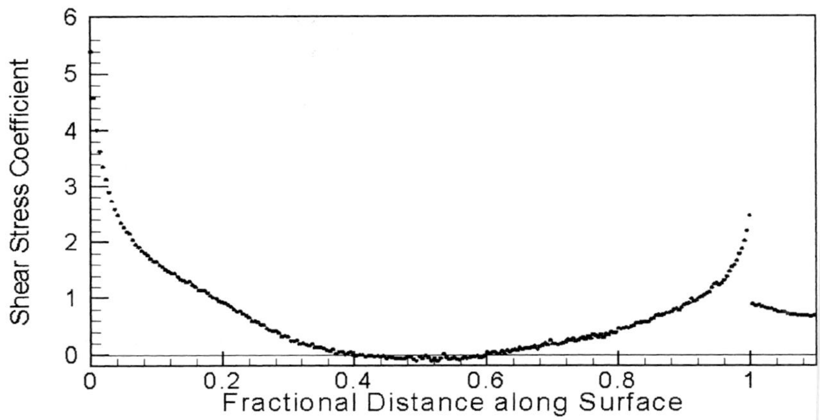

Figure 8 The shear stress distribution in the presence of flow disturbances.

4 Flows with a Streamwise Oscillation of the Model

The DSMC "wind tunnel" program DSWT program was extended to include a simple harmonic oscillation of the model in the streamwise direction. The resulting WTOM program actually employed a change in the frame of reference such that the model remained stationary while the oscillation was applied to all the simulated molecules. This is an interactive program and the frequency and amplitude of the oscillation were set by on-screen slider controls.

The oscillation affects the shear stress because the molecular flux is larger when the surface is moving against the flow than when it is moving with it. This can cause the shear stress to be positive even when the flow is separated and the average flow above the surface is in the negative direction. This meant that the extent of the separated region had to be determined from the streamlines rather than the shear stress.

The results are summarized in Table 1. The flow properties were sampled through a time average over a large number of oscillations and, as noted in the Introduction, this is sensible only if the period of the oscillation is small in comparison with the flow transit time. The frequency is expressed in Table 1 as the ratio of the period of the oscillation to the flow transit time. The amplitude of the oscillation is normalized by the width of the computational region. The magnitude of the oscillation is also indicated by the ratio of the maximum oscillation speed to the freestream speed.

Case 0 is the reference case with the 25°/50°double wedge at a Mach number of 5 and a Knudsen number of 0.0168. There is no significant change in the extent of the separated flow region until the maximum speed of the oscillation reaches five or ten percent of the freestream speed. When this ratio reaches about one third, the separation increases rapidly and it can be seen from Figure 9 that a separation ratio of 0.6 effectively converts the double wedge to a single wedge with an attached oblique shock wave.

Table 1 The effect of the oscillation on the extent of the separated flow region.

Case	Ratio of Period to Transit Time	Ratio of Amplitude to Flow Width	Ratio of Extent of Separation to Surface Length	Ratio of Max. Oscillation Speed to the Stream Speed
0	-	-	0.24	-
1	0.2	0.0005	0.26	0.045
2	0.2	0.001	0.29	0.09
3	0.2	0.002	0.32	0.18
4	0.2	0.003	0.64	0.27
5	0.2	0.004	0.82	0.36
6	0.2	0.005	0.97	0.45
7	0.08	0.0025	0.24	0.09
8	0.08	0.005	0.33	0.18
9	0.08	0.01	0.59	0.36

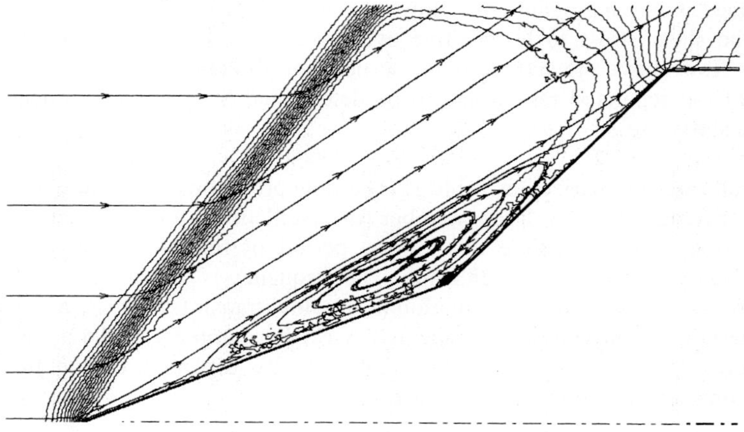

Figure 9 Streamlines and contours of constant pressure for Case 9.

5 Conclusions

It is not possible to draw general conclusions from a restricted set of computations, particularly when these are near the low Reynolds boundary of the continuum flow region. However, the results indicate that, to have any effect on the extent of separated flow regions, the disturbances have to be much stronger than those that can reasonably be expected to be present in hypersomic wind or shock tunnels.

Acknowledgement

This work was supported by the NASA Langley Research Center under contract L 13705 to GAB Consulting Pty. Ltd.

References

[1] M.S. Holden and T.P.Wadhams, "Code Validation Study of Laminar Shock/Boundary Layer and Shock/Shock Interactions in Hypersonic Flow, Part A: Experimental Measurements", AIAA Paper 2001-1031, 2001.

[2] G.A. Bird, "Molecular Gas Dynamics and the Direct Simulation of Gas Flows", Oxford University Press, 1994.

[3] Program manuals may be downloaded from gab.com.au.

Computational Bluff Body Fluid Dynamics and Aeroelasticity

H. M. Blackburn
CSIRO Building Construction and Engineering,
PO Box 56, Highett 3190, Australia
hugh.blackburn@csiro.au

Summary

Computational approaches are now making a substantial contribution to investigations in bluff body fluid mechanics, which were previously the sole domain of experimentalists. Stability analysis, direct numerical simulation and, increasingly, large eddy simulation, is being used to study fundamental aspects of the phenomena involved. Accelerating reference frame solution techniques have recently been developed and employed in the study of bluff body aeroelasticity: these are described and applied to the study of flows past slender cylinders with prescribed motion, and with aeroelastic fluid–structure interaction.

1 Introduction

Bluff body fluid dynamics and aeroelasticity differ from more classical studies in aeroelasticity primarily because the extensive regions of separated flow inherent in the problems under study defy compact theoretical treatment. For this reason, progress is heavily dependent on experimental approaches, both physical and numerical. As noted by the editors of a recent collection of works in the area [1], computational fluid dynamics is now making a significant contribution to fundamental understanding in bluff body fluid dynamics, owing to the ability to control and observe independent and dependent variables in ways that are difficult or impossible to achieve in physical experiments.

In this paper we describe 'accelerating reference frame', or ARF, techniques — a class of methods that recently has been successfully applied to studies of bluff body fluid–structure interaction problems. Example results from computational studies of interaction between vortex street wakes and circular cylinders are reviewed. Finally, we outline prospects for future work in the area.

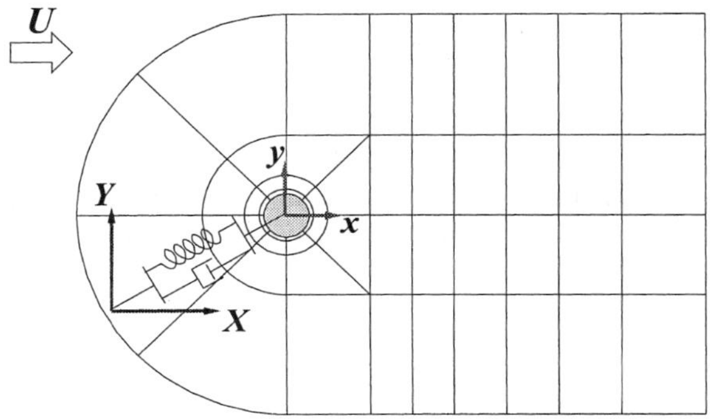

Figure 1 Schematic illustrating the basis of accelerating reference frame (ARF) techniques. Flows are solved in an ARF attached to the structure: the mesh moves with the structure, does not distort with time, and is coupled through motion-dependent forces to a global reference frame in which the far-field flow $\boldsymbol{U}$ is prescribed.

2 Computational Methods

An elastic structure will distort in response to loads exerted on it by a flowing fluid, and when the motion of the structure can feed back to affect the flow, an aeroelastic coupling arises. In analysis techniques, separate discretisations of the fluid and the structure are typically employed, and the problems are coupled through boundary conditions. Here we will focus on the fluid-mechanical part of the problem and assume that the dynamic structural motion can be obtained through any appropriate technique.

Two basic methods of approach can be identified for mesh-based numerical discretisations of the flow problems. The most general approach allows the computational mesh local to the structure to distort continuously in time as the structure moves. In this case, the far-field boundaries of the mesh, and the associated boundary conditions, are usually kept fixed in time. This is the 'arbitrary Lagrangian–Eulerian', or ALE approach. However, the method has significant computational overheads associated with the temporal changes of the mesh interpolation functions. An alternative is leave the mesh unchanged in time, but fix it to the structure and allow it to move in space, adjusting the momentum equations and boundary conditions as appropriate. This is the 'accelerating reference frame' or ARF, approach, as illustrated schematically in Figure 1. For open flows past an isolated structure, ARF methods can be just as appropriate as ALE methods, and avoid computational overheads associated with mesh distortion.

2.1 ARF Techniques for Fluid–Structure Interaction

Consider the interaction between an essentially two-dimensional slender elastic structure and the (possibly three-dimensional) flow past it. The structure is characterised by its mass per unit length m, natural frequency $f_\mathrm{n} = \omega_\mathrm{n}/2\pi$ and dimensionless structural damping ζ, and it carries a set of reference frame axes. In an inertial, global, reference frame, the structure moves in response to the force per unit length $\boldsymbol{f}$ exerted on it by the fluid, according to

$$\ddot{\boldsymbol{x}} + 2\zeta\omega_\mathrm{n}\dot{\boldsymbol{x}} + \omega_\mathrm{n}^2\boldsymbol{x} = \boldsymbol{f}/m \tag{1}$$

or as a set of first-order ODEs, with $\boldsymbol{v} = \dot{\boldsymbol{x}}$

$$\dot{\boldsymbol{v}} = \boldsymbol{f}/m - 2\zeta\omega_n\boldsymbol{v} - \omega_\mathrm{n}^2\boldsymbol{x}, \qquad \dot{\boldsymbol{x}} = \boldsymbol{v}. \tag{2}$$

The equations that describe the *relative* motion of incompressible fluid of density ρ in the reference frame attached to the structure are

$$\partial_t\boldsymbol{u} + \boldsymbol{N}(\boldsymbol{u}) = -\rho^{-1}\nabla p + \nu\nabla^2\boldsymbol{u} - \nabla\cdot\boldsymbol{\tau} - \ddot{\boldsymbol{x}}, \qquad \nabla\cdot\boldsymbol{u} = 0, \tag{3}$$

where $\boldsymbol{u}$ is the velocity field, $\boldsymbol{N}(\boldsymbol{u})$ represents nonlinear advection terms, and $\boldsymbol{\tau}$ is a possible sub-grid scale stress in the case of a large eddy simulation. Note the direct coupling of (3) to (1) through the frame acceleration $\ddot{\boldsymbol{x}}$.

At domain boundaries where far-field velocity boundary conditions, $\boldsymbol{U}$, are prescribed in the inertial reference frame, the appropriate conditions in the ARF are

$$\boldsymbol{u} = \boldsymbol{U} - \boldsymbol{v}. \tag{4}$$

On the surface of the structure, the velocity boundary condition is typically $\boldsymbol{u} = 0$. Pressure boundary conditions, if required, are obtained by dotting the domain unit outward normal $\boldsymbol{n}$ into (3); employing also the vector identity $\nabla^2\boldsymbol{u} = \nabla(\nabla\cdot\boldsymbol{u}) - \nabla\times\nabla\times\boldsymbol{u}$ provides

$$\partial_n p = \rho\boldsymbol{n}\cdot\left[-\boldsymbol{N}(\boldsymbol{u}) - \nu\nabla\times\nabla\times\boldsymbol{u} - \nabla\cdot\boldsymbol{\tau} - \ddot{\boldsymbol{x}} - \partial_t\boldsymbol{u}\right]. \tag{5}$$

Two special cases of (5) occur: on the far-field boundary, $\partial_t\boldsymbol{u} = -\ddot{\boldsymbol{x}}$, while, on the surface of the structure, often $\partial_t\boldsymbol{u} = \boldsymbol{u} = 0$.

Coupling of (1) back to (3) occurs through the force per unit length exerted by the fluid on the structure through pressure and viscous stress

$$\boldsymbol{f} = \oint p\boldsymbol{n}\,\mathrm{d}s - \oint \mu\boldsymbol{n}\cdot\left[\nabla\boldsymbol{u} + (\nabla\boldsymbol{u})^T\right]\mathrm{d}s, \tag{6}$$

where the integrals are taken around the surface perimeter of the structure.

These techniques can also be applied to cases where the structure has prescribed motion, in which case $\dot{\boldsymbol{x}}$ and $\ddot{\boldsymbol{x}}$ are explicitly supplied.

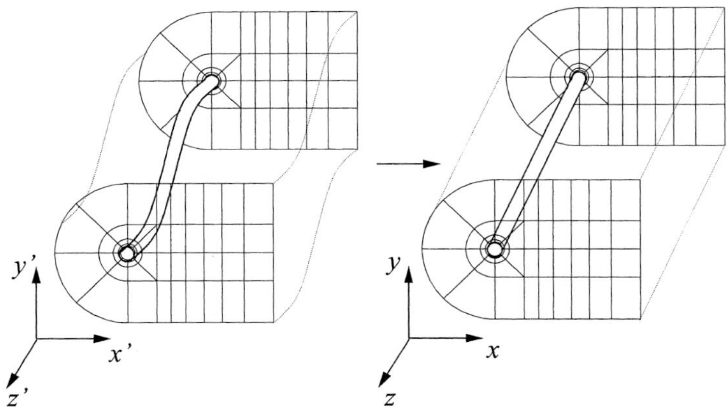

Figure 2 Schematic used to illustrate the application of ARF techniques to cases in which the structure deforms in directions normal to its axis. Coupling terms are computed on the basis of the mapping $\boldsymbol{x} \to \boldsymbol{x}'$.

2.2 Generalised ARF Techniques

ARF techniques as described above were first applied to the study of vortex-induced vibration of circular cylinders [2, 3, 4]. An extension allows the same technique to be applied in cases where the flexible structure has a rotational degree of freedom in addition to translational freedoms [5].

The method has also been generalised to the case where the displacement of the structure varies in the third dimension, as illustrated in Figure 2 [6]. If the mapping that takes the position of a point on the structure, $\boldsymbol{x}'$, into a system of coordinates in which the structure is undistorted, $\boldsymbol{x}$, is given by

$$x = x' - \zeta(z,t), \quad y = y' - \eta(z,t), \quad z = z', \tag{7}$$

then the corresponding velocity components and the pressure are related by

$$u = u' - \partial_t\zeta - w\partial_z\zeta, \quad v = v' - \partial_t\eta - w\partial_z\eta, \quad w = w', \quad p = p'. \tag{8}$$

This transformation supplies additional pressure and viscous coupling terms to (3), which becomes

$$\partial_t\boldsymbol{u} + \boldsymbol{N}(\boldsymbol{u}) = -\rho^{-1}\nabla p + \nu\nabla^2\boldsymbol{u} + \boldsymbol{A}(\boldsymbol{u}, p, t), \tag{9}$$

with

$$A_x = -\mathrm{d}_{tt}\zeta + \nu\left[\partial_{z'z'}\left(u + w\partial_z\zeta\right) - \partial_{zz}u + \partial_z\zeta\nabla^2_{xy}w + \partial_{tzz}\zeta\right], \tag{10a}$$
$$A_y = -\mathrm{d}_{tt}\eta + \nu\left[\partial_{z'z'}\left(v + w\partial_z\eta\right) - \partial_{zz}v + \partial_z\eta\nabla^2_{xy}w + \partial_{tzz}\eta\right], \tag{10b}$$
$$A_z = \partial_z\zeta\partial_x p + \partial_z\eta\partial_y p + \nu\left[\partial_{z'z'}w - \partial_{zz}w\right], \tag{10c}$$

where the operators

$$\mathrm{d}_t \equiv \partial_t + u\partial_x + v\partial_y + w\partial_z, \tag{11a}$$

$$\partial_{z'} \equiv \partial_z - \partial_z\zeta\partial_x - \partial_z\eta\partial_y, \tag{11b}$$

$$\nabla^2_{xy} \equiv \partial_{xx} + \partial_{yy}. \tag{11c}$$

Now that the structure has a third degree of freedom, partial differential equations, derived e.g. from beam theory, replace the ODEs (1).

2.3 Discretisations

ARF techniques can be allied with any suitable spatial discretisation or temporal integration methods. In the applications to be described here, spectral elements have been used for spatial discretisations, in conjunction with mixed explicit–implicit time integration [7], usually of second order.

3 Prescribed Motion

First we review some studies of flows either past or generated by circular cylinders with prescribed oscillation. In each case the focus of the study has been to examine flow phenomena through numerical investigations that would be difficult or impossible to replicate through physical experiments.

3.1 Phase Switching in the Wake of an Oscillating Cylinder

The purpose of this study [8] was to investigate the mechanism underlying 'phase switching' that occurs in the wake of a circular cylinder forced to oscillate cross flow. In the primary synchronisation regime, when the Kármán wake oscillation frequency is entrained by that of cross flow oscillation, it is known from experimental studies that a rapid variation in the phase angle between cylinder motion and lift force occurs over a narrow range of cylinder motion frequencies. This phase switch is accompanied by a change in sign of the time-average mechanical energy transfer between the structure and fluid.

The mechanical energy transferred from the flowing fluid, freestream speed U, to the oscillating cylinder, diameter D, per motion cycle, period T, can be written in dimensionless form as

$$E = \frac{2}{\rho U^2 D^2} \int_0^T \dot{y} F_l \,\mathrm{d}t = \int_0^T \dot{\alpha} C_l \,\mathrm{d}t = \frac{1}{2} \oint \left(C_l \,\mathrm{d}\alpha + \alpha \,\mathrm{d}C_l \right), \tag{12}$$

where $\alpha = y/D$ is the dimensionless cross flow displacement, F_l is the cross-flow force per unit length and C_l the corresponding lift coefficient. The quantity E is positive when work is done on the cylinder, negative when work is done on the fluid. The last term in (12) assumes that a time-periodic solution has been obtained; the path integrals are taken around a limit cycle

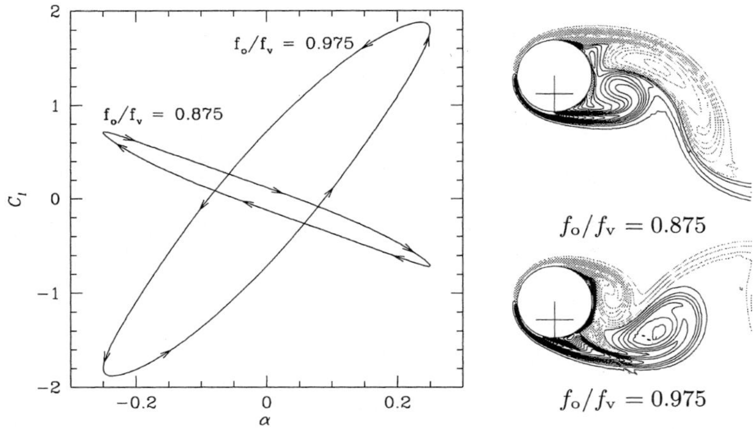

Figure 3 Limit cycles of coefficient of lift vs. dimensionless cross flow displacement for two-dimensional flow past an oscillating cylinder, $Re = 500$, $\alpha_{\max} = 0.25$ [8]. The dimensionless energy transfer between the cylinder and the fluid per motion cycle (E) is given by the area enclosed by the cycle, and its sign by the orientation of traverse. E is positive for $f_o/f_v = 0.875$, negative for $f_o/f_v = 0.975$. Also shown are the corresponding vorticity contours, displayed at the point of maximum cylinder displacement.

in (α, C_l) space. Figure 3 shows two of these limit cycles, obtained from two-dimensional simulations at $Re = 500$ and an amplitude ratio $\alpha_{\max} = 0.25$, and at two different frequency ratios $f_o/f_v = 0.875$ and 0.975, where f_o is the imposed cross flow oscillation frequency and f_v is the fixed-cylinder vortex shedding frequency. The orientation of traverse around the limit cycles, and hence the sign of E, is opposite in the two cases. Also shown in Figure 3 are contours of vorticity for these two frequency ratios, computed at the same phase in the cylinder motion cycle, illustrating the fact that the timing of vortex shedding is also changed dramatically by variation in f_o/f_v.

An hypothesis was advanced that the change in flow structure between the two types of solution illustrated in Figure 3 resulted from a change in balance between two different vorticity production mechanisms, one associated with surface-tangential pressure gradients, the other with the tangential component of cylinder acceleration. This hypothesis was investigated by numerically manipulating the local tangential motion of the cylinder boundary to reduce the motion-induced vorticity production progressively to zero. As this was done, the solution branch associated with the flow shown for $f_o/f_v = 0.975$ became smaller in extent, and eventually disappeared altogether when the motion-induced vorticity production was zero, while the extent of the other solution branch was left comparatively unchanged. The outcome of this numerical experiment thus appears to support the hypothesis.

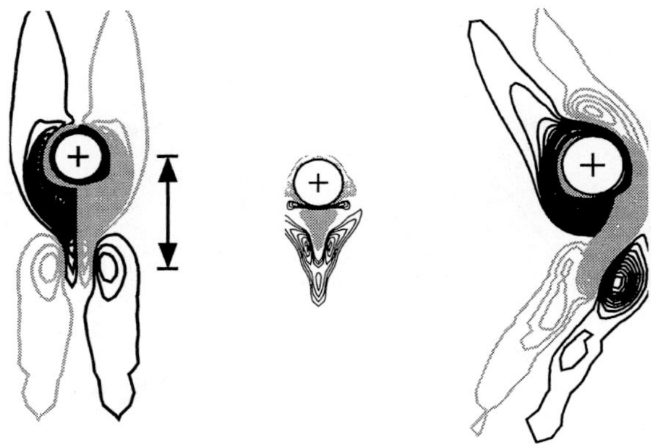

(a) (b) (c)

Figure 4 Vorticity contours for the two-dimensional flows produced by a cylinder in vertical oscillation of indicated amplitude, $KC = 7$. For $\beta = 13.5$, the two-dimensional periodic flow shown in (a) is linearly unstable to the two-dimensional symmetry breaking Floquet mode shown in (b), leading to another periodic two-dimensional flow, but with broken reflection symmetry, illustrated in (c). The two-dimensional asymmetric periodic flows are further unstable to three-dimensional Floquet modes [9]. The symmetric base flow, its unstable Floquet mode, and the resulting asymptotic flow are all presented at the phase of maximum oscillation amplitude, $\alpha_{\max}$.

3.2 Instability of the Flow Generated by an Oscillating Cylinder

The oscillatory rectilinear translation of a long circular cylinder normal to its axis in a quiescent body of fluid at low motion amplitudes and frequencies generates a time-mean streaming flow which is inwards towards the low-pressure shoulders of the moving cylinder and outwards along the motion axis. The two dimensionless groups that describe the problem are the Keulegan–Carpenter number $KC = 2\pi\alpha_{\max}$ and Stokes number $\beta = f_o D^2/\nu$. At low values of KC and β, the flow is time-periodic, two-dimensional and has reflection symmetry about the translation axis, as shown in Figure 4 (a). Increasing either KC or β can lead to three-dimensional flows through either of two instability mechanisms [9]. As the two-dimensional flow is time-periodic, Floquet analysis is the appropriate tool with which to study the problem.

At low values of β, increasing KC leads to instability of a two-dimensional Floquet mode that breaks the reflection symmetry about the translation axis (Figure 4 b); the subsequent asymmetric periodic flows (Figure 4 c) are further unstable to three-dimensional Floquet modes. At low values of KC, increasing β produces instability of three-dimensional Floquet modes that break the cylinder-axis translation symmetry, leading to a three-dimensional flow that retains reflection symmetry in the spanwise average.

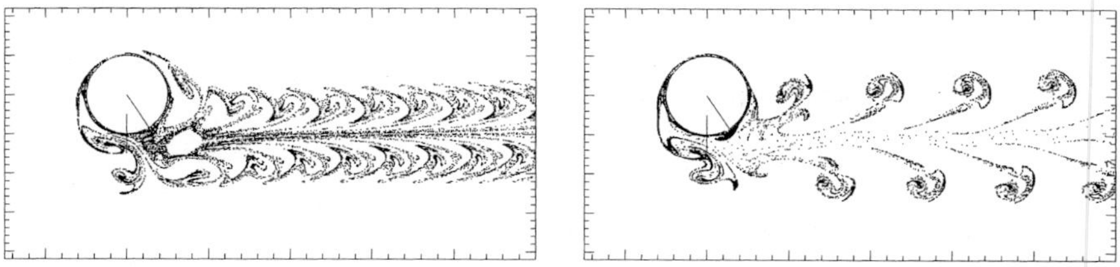

(a) (b)

Figure 5 Particle-tracking visualisations of the two-dimensional flows produced by a circular cylinder which has a vertical translational oscillation combined with a rotational oscillation [10]. Flows generated (a), when the cylinder does not translate in the horizontal direction; (b), with the cylinder swimming to left at its terminal speed. $Re_{\mathrm{rms}} = 200$, based on vertical oscillation velocity.

3.3 Propulsion Produced by Oscillatory Rotation and Translation

In [10], it was reported that when a oscillatory rotation of the cylinder is added to the rectilinear oscillatory translation (as in § 3.2), a new streaming flow can result. If the periods of the two motions are the same, and the phase angle between them set appropriately, the time-mean streaming flow is directed outwards along a line *perpendicular* to the cylinder translation axis, as seen in Figure 5 (a). Thus results a time-mean force in a direction normal to the cylinder translation axis. If the cylinder is left free to accelerate in the direction normal to the imposed oscillatory motion, it can gain speed and eventually will swim at a terminal mean velocity, producing the flow illustrated in Figure 5 (b). For two-dimensional flows, the asymptotic state is time-periodic and again amenable to Floquet analysis, which shows three-dimensionally-unstable Floquet modes.

4 Coupled Fluid–Structure Problems

4.1 Mass-Damping Effects

An early application of the ARF method in aeroelastic studies was to two-dimensional flow past a flexibly mounted circular cylinder at $Re = 200$ [2]. The structural oscillation frequency f_{n} was set to match the vortex shedding frequency for the fixed cylinder. The remaining dimensionless groups are then the structural damping ratio ζ and the density ratio $m/\rho D^2$. These may be combined into a single mass-damping parameter of the form $m\zeta/\rho D^2$, often used to correlate the observed maximum amplitude of cross flow response, as in Griffin's compilation of experimental results [11], shown in Figure 6. As the damping $\zeta \to 0$, the peak amplitudes asymptote to maximum values, and the simulation results, also plotted in Figure 6, show that these are only weakly dependent on the density ratio.

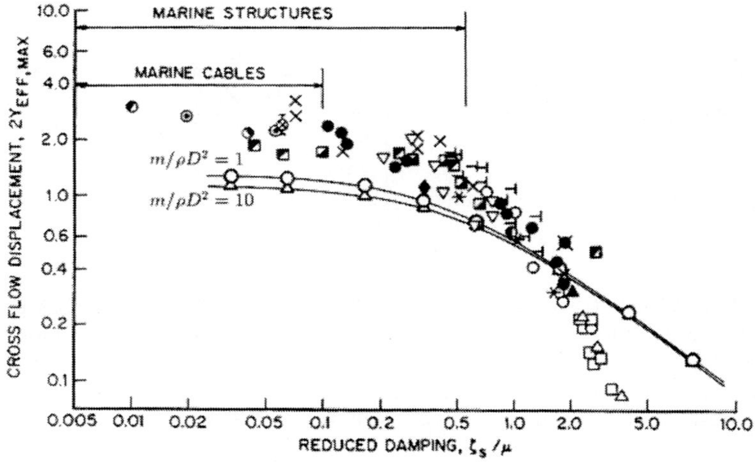

Figure 6 Maximum values of asymptotic peak-to-peak free vibration oscillation amplitudes as functions of the mass-damping product $\zeta_s/\mu = 8\pi^2 St^2 m\zeta/\rho D^2$ [2]. A comparison of computed values for density ratios $m/\rho D^2 = 1$ (○) and 10 (△), $Re = 200$, with a compilation of experimental values [11]

4.2 Lock-In Effects

In order to study the effect of variation of the ratio of structural natural frequency to fixed-cylinder vortex shedding frequency, another set of two-dimensional simulations was carried out [3], in this case with the damping and density ratios set to $\zeta = 0.01$ and $m/\rho D^2 = 10$ respectively. For these simulations, the Reynolds number was fixed at $Re = 250$. The results are summarised in Figure 7.

For all simulations, cylinder cross-flow oscillation and lift frequencies were found to coincide in the asymptotic state, however the lift/oscillation frequency f_o was nearly the same as f_v away from $f_n/f_v = 1$. Near $f_n/f_v = 1$ both frequencies changed together to fall near (but not *exactly* on) f_n: exact coincidence with f_n is indicated by the slanted thin line near the centre of the figure. This change in vortex shedding frequency to nearly match the cylinder natural frequency is the lock-in phenomenon. During lock-in, amplitudes of cross-flow oscillation increased markedly: the largest steady-state values correspond to peak-to-peak oscillation amplitudes of approximately $0.9\,D$ (rms values of α are presented in Fig. 7). The lack of exact coincidence of f_o with f_n during lock-in can be accounted for by a phase difference between cylinder motion and lift forces, which is known to be a function of f_n/f_v.

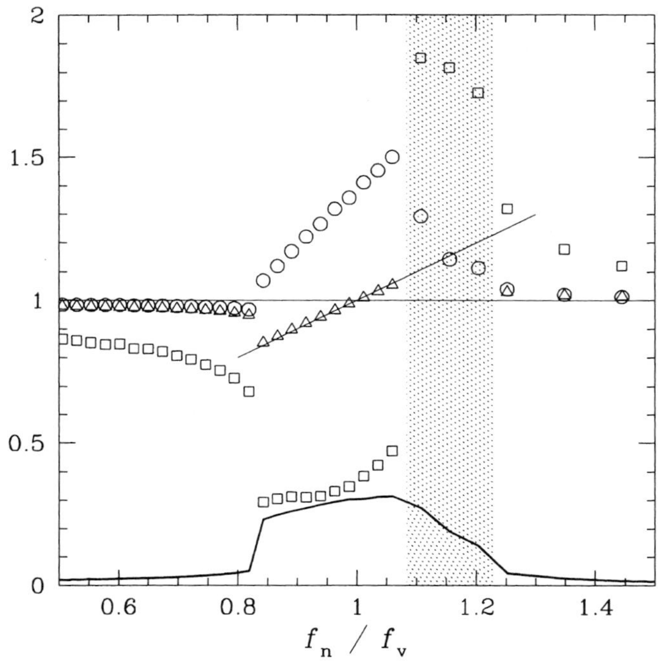

Figure 7 Cylinder response diagram for two-dimensional simulations at $Re = 250$ [3]. Abscissa values give the ratio of cylinder *in vacuo* natural frequency f_{n} to Strouhal frequency for the fixed cylinder f_{v}. □, ratio of fluctuating lift coefficients $C_{\mathrm{l_{rms}}}/C_{\mathrm{l_{rms_0}}}$; ○, ratio of mean drag coefficients $C_{\mathrm{d}}/C_{\mathrm{d_0}}$; △, ratio of oscillation frequency to fixed-cylinder Strouhal frequency $f_{\mathrm{o}}/f_{\mathrm{v}}$; ——, cylinder cross-flow response amplitude α_{rms}. Shaded region indicates a chaotic response regime.

4.3 Three-Dimensional Wake Effects

The simulations in §§ 4.1 and 4.2 were for two-dimensional flows, whereas since fixed-cylinder wakes become three-dimensional for Reynolds numbers above approximately 190, effectively all real-world occurrences of vortex-induced vibrations of circular cylinders must involve three-dimensional wake flows. Two-dimensional simulations, while significant, can potentially fail to reveal effects of dynamically important flow physics. With increases in computer capacity, full three-dimensional DNS and LES studies can now be undertaken, however these are still expensive, owing to the long integration times involved.

Figure 8, reproduced from [4], compares cylinder cross flow amplitude responses obtained using physical experiments with those obtained using both two-dimensional and three-dimensional simulations, where all relevant pa-

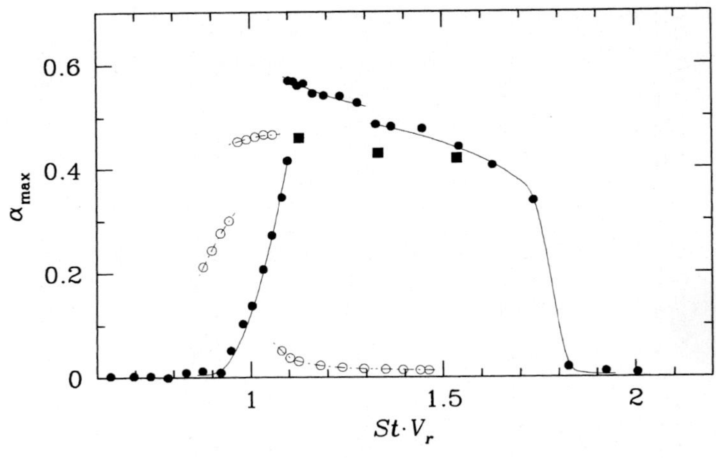

Figure 8 Dimensionless average peak response amplitude $\alpha_{\max}$ in vortex-induced vibration of a circular cylinder as a function of dimensionless flow speed $St \cdot V_{\mathrm{r}}$ for $Re \simeq 750$ [4]: •, experimental results; ■, three-dimensional simulations; ∘, two-dimensional simulations.

rameter values (Re, $m/\rho D^2$, ζ) were matched. The flow speed is given dimensionlessly as $St \cdot V_{\mathrm{r}}$, where $St = f_{\mathrm{v}}D/U$ is the fixed-cylinder Strouhal number and $V_{\mathrm{r}} = U/f_{\mathrm{n}}D$ is the reduced velocity. It is clear that the nature of the responses for two-dimensional simulations and experiment are very different, in terms of peak amplitudes achieved, but particularly in the number and extent of observed solution branches. The three comparable values obtained using three-dimensional DNS reveal much better agreement with experiment in terms of the extent and position of the response amplitude envelope.

The comparison of response amplitudes obtained with three-dimensional DNS with those obtained experimentally is encouraging, although there are significant differences and a more extensive study is desirable. However, another important point of comparison is of the coherent structures of the wake flows. No two-dimensional DNS result has been able to reproduce the experimentally observed '2P' wake mode, where two counter-rotating pairs of vortices are formed in the near wake for each cylinder motion cycle. Figure 9 shows a comparison of phase-averaged contours of the spanwise vorticity component in the near wake, obtained from DNS and from physical experiments (though at a slightly higher Re). It is clear that the simulations reproduce the '2P' wake mode observed in the experiment.

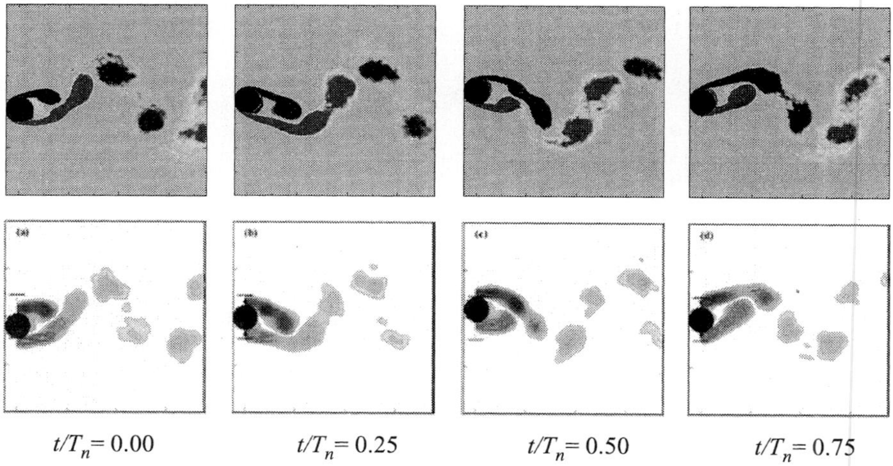

Figure 9 Simulation results showing phase and spanwise average contours of spanwise vorticity for four phases of the cylinder motion cycle at $Re \simeq 750$ (upper plots), compared to (lower) similar (two-dimensional) phase averages from experiments at a slightly higher Reynolds number [4].

Table 1 Global flow parameters from LES of flow past a fixed cylinder at $Re = 3900$ compared to experimental values at $Re = 3000$.

Source	St	C_{pb}	C_d
Norberg [13]	0.210	-0.88	0.99
Spectral element LES	0.218	-0.93	1.01

5 Stationary Bluff Body Flows

We complete our survey of computational bluff-body fluid dynamics by noting that while most of the recent fundamental computational investigations in the area have used direct numerical simulation techniques, progress to more applicable Reynolds numbers requires some form of turbulence modelling. In view of the relative infancy of turbulence models applied to bluff-body flows, large eddy simulations would appear to be the most promising avenue for extension in this area. Results from eddy-viscosity based LES of flow past a stationary circular cylinder, $Re = 3900$ [12], are presented in Table 1 and Figure 10. The relatively good agreement with the available validation data encourages further development — but more detailed comparisons also serve to highlight the need for more extensive validation experiments.

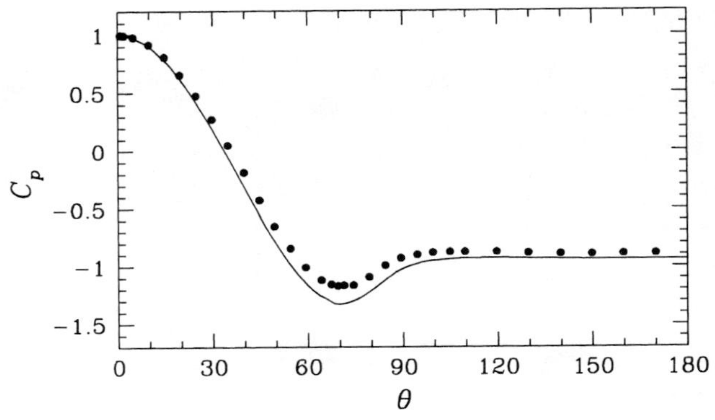

Figure 10 Time-mean pressure coefficient on the surface of a stationary cylinder as a function of angular position, LES results for $Re = 3900$ [12]. —, spectral element simulation; •, experimental results for $Re = 3000$ [13].

6 Discussion and Prospects

The immediate prospects for application of ARF methods are to moderate Reynolds number flows through adoption of LES approximations in the flow solver. This will allow more direct comparison with experimental results, and after this validation to better understanding of bluff-body vortex-induced vibration. The path is also open to better understanding of buffeting and quasi-steady aeroelastic effects such as stall-flutter and galloping.

While the applications of ARF techniques here have all been to separated flows, they can equally be applied to DNS or LES of aeroelastic fluid–structure interaction of attached flows. We can also anticipate extensions of the method to deal with fully three-dimensional structures. Finally, the ARF approach can be applied to any problem where the flow can be described in an accelerating reference frame. An intriguing example presently under study is the violent overturning instability in a spinning, precessing container of fluid [14].

Acknowledgments

Research in this area has been made more pleasurable through interaction with my colleagues. I gratefully acknowledge the contribution of my collaborators in the work described here: John Elston, Raghuraman Govardhan, Ron Henderson, George Karniadakis, Richard Manasseh, Stefan Schmidt, John Sheridan and Charles Williamson. The support of APAC and CSIRO HPCCC computing facilities has been essential for this work, and I thank their staff for continued support.

References

[1] T. Leweke, P. W. Bearman, and C. H. K. Williamson, "Preface — Special Issue on Bluff Body Wakes and Vortex-Induced Vibrations," J. Fluids & Structures **15,** 377–8 (2001).

[2] H. M. Blackburn and G. E. Karniadakis, "Two- and Three-Dimensional Simulations of Vortex-Induced Vibration of a Circular Cylinder," In *3rd Intnl Offshore & Polar Engng Conf.*, **III,** 715–720 (Singapore, 1993).

[3] H. M. Blackburn and R. D. Henderson, "Lock-In Behaviour in Simulated Vortex-Induced Vibration," Exptl Thermal & Fluid Sci. **12,** 184–189 (1996).

[4] H. M. Blackburn, R. N. Govardhan, and C. H. K. Williamson, "A Complementary Numerical and Physical Investigation of Vortex-Induced Vibration," J. Fluids & Structures **15,** 481–488 (2001).

[5] L. Li, S. J. Sherwin, and P. W. Bearman, "A Moving Frame of Reference Algorithm for Fluid/Structure Interaction of Rotating and Translating Bodies," Intnl J. Num. Meth. Fluids **38,** 187–206 (2002).

[6] D. J. Newman and G. E. Karniadakis, "A Direct Numerical Simulation of Flow Past a Freely Vibrating Cable," J. Fluid Mech. **344,** 95–136 (1997).

[7] G. E. Karniadakis, M. Israeli, and S. A. Orszag, "High-Order Splitting Methods for the Incompressible Navier–Stokes Equations," J. Comput. Phys. **97,** 414–443 (1991).

[8] H. M. Blackburn and R. D. Henderson, "A Study of Two-Dimensional Flow Past an Oscillating Cylinder," J. Fluid Mech. **385,** 255–286 (1999).

[9] J. R. Elston, J. Sheridan, and H. M. Blackburn, "The Transition to Three-Dimensionality in the Flow Produced by an Oscillating Circular Cylinder," In *14th Australasian Fluid Mechanics Conference*, (Adelaide, 2001).

[10] H. M. Blackburn, J. R. Elston, and J. Sheridan, "Bluff Body Propulsion Produced by Combined Rotary and Translational Oscillation," Phys. Fluids **11,** 4–6 (1999).

[11] O. M. Griffin, J. H. Pattison, R. A. Skop, S. E. Ramberg, and D. J. Meggitt, "Vortex-Excited Vibrations of Marine Cables," ASCE J. Wat. Port, Coastal Ocean. Div. **WW2,** 183–204 (1980).

[12] H. M. Blackburn and S. Schmidt, "Large Eddy Simulation of Flow Past a Circular Cylinder," In *14th Australasian Fluid Mechanics Conference*, (Adelaide, 2001).

[13] C. Norberg, "Effects of Reynolds Number and Low-Intensity Freestream Turbulence on the Flow Around a Circular Cylinder," Technical Report No. 87/2, Dept Appl. Thermodyn., Chalmers Univ. Tech. (1987) .

[14] R. Manasseh, "Breakdown Regimes of Inertia Waves in a Precessing Cylinder," J. Fluid Mech. **243,** 261–296 (1992).

Dynamics and Control of Flames Interacting with Pressure Waves

S. CANDEL[1,2], S. DUCRUIX[1], T. SCHULLER[1] and D. DUROX[1]

[1]Laboratoire EM2C, CNRS UPR 288, Ecole Centrale Paris
92295 Châtenay-Malabry, France
[2]and Institut Universitaire de France

Summary

The coupling between pressure waves and combustion gives rise to dynamical phenomena which constitute some of the most challenging problems in combustion research. A number of related issues have been investigated over the last decades. Studies have concerned fundamental aspects and practical implications. Important progress has been accomplished in the understanding, modeling and simulation of combustion dynamics. Detailed experiments have provided a wealth of information on elementary dynamical processes involving vortex interactions, mutual flame annihilation, flame collisions with boundaries, and coupling between flames and acoustics. The mechanisms driving instabilities and the coupling between pressure waves and combustion has been extensively investigated in gas turbine model combustors. Progress in numerical modeling has allowed simulations of dynamical flames interacting with pressure perturbations. This has allowed advances in prediction methods for combustion instabilities. Efforts have also concerned the development of the related subject of combustion control. A number of experiments on laboratory scale combustors have shown that the amplitude of combustion instabilities could be reduced by applying control principles and some applications of these principles have been made in full scale terrestrial gas turbine systems. Research has focused on algorithms, actuators, sensors and systems integration. In recent years, scaling from laboratory experiments to practical devices has been demonstrated with some success but limitations have also been revealed. After a brief review of the state of the art of these topics, this article describes some of our recent work on (1) Elementary processes in flame/acoustics coupling, (2) Simulation of combustion dynamics, (3) Multi-dimensional simulation of active control. Experiments on premixed flames responding to an acoustic modulation imposed on the upstream flow are used to illustrate fundamental interactions. It is shown that a flame impinging on a solid boundary and perturbed from upstream may induce a considerable amplification of the sound radiated by the system. The simulation of turbulent flames coupled to plane acoustic modes is then considered. The last topic is concerned with the numerical simulation of active control. This involves the coupling of a flow solver with a control algorithm. It is indicated that this coupling requires special precautions.

1 Introduction and Background

There are numerous coupling problems in combustion. Burning itself is a complicated process involving fluid mechanics, transport phenomena of heat and mass, chemical kinetics and multispecies and often multiphase thermodynamics. The flame appears as a complex combination of interacting mechanisms. The structure of the flame is governed by a competition between convection, diffusion and reaction involving a large number of chemical species. The situation already complicated when the flow is laminar becomes more intricate when combustion takes place in a turbulent stream (which is the case in practical devices like those used in aeronautical and space propulsion). A further complication is introduced by the coupling of combustion with pressure waves which is specifically effective when the flame is confined as is the case in aircraft engine combustors and afterburners, in rocket motors or energy production powerplants. This confinement allows resonant interactions which in turn give rise to powerful dynamical phenomena. These have many serious consequences in the form of "combustion instabilities".

The coupling of combustion and pressure waves is a topic of continued interest which has evolved over the past fifty years from phenomenology to a situation where it is amenable to simulation (see [1-3] for recent reviews). The related area of control has also evolved during this period from early concepts of feedback stabilization of the combustion process to more recent demonstrations in laboratory experiments of active control and full scale application in gas turbine combustors. Control of combustion is made difficult because the process is complex, nonlinear, involves a variety of coupled mechanisms and is not easily described mathematically. This is compounded with difficulties associated with the harsh conditions prevailing in combustors and the difficulties of integrating sensors and actuators in practical devices. One important design objective is to keep the combustion process under control and avoid possible flashback, blow-out and resonant instability. At this point, it is worth listing some of the reasons why flames are so susceptible to instabilities coupled by pressure waves :

- The energy density associated to combustion is quite sizable. A small fraction of this energy is sufficient to drive the oscillation.
- The combustion process involves time lags. Reactants introduced in the combustor at one instant are converted into burnt gases at a later time. Systems with delays are known to be more readily unstable.
- Resonant interactions may readily occur in the confined and weakly damped geometries used in most practical combustors. Among the possible coupling modes acoustics is often dominant. Resonance is sharp because the losses are reduced.
- Numerous mechanisms may feed energy in the perturbed motion. A gain is obtained if according to the Rayleigh criterion the heat release fluctuation is in phase with pressure or more exactly if $\int p'q'dt > 0$.

To avoid strong pressure wave/combustion interactions one tries to change the balance between gain and losses. This is usually achieved by passive methods involving a removal of energy from the oscillation or a modified design to avoid certain modes of oscillation. It is now possible to envisage active control solutions using feedback concepts. Such proposals already made during the 1950's [4-6], were only demonstrated in the 1980's and full scale application is only recent.
Because the process is so complex design tools are not available. However, the situation changes rapidly and current efforts focus on the development of detailed numerical models for combustion dynamics and control. This will rely on the considerable amount of experimental data and insight gained on the process during the previous years. There are many complexities which cannot be covered here. Three aspects are considered to describe some current efforts. Elementary processes in combustion/acoustics coupling are illustrated in Section 2 with experiments on premixed laminar flames. The simulation of combustion dynamics is envisaged in Section 3 and exemplified in Section 4 with calculations of a premixed turbulent flame in an unstable combustion process. Simulations of active control are briefly discussed in Section 5.

2 Perturbed Flames Impinging on a Solid Plate

Studies of the response of laminar flames to acoustic modulations are useful for many reasons : (1) The laminar case provides a simple well controlled configuration for detailed observations and it may be used to uncover novel coupling phenomena. (2) The interaction may be modeled from basic principles and does not depend on closure schemes like those required for descriptions of turbulent transport and combustion. (3) Experimental data may be used to validate computational tools for combustion dynamics. There are also many practical applications in which the flame is quasi-laminar (domestic boilers, radiant heaters and industrial dryers). Systematic experiments on laminar flame dynamics have been carried out in the case of a premixed flame anchored on a burner (see [7]). Recent studies in which the flame impinges on a flat plate have revealed some remarkable interactions [8]. A short account is given below.

A premixed flame is anchored on an axisymmetric burner comprising a cylindrical tube, 16 cm in length, followed by a convergent nozzle. The outlet diameter is 22 mm(Figure 1). The flow is modulated around its mean value $\bar{U}$ by an actuator set at the bottom of the tube. This produces a harmonic velocity perturbation u' at the burner outlet which wrinkles the surface of the flame front $A(t)$. Figure 2 shows the overall pressure level radiated 25 cm away from the flame as a function of the modulation frequency f_e. The sound produced in the case of a perturbed impinging jet flame is always 10 to 20 dB louder than the sound of the same upstream perturbation but without combustion or without the plate. This amplification is due to combustion and only occurs when the flame interacts with the plate [8]. Phase locked images (not shown) indicate that the flame front is strongly modulated. The

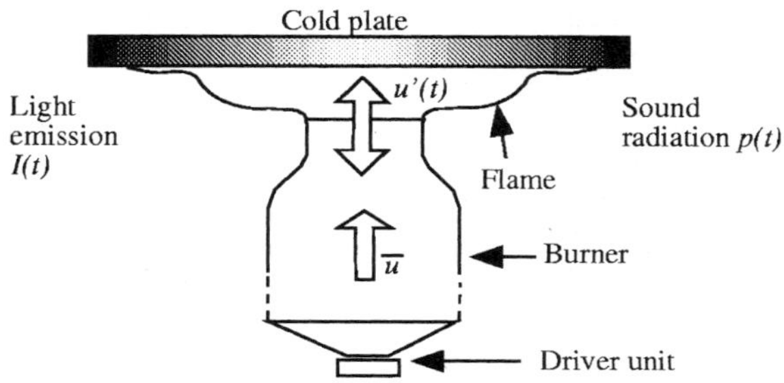

Figure 1. Experimental setup. $\bar{U}$ is the mean bulk velocity and u' the fluctuating velocity component at the burner outlet.

perturbations are convected along the flame front away from the burner towards the plate. When they reach the plate a large portion of the flame surface vanishes in a short period of time [8]. This process induces a source of sound located where the flame interacts with the plate. The corresponding far-field pressure p_∞ is given by the time retarded rate of change of the flame surface area $A(t)$. It is possible to relate the sound radiated to the rate of change of flame surface [9-11]:

$$p_\infty(r,t) = \frac{\rho_\infty}{4\pi r}\left(\frac{\rho_u}{\rho_b} - 1\right) S_L \left[\frac{dA}{dt}\right]_{t-\tau} \tag{1}$$

In this equation, $\rho_\infty, \rho_u, \rho_b$ are respectively the far-field air, the reactant gas and the burned gas densities. τ is the travel time over a distance r from the sources to the detector. In a first approximation the chemiluminescence of the flame $I(t)$ can be considered to be proportional to the volumetric rate of consumption of gas reactants $Q(t)$ [12]. Assuming a constant flame speed S_L, $I(t)$ is also proportional to the total flame surface area $A(t)$ [8]. The far-field pressure signal p_∞ should then be proportional to the rate of change of $I(t)$. Light emission $I(t)$, direct estimations of the flame surface area $A(t)$ and measurements of the farfield pressure signals p_∞ are correlated in Figure 3. The fast rate of extinction of the flame surface at the cold boundary corresponds to the significant acoustic pressure drop measured by the microphone. The periodic flame-wall quenching constitutes a strong source of sound which in a confined situation could feed acoustic energy in the natural modes of oscillation of the system. This will lead to a resonant feedback process driven by the flame-wall interaction – a remarkable example of the coupling between combustion, acoustics and solid boundaries [13].

3 Modeling of Combustion Dynamics

The theoretical description of combustion dynamics is usually approached by emphasizing one of the processes in this closely coupled phenomenon :

- Considering acoustics as the central mechanism one may write a wave equation for the reacting flow and derive a unified framework for the analysis of

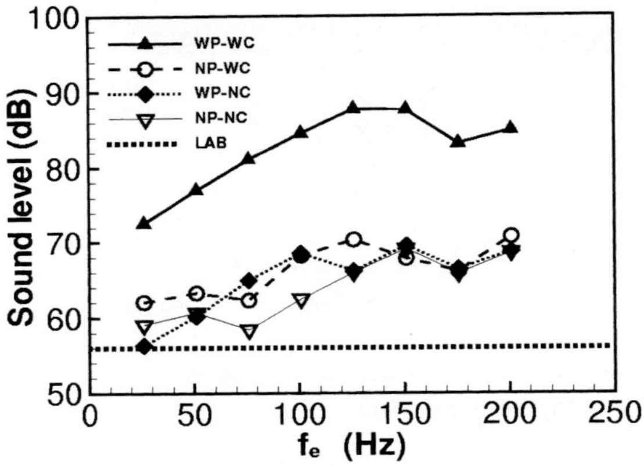

Figure 2. Overall sound level as a function of the driving frequency. Mean axial velocity: $\bar{U} = 1.44$ m/s. Equivalence ratio: $\Phi = 0.95$. Nozzle-to-plate distance $z = 7.6$ mm. The convention adopted in the caption is WP: with plate, NP: without plate, WC: with combustion, NC: without combustion, and LAB: mean background noise in the laboratory.

combustion oscillations. While this is an attractive approach it tends to hide the difficult problem of describing the response of the flame to the wave motion.

- Considering combustion as the central process the modeling is aimed at the description of the flame dynamical response to acoustic waves. The analysis emphasizes the fluid mechanics, the flame motion in a field of perturbations, and the differential response of the cold and hot streams of gases.

The first viewpoint is exemplified in the analytical framework established by Culick and his co-workers [3]. This uses a projection method equivalent to the Galerkin procedure introduced in this field by Zinn [14].

The starting point is the formulation of a boundary value problem for acoustic perturbations in a reactive medium

$$\nabla^2 p' - \frac{1}{c^2}\frac{\partial^2 p'}{\partial t^2} = h(\overline{\boldsymbol{v}}, \overline{p}, \boldsymbol{v}', p'; \dots), \quad \text{with} \quad \boldsymbol{n} \cdot \nabla p' = -f \tag{2}$$

Inhomogeneous Neumann boundary conditions are used in this formulation but other conditions could be used as well.

The acoustic field is then expanded in a series of normal modes :

$$p'(\boldsymbol{x}, t) = \overline{p} \sum_{n=1}^{\infty} \eta_n(t) \psi_n(\boldsymbol{x}) \tag{3}$$

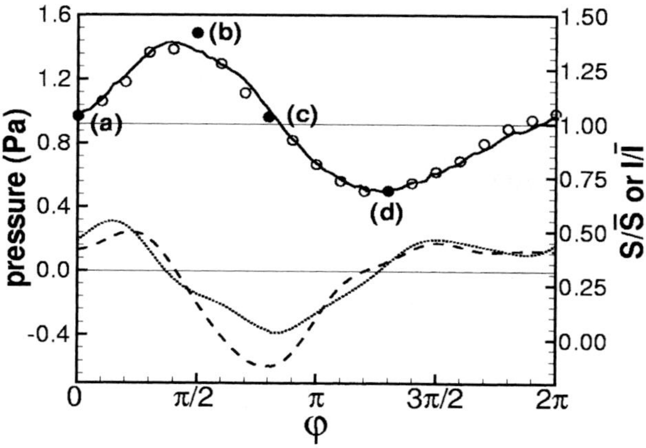

Figure 3. Evolution of the flame properties over a cycle of excitation. Flame light: $I/\bar{I}$ (full line). Flame surface area: $S/\bar{S}$ (circles), filled black circles indicate phases φ at which images are recorded. Pressure measured by the microphone: p_∞ (dash lines). Rate of change of $I(t)$: $dI/dt_{t-\tau}$ (dots line). $f_e = 101$ Hz, $\bar{U} = 1.20$ m/s, $\Phi = 1.05$, $z = 7.6$ mm.

The eigenfunctions $\psi_n(\boldsymbol{x})$ satisfy a homogeneous Helmholtz equation and corresponding homogeneous boundary conditions :

$$\boldsymbol{\nabla}^2\psi_n + (\omega_n^2/c^2)\psi_n = 0, \quad \text{with} \quad \boldsymbol{n}\cdot\boldsymbol{\nabla}\psi_n = 0 \tag{4}$$

The amplitudes of the normal mode expansion then satisfy a set of second order differential equations :

$$\frac{d^2\eta_n}{dt^2} + \omega_n^2\eta_n = F_n \quad \text{where} \quad F_n = -\frac{c^2}{\overline{p}E_n^2}\left[\int_V h\psi_n dV + \int_A f\psi_n dA\right] \tag{5}$$

This formulation emphasizes the wave motion and may be used to determine the modal amplitudes η_n. One difficulty is to specify the source terms arising from the distributed coupling process between acoustics and combustion and formally represented in the previous equations by the functions $h(\overline{\boldsymbol{v}}, \overline{p}, \boldsymbol{v}', p'; \dots)$ and f. Another difficulty is encountered in developing a suitable representation of source terms associated with control. Because the formulation is linear, these terms generally appear as additive sources, h becomes $h + h_c$ and f becomes $f + f_c$ but these linear forms do not describe the subtle interaction which takes place between control sources and the unstable motion. This framework is however valuable and may be used to devise "low order models" of combustion dynamics and control.

Adopting the second viewpoint which emphasizes the flow and combustion dynamics, there is no unique framework but one may identify some modeling levels : (1)

Direct simulation. The flame motion is calculated with a DNS method, the acoustics are in the form of characteristic boundary conditions possibly involving time domain impedance models ... (2) Large eddy simulation of the flame motion, acoustics and corresponding boundary conditions. (3) Large eddy simulation for the flame motion, analytical or linearized Euler description of the wave motion outside the flame zone. (4) Low order models for the flame zone, analytical or linearized Euler description of waves.

In most practical cases the flame is turbulent and the LES approach is the most suitable. There are now well established large eddy simulations (LES) and subgrid scale modeling (SSM) methodologies for nonreactive flows. Combustion LES is less well covered and constitutes a relatively new field not much older than about a dozen years. An important motivation for the development of LES is that it may provide a route to an improved modeling of combustion and is specifically well suited to the description of unsteady dynamical phenomena which are the subject of this lecture. Important advances have been made in the numerical modeling of flame dynamics using large-eddy simulations ([15-23]). Among the many problems encountered in the application of LES to combustion dynamics, the most prominent are the description of the flame motion and the associated subgrid scale modeling problem, the reduction of numerical dissipation and dispersion to acceptable levels and the treatment of boundary conditions.

One difficulty with combustion LES stems from the fact that chemical conversion of reactants takes place in thin layers which cannot be captured on the relatively coarse grid used for LES. It is therefore necessary to adapt the flame description to the LES framework. In the premixed case, this has been achieved by following two distinct routes : (1) The flame is represented as a thin front and its motion is calculated by a front tracking technique or with a G-equation formulation. (2) The flame is artificially thickened while preserving its displacement speed.

If the system is premixed a rather general method consists in describing the flame as a thin front. Most studies rely on a kinematic description of the motion of this front based on the G-equation. The flame corresponds to an iso-level $G = G_0$ of a function G describing the propagation of a front featuring a normal velocity S_d. One considers for example that $G = 0$ in the fresh gases and $G = 1$ in the reacted mixture and one has to solve the first order equation :

$$\frac{\partial G}{\partial t} + \boldsymbol{v} \cdot \boldsymbol{\nabla} G = S_d \, |\boldsymbol{\nabla} G| \tag{6}$$

This method based on the theoretical work of [24], was explored by [15]. In most analytical studies using this formulation the change in density accross the flame front is neglected. Various arguments are used to justify an assumption which is obviously not quite right but is often made in combustion and has the advantage of simplifying the analysis. Density changes across the flame may be included in this

framework as shown for example in [16-17].
The second representation of premixed flames in the LES framework, consists in artificially thickening the reaction zone, while keeping its propagative properties unchanged. The idea originally developed by [25-26] for the computation of laminar flames in complex geometries. has been applied in many recent large eddy simulations of premixed combustion (for example, [19], [20], [22]). Details may be found in these references and will not be reproduced here. Fine tuning of the thickened flame model have been devised to account for combustion taking place in the small scales [22], [27]. An efficiency function E is used to incorporate the effects of the unresolved scales on the resolved rate of reaction.

Another issue of considerable importance in LES for combustion dynamics is related to boundary conditions. It is known from computational fluid dynamics that the treatment of inflow, outflow conditions together with conditions at the walls determines to a great extent the quality of the solution (see for example [28]). In combustion instability simulations one has to establish boundary conditions for the fluid flow but simultaneously impose conditions on the various waves involved in the process (see for example [29]).

4 Turbulent Flame Dynamics

Turbulent combustion dynamics is illustrated in this section with unsteady flame calculations in a realistic geometry [30]. The premixed combustor developed by the University of Cambridge is a model scale version of a gas turbine unit. The airflow enters the injector through an annular section and is then divided into two streams which flow through concentric channels (Figure 4). Vanes are placed in these channels to create a rotating velocity component ("swirl") with an opposite sign. Gaseous fuel is injected in the airflow through cylindrical bars fitted with two 0.45 mm diameter holes. The swirler unit is plugged on a 70 mm in diameter quartz tube. Burnt gases are exhausted from this tube at atmospheric pressure. Simulations of the complete system (air supply, fuel supply, swirler unit and combustion chamber) would require an excessive amount of computer resources. It is therefore assumed that (1) the flow beyond the injection vanes is perfectly premixed, (2) the rotating component u_θ of the flow created by the vanes is proportional to the axial flow velocity and to the tangent of the vane angle and (3) u_θ is a constant in a given section. The system is considered to be axisymmetric and the calculations are carried out over a sector corresponding to a 10° angle around the symmetry axis.
The artificial thickening approach described in the previous section is implemented in AVBP, a LES flow solver (see [31]). The thickening factor F must be large enough to resolve the flame front on the LES mesh, but small enough to let the flame be wrinkled by the vortices calculated in the simulation. A thickening factor $F = 8$ meets both requirements. Calculations correspond to an equivalence ratio $\Phi = 0.7$. The flow velocity is equal to 15 m s^{-1}, corresponding to an airflow rate of

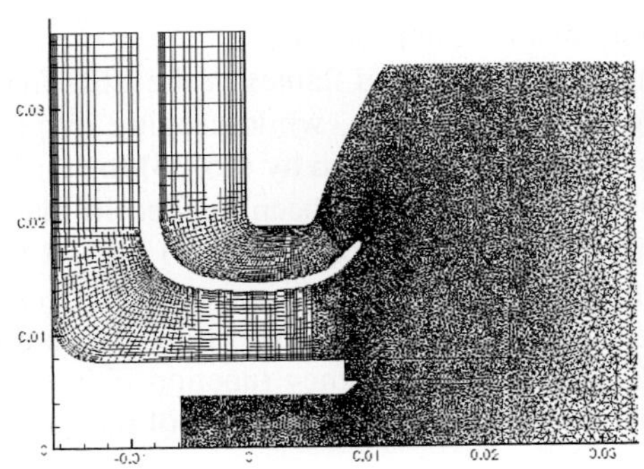

Figure 4. Enlargement of the mesh at the outlet of the swirler unit.

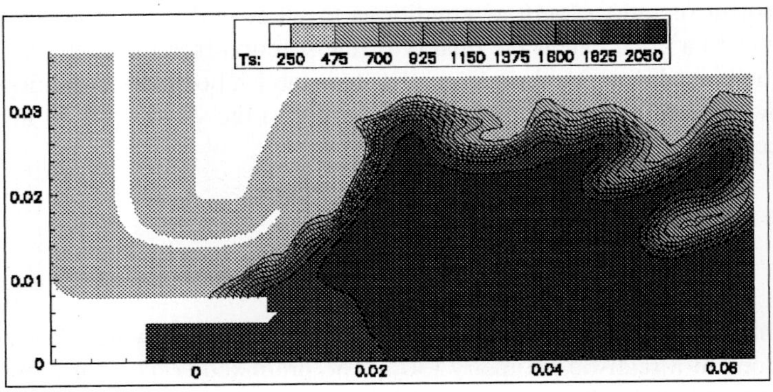

Figure 5. Snapshot of the field of temperature near the injector outlet.

60 g s^{-1}. Figure 5 shows the temperature in the swirler unit near field. A central recirculation zone serves to stabilize the flame. Burned gases in contact with fresh gases, provide a continuous source of ignition. The stabilization region is near the lower border of the internal channel in a region of reduced flow velocities. Snapshots of the temperature field show that vortices develop and the flame stabilization position moves when these structures are shed. This causes fluctuations which may lead to combustion instabilities. At certain instants the flame propagates inwards along the internal channel separation. This could eventually lead to flashback. Vortices created in the region between the outer jet and the chamber wall (Figure 5) are convected in the chamber and wrinkle the flame front producing large pockets of fresh gases which are conveyed into the recirculation region and destabilize the flame. Pockets of fresh gases are also created near the flame extremity. These pockets are convected and burned with a certain delay. Depending on the delay, this phenomenon may participate in the development of acoustic instabilities. The flame length varies between 8 and 12 cm during the calculation. In the first part of the

flame front ($0.1 < x < 0.3$), the amplitude of the motion is relatively small and the flame remains close to its mean position, whereas large movements can be observed further downstream. The flame is slightly lifted with respect to the injector channel, even though a small reaction rate exists near the separation wall. It is interesting to note that the stabilization position is not fixed in time. Figure 6 shows the temporal

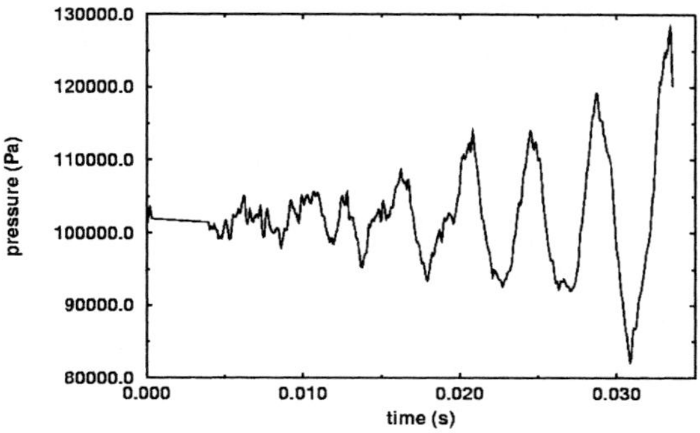

Figure 6. Temporal evolution of the pressure at a point in the middle of the chamber.

evolution of the pressure in the chamber for a time period of 35 ms. After a transition period, the signal is periodic. The growth rate is still positive at the end of the simulation indicating that no limit cycle has been reached at that time. Spectral analysis indicates that the signal frequency is close to 250 Hz while experimental observations indicate an instability around 200 Hz. These calculations indicate that it is possible to simulate the dynamical behaviour of turbulent flames in a gas turbine combustor. Results are qualitatively close to observations but validations are needed to see if the method may be extended to predict instability boundaries.

5 Simulation of Active Control

Instability control is a central design issue in combustion engineering. One may distinguish passive control schemes, which require a physical understanding of the phenomenon and involve careful modifications of the combustion system and active methods which rely on feedback concepts. The principles of active control of instabilities were described in some early studies but practical demonstrations were only achieved in the recent past (see [32] for a review).

Developments of active control strategies have relied on experiments with a limited number of studies directed at the full simulation of control with the objective of testing and improving control algorithms ([33],[34]). Simulation complements experimentation and it has potential in the development of flow control methods.

Many problems arise when one wishes to couple a flow solver with a control algorithm. This is illustrated in [34, 35]. This investigation is directed at the control of vortex driven instabilities found in solid segmented rocket motors. An adaptive controller is employed in this simulation and the conceptual principle is shown in Figure 7.

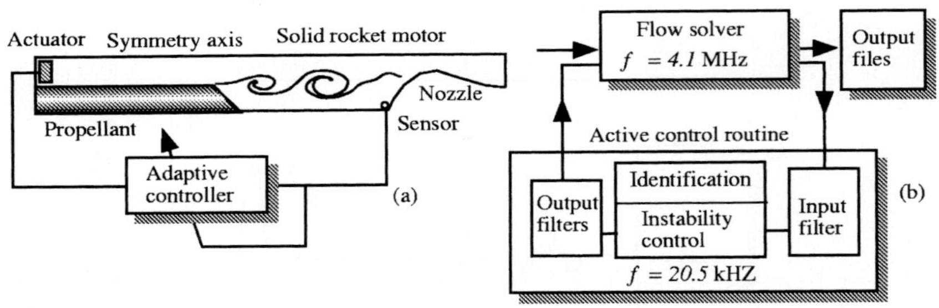

Figure 7. Active control of a solid rocket motor. (a) Principle, (b) Coupling between the flow solver Sierra and the adaptive controller RAC. (Adapted from [35].)

One aspect which deserves special care is the definition of the "software actuator". Problems related to the coupling of the flow simulation module with the control algorithm must be treated next because there is a large mismatch between the timestepping of the flow solver and the sampling period of the controller. It becomes immediately apparent that coupling requires filtering of the input and output of the controller (see [34], [35] for details).

6 Conclusions and Perspectives

This article focuses on dynamics and control of combustion. These topics are technologically important and give rise to many scientific problems. Fundamental aspects have been studied in the past and much has been learned from well controlled laboratory experiments. The numerical modeling of combustion dynamics has also progressed with the development of LES methods but prediction of instability still poses challenges. One objective for the future will be to advance first generation subgrid models for combustion LES. Combined studies are needed in modeling, simulation and experimentation oriented towards LES needs. Progress is also required in the treatment of boundary conditions and in the integration of computational strategies describing the mean flow as well as the wave motion. Active control of combustion instability has also progressed in many ways. While initial demonstrations were carried out on small scale systems, control has also been implemented in some full scale gas turbines. A considerable amount of research has been made to study the problem with modern control tools. Theoretical instability models have been devised and control strategies have been explored using simple low order and essentially linear instability models. This research has been mostly theoretical with limited demonstrations on practical devices. One challenge for the future will be to apply these strategies (state feedback, robust control, model based adaptive control ...) to real systems. The multi-dimensional simulation (MDS) of active control has become an important area of research. The objective is to allow realistic software studies of control concepts. This problem combines CFD difficulties with those of coupling a flow solver with a control scheme. Early developments of such a capability were limited. It has been found from recent studies that a software description of the actuator is essential and that coupling of the flow solver and controller needs to be treated with care.

This work is supported by DGA, CNES and the Brite Euram project Aciacoc.

References

[1] S. Candel. "Combustion instabilities coupled by pressure waves and their active control". *24th Int. Symp. on Combustion*, The Combustion Institute, Pittsburgh, 1992, pp. 1277-1296.

[2] S. Candel, C. Huynh Huu and T. Poinsot. "Some modeling methods of combustion instabilities". In *Unsteady Combustion* edited by F.E.C. Culick, M.V. Heitor and J.H. Whitelaw, Kluwer, Dordrecht, 1996, pp. 83-112.

[3] F.E.C. Culick. "Combustion instabilities in propulsion systems". In *Unsteady combustion* edited by F.E.C. Culick, M.V. Heitor and J.H. Whitelaw, Kluwer, Dordrecht, 1996, pp. 173-241.

[4] L. Crocco and S.L. Cheng. *Theory of combustion instability in liquid propellant rocket motors* Butterworths Science Publication, London, 1956.

[5] H.S. Tsien. "Servo-stabilization of combustion in rocket motors". *Am. Rocket Soc. J.*, Vol. 22 , 1952, pp. 256-263.

[6] F.E. Marble and D.W. Cox, jr. "Servo-stabilization of low-frequency oscillations in a liquid bipropellant rocket motor". *Am. Rocket Soc. Journal*, Vol. 23, 1953, pp. 63-81.

[7] S. Ducruix, D. Durox and S. Candel. "Theoretical and experimental determination of the transfer function of a laminar premixed flame". *Proc. Comb. Inst.*, Vol. 28 , 2000, pp. 765-773.

[8] T. Schuller, D. Durox and S. Candel. "Dynamics of and noise radiation from a perturbed impinging premixed jet flame". *Comb. & Flame*, Vol. 128, 2001, pp. 88-110.

[9] S.L. Bragg. "Combustion noise". *Journal of the Institute of Fuel*, Vol. 36 , 1963, pp. 12-16.

[10] D.I. Abugov and O.I. Obrezkov. "Acoustic noise in turbulent flames". *Combustion Explosions and Shock Waves*, Vol. 14 , 1978, pp. 606-612.

[11] P. Clavin and E.D. Siggia. "Turbulent premixed flames and sound generation". *Combust. Si. Tech.*, Vol. 78 , 1991, pp. 147-155.

[12] R.B. Price, I.R. Hurle and T.M. Sudgen. "Optical studies of the generation of noise in turbulent flames". *12th Int. Symp. on Combustion*, The Combustion Institute, Pittsburgh, 1969, pp. 1093-1102.

[13] D. Durox, T. Schuller and S. Candel. "Self-induced instabilities of a premixed jet flame impinging on a plate". *Proc. of the Comb. Inst.*, Vol. 29 , 2002.

[14] B.T. Zinn and E.M. Lores. "Application of the Galerkin method in the solution of nonlinear axial combustion instability problems in liquid rocket". *Combust. Sci. Tech.*, Vol. 4 , 1972,

pp. 269-278.

[15] S. Menon and W. Jou. "Large-eddy simulations of combustion instabilities in an axisymmetric ramjet combustor". *Combust. Sci. Tech.*, Vol. 75 , 1991, pp. 53-72.

[16] J. Piana, D. Veynante, S. Candel and T. Poinsot. "Direct numerical simulation analysis of the G-equation in premixed combustion," *Proc. of the Second ERCOFTAC workshop on direct and large-eddy simulation* (Grenoble), 1996.

[17] J. Piana, F. Ducros and D. Veynante. "Large eddy simulations of turbulent premixed flames based on the G-equation and a flame wrinkling description". In *11th Symposium on Turbulent Shear Flows* F. Durst, N. Kasagi, B.E. Launder, F.W. Schmidt and K. Suzuki, eds. Springer-Verlag, Berlin, 1997, pp. 21.13-21.18.

[18] D. Veynante and T. Poinsot. "Reynolds average and large eddy simulation modeling for turbulent combustion". In *New Tools in Turbulence Modelling* edited by O. Metais and J. Ferziger, Springer Verlag, New York, 1997, pp. 105-140.

[19] D. Thibaut and S. Candel. "Numerical study of unsteady turbulent premixed combustion. Application to flashback simulation". *Comb. & Flame*, Vol. 113 , 1998, pp. 51-65.

[20] C. Angelberger, D. Veynante and F. Egolfopoulos. "LES of chemical and acoustic forcing of a premixed dump combustor". *Flow, Turb. Comb.*, Vol. 65 , 2001, pp. 205-222.

[21] H.G. Weller, G. Tabor, A.D. Gosman and C. Fureby. "Application of flame wrinkling LES combustion model to a turbulent mixing layer". *Proc. Comb. Inst.*, Vol. 27 , 1998, pp. 899-907.

[22] O. Colin, F. Ducros, D. Veynante and T. Poinsot. "A thickened flame model for large eddy simulations of turbulent permixed combustion". *Phys. Fluids*, Vol. 12 , 2000, pp. 1-21.

[23] E.R. Hawkes and R.S. Cant. "A flame surface density approach to large-eddy simulation of premixed turbulent combustion". *Proc. Comb. Inst.*, Vol. 28 , 2000, pp. 51-58.

[24] A.R. Kerstein, W.T. Ashurst and F.A. Williams. "Field equation for interface propagation in an unsteady hopmogeneous flowfield". *Phys. Rev.*, Vol. A37 , 1988, pp. 2728-2731.

[25] T.D. Butler and P.J. O'Rourke. "A numerical method for two dimensional unsteady reacting flows". *16th Int. Symp. on Combustion*, The Combustion Institute, Pittsburgh, 1976, pp. 1503-1515.

[26] P.J. O'Rourke and F.V. Bracco. "Two scaling transformations for the numerical computation of multidimensional unsteady laminar flames". *J. of Computat. Phys.*, Vol. 33 , 1979, pp. 185-203.

[27] C. Nottin, R. Knikker, M. Boger and D. Veynante. "Large eddy simulations of an acoustically excited turbulent premixed flame". *Proc. Comb. Inst.*, Vol. 28 , 2000, pp. 67-73.

[28] T.J. Poinsot and S.K. Lele. "Boundary conditions for direct simulation of compressible viscous flows". *J. Computat. Phys.*, Vol. 101 , 1992, pp. 104-129.

[29] S. Ducruix, S. Candel, F. Nicoud and T. Poinsot. "Conditions aux limites permettant l'injection d'ondes acoustiques dans une simulation numerique," *Proceedings of the 14 ème Congrès Français de Mécanique* (Toulouse), 1999.

[30] S. Ducruix, T. Poinsot and S. Candel. "Large eddy simulations of combustion instabilities in a swirled combustor," *Proc. of the IUTAM Colloquium on Turbulent Mixing and Combustion* (Kingston, Canada), 2001.

[31] T. Schonfeld, C. Angelberger, J.P. Legier and S. Ducruix. "Numerical simulation of compressible reactive flows on unstructured grids". AIAA paper 99-, 1999.

[32] K.R. McManus, T. Poinsot and S. Candel. "A review of active control of combustion instabilities". *Prog. Energy Combust. Sci.*, Vol. 19 , 1993, pp. 1-29.

[33] S. Menon. "Active combustion control in a ramjet using large-eddy simulations". *Combust. Sci. Tech.*, Vol. 84 , 1992, pp. 51-79.

[34] M. Mettenleiter, F. Vuillot and S. Candel. "Numerical simulation of active control in unstable solid rocket motors," *Proceedings of the RTO (Agard) Symposium on Active Control* (Braunschweig, Germany), RTO, 2000.

[35] M. Mettenleiter, F. Vuillot and S. Candel. "Numerical simulation of active control : Application to unstable solid rocket motors," *AIAA Journal*, Vol. 40, 2002, pp. 860-868.

A General Methodology for Studying the Hydrodynamic Stability of Flows in Enclosures

PATRICK LE QUÉRÉ[1], EMILIE GADOIN[1,2], OLIVIER DAUBE[1,3], SHIHE XIN[1,4], and LAURETTE TUCKERMAN[1]

[1] LIMSI-CNRS, BP 133, F-91403 Orsay Cedex, France
[2] Ecole Polytechnique Université de Nantes
[3] Université d'Evry
[4] Université Paris-Sud
plq@limsi.fr,
WWW home page: http://www.limsi.fr

Summary

We present a general methodology for studying the hydrodynamic stability of flows enclosed in cavities of complex geometry. To this aim, different tools consisting of time integration of the unsteady equations, steady state solving, computation of the most unstable eigenmodes of the Jacobian and its adjoint have been developed. The methodology is applied to the classical differentially heated cavity, where the steady solution branch is followed for very large values of the Rayleigh number and most unstable eigenmodes are computed at selected Rayleigh values. Its effectiveness for complex geometries is illustrated on a configuration consisting of a cavity with internal heated partitions. We finally propose to reduce the Navier-Stokes equations to a low order differential system by expanding the unsteady solution as the sum of the steady state solution and of a linear combination of the leading eigenmodes. The principle of the method, which requires computing the eigenmodes of the adjoint of the jacobian, is exposed and preliminary results are presented.

1 Introduction

Computation of flow instabilities is certainly one of most fascinating and challenging aspects of present computational fluid dynamics. The aim of hydrodynamic instability theory is to predict the values of the parameters at which instabilities occur, the nature of the instability (steady or unsteady, supercritical or subcritical), the type of solution that results from the instability, the modification of the base flow solution that results from the non-linear interactions of the unstable modes and the subsequent route to chaos. Mastering or promoting these instabilities, one aspect of active fluid mechanics, is also a very challenging prospect.

As soon as the geometry involves more that one inhomogeneous spatial direction, one generally lacks analytically-known base flow solutions and therefore the stability analysis has to be carried out numerically by resolution of the governing equations. This is usually done by direct integration of the unsteady Navier-Stokes equations. This methodology places severe constraints on the numerical algorithms which must be both spatially and time accurate. This has in particular promoted the development of spectral and pseudo-spectral spatial Chebyshev approximations in space coupled to second order timestepping schemes. It is indeed true that these unsteady methods have proven extremely successful in predicting loss of stability of several classes of flows, although like any other method, they have to be used with sufficient grid resolution. While these methods are still heavily used in this context and also to perform Direct Numerical Simulations of flows in simple geometries, they have also shown their limitations in establishing bifurcation diagrams and for complex geometries.

For the purpose of investigating the stability of flows in general, it has been recognized [1] that it is highly desirable to have efficient solvers to follow the path of steady solution branches even when they become unstable. Designing a solver to compute the unstable steady solutions is not a trivial matter since it is almost mandatory to use Newton's iterations, coupled in general to arclength continuation. Newton's iterative technique makes use of the Jacobian of the equations, which is on the order of 3 to 4 times the number of discretization points in two spatial dimensions. When the instability takes place at large values of the Reynolds or Rayleigh number as is the case for flows between rotating disks or for natural convection in differentially heated cavities, the number of discretization points needed to approximate the solution can become very large. Building the corresponding Jacobian matrix may also become difficult, in particular in the context of finite volume staggered meshes[1], although it is very sparse. Further, the incompressibility constraint in velocity-pressure formulation makes this matrix not easily invertible. Matrix free methods are therefore needed, *i.e.* a method which requires only the evaluation of the action of the Jacobian on a vector field to produce a solenoidal vector field, but this is not an easy task either due to the bad conditioning of the matrix.

An elegant solution to this problem was proposed by Tuckerman [2],[3] based on the fact that a family of efficient timestepping algorithms have been developed for integrating the unsteady equations. These are generally based on an implicit treatment of the viscous terms and an explicit treatment of the non-linear terms. She noted that the operator consisting of the difference between the solutions at two consecutive timesteps has the same roots as the steady state operator, a simple but far-reaching observation indeed, since it turns out that the Jacobian of the incremental operator is a straightforward

[1] This objection does not hold for the finite element community which has usually employed the true Newton's method, as it did not benefit from the natural separability of the operators, even on structured meshes

modification of the initial timestepping scheme and can thus be used very efficiently in the Newton's iterations. Up to date, most implementations have been performed for simple geometries in which the unsteady Stokes problem is solved by direct algorithms such as implicit influence matrix methods either with finite differences or spectral methods [2],[3],[4],[5]. Our experience with this algorithm has shown us that a key point in the efficiency and global performance of the method seems related to the fact that the timestepping solver simultaneously enforces the discrete momentum and continuity equations although there have been successful implementations with fractional step algorithms [6],[7]. One of our objectives in this work is to propose an extension of the method applicable to convection dominated flows in moderately complex geometries such as, for example, natural convection in cavities with internal partitions. To this aim we have implemented an iterative solver of the unsteady Stokes problem discretized on a classical staggered grid based on Vanka's multigrid scheme [8].

Once a steady solution is obtained, its linear stability can be investigated in several ways. One possibility is to perform time integration of the linearized equations governing the fluctuations. This is most efficiently done by adapting an unsteady code such as a fractional step method for the integration of the equations governing the fluctuating quantities with the quadratic non-linearities and replaced by linearized quantities about the steady solution. Integrating these unsteady equations will then eventually "polarize" the fluctuating solution along the most unstable (or least stable) eigenmode. This method works generally well if the few most unstable eigenmodes are sufficiently far apart although it is necessary to use small enough timesteps so that it may require integrating over several tens or hundreds of thousands of time steps in order for the solution to polarize. It can also be highly non monotonic in the sense that, even if the base flow solution is stable, the energy of the fluctuations can grow substantially before reaching its asymptotical decaying rate. The amplitude of this energy growth is related to the non-normality of the Jacobian [9], which increases as the relative size of the viscous terms to the convective terms decreases.

Another technique is to compute the spectrum of the Jacobian which gives access to the Jacobian eigenvalues. The solution will be linearly stable if all eigenvalues are of negative real part, unstable otherwise. Some implementations of this methodology have been based on an explicit representation of the Jacobian and on the use of generalized eigensolvers such as QR algorithms [10]. This approach seems limited to configurations which undergo instability at small Reynolds or Rayleigh number values, thus amenable to small or moderate spatial resolutions. When this is not the case, the memory requirements for storing the Jacobian can become prohibitively large. Further it is even useless to compute the whole spectrum since it is only those eigenvalues of the Jacobian of largest real part which are of interest. These two remarks have led to the development of Arnoldi-Krylov algorithms which allow for the computation of a selected part of the spectrum using matrix free methods. An invariant low dimensional subspace is determined by the repeated action of the Jacobian operator on an initial iterate; the corresponding eigenmodes are then computed in this space. This was done in this work by using the ARPACK library [11].

Our last objective is to establish a methodology to reduce the dynamics of the corresponding flows to a low dimensional dynamical system. We propose to expand the unsteady solution in a basis consisting of some of the most unstable eigenmodes. After substitution of this expansion in the governing equations, obtaining the dynamical system requires taking the scalar product with the orthogonal functions which are the eigenmodes of the adjoint of the Jacobian. It is also well known that these adjoint eigenmodes are instrumental for the purpose of controlling the flow [12].

This paper is organized as follows : we will present in the next section the equations that govern the flows and configurations in which we are interested. The following three sections will be devoted to the description of the numerical algorithms needed throughout this work. First the computation of branches of steady solutions will be addressed and its efficiency will be illustrated by computing the solutions on the steady branch for the differentially heated cavity of aspect ratio 4 for values of the Rayleigh number more than one order of magnitude above the critical value and for a cavity with internal heated partitions representative of cooling of electronic equipment. The following section will address the determination of the spectrum of the Jacobian of the Navier-Stokes equations linearized about the solutions on the steady branch. Illustrative examples will be given for the corresponding configurations. Lastly, we will address the determination of the adjoint Jacobian spectrum and the principle of generating the dynamical system will be discussed.

2 Governing Equations

The class of flows in which we are interested are modelled with the Navier-Stokes equations in the Boussinesq approximation. The fluid is characterized by constant kinematic and thermal diffusivities ν and κ respectively, assumed to be constant. It is enclosed in a two-dimensional cavity of height H and width ℓ, corresponding to a vertical ratio aspect $\mathrm{R}_F = \frac{H}{\ell}$. The corresponding equations are made dimensionless by introducing $L_{ref} = H$, $\mathrm{V}_{ref} = \frac{\kappa}{H} Ra^{1/2}$ and $\mathrm{t}_{ref} = \frac{H^2}{\kappa} Ra^{-1/2}$ as reference quantities for length, velocity and time where Ra is the Rayleigh number based on $H (= \frac{g\beta H^3 \Delta T}{\nu\kappa})$ and $Pr (= \frac{\nu}{\kappa})$ is the fluid Prandtl number. The dimensionless temperature is defined as $\Theta = \frac{T - T_1}{T_2 - T_1}$ where T_2 and T_1 are the hot and cold temperatures applied at cavity walls. The governing equations can then be cast in the following non-dimensional form:

$$\frac{\partial u}{\partial x}+\frac{\partial w}{\partial z}=0 \tag{1}$$

$$\frac{\partial u}{\partial t}+u\frac{\partial u}{\partial x}+w\frac{\partial u}{\partial z}=-\frac{\partial P}{\partial x}+\frac{Pr}{\sqrt{Ra}}\nabla^2 u \tag{2}$$

$$\frac{\partial w}{\partial t}+u\frac{\partial w}{\partial x}+w\frac{\partial w}{\partial z}=-\frac{\partial P}{\partial z}+\frac{Pr}{\sqrt{Ra}}\nabla^2 w+Pr\Theta \tag{3}$$

$$\frac{\partial \Theta}{\partial t}+u\frac{\partial \Theta}{\partial x}+w\frac{\partial \Theta}{\partial z}=\frac{1}{\sqrt{Ra}}\nabla^2 \Theta \tag{4}$$

where t is the time, u and w are the velocity components in the x (horizontal) and z directions respectively and P the dimensionless pressure. In these equations, ∇^2 is the laplacian operator $(\frac{\partial^2}{\partial x^2}+\frac{\partial^2}{\partial z^2})$.

We have considered two geometrical configurations. The first one is the classical differentially heated cavity, which consists of an enclosure whose left wall is maintained at a hot temperature, the right one at a cold temperature whereas the walls at top and bottom are adiabatic. We will consider a cavity of aspect ratio $\frac{H}{\ell}=4$ which was investigated in detail in the past [13]. This will help demonstrate the efficiency of the methods described hereafter. The second geometrical configuration is representative of problems met in passive cooling of electronic equipment. It consists of a square cavity with four internal heated vertical plates featuring electronic boards. The internal plates are heated while the outer vertical walls are cold, the top and bottom walls being adiabatic. This will help illustrate the capability of the methodology for (relatively) complex geometries. For brevity, this configuration will not be discussed here. Some representative results are available in [14].

3 Numerical Algorithms

As said in the introduction, investigating the stability of the solutions of the Navier-Stokes equations require the joint use of several algorithms either capable of integrating the unsteady equations or of solving the steady state equations.

3.1 Steady State Solving

Algorithm Description The steady state solutions satisfy equations (1)-(4) with $\frac{\partial}{\partial t}=0$. Let us call $U_0=(\overline{u},\overline{w},\overline{P},\overline{\Theta})$ such a steady state. In short-hand notation U_0 satisfies :

$$(L-N(U_0)).U_0=0 \tag{5}$$

with the constraint that the velocity field $(\overline{u},\overline{w})$ be divergence free. Solving these equations directly is not easy even when the corresponding solution is stable. Because they are nonlinear, iterative techniques must be used. Several algorithms have been developed for this purpose although the convergence of most of them degrades as the Rayleigh number increases *i.e.*when the equations become increasingly convection-dominated. Solving for an unstable steady state solution is far more difficult because iterative techniques based on pointwise relaxation generally fail to converge when the relaxation operator is not definite, and Newton's methods, long used in the finite element community but not so much in the context of finite differences or finite volumes discretizations, seem preferable. Starting from an initial guess U^0, one Newton step consists of computing an increment $\Delta U^k = U^{k+1} - U^k$ which is the solution of

$$(L - DN_{U^k}).\Delta U^k = -(L - N(U^k)).U^k \tag{6}$$

where $(L - DN_{U^k}) = \mathcal{L}_{U^k}$ is the Jacobian of the nonlinear operator about the current iterate U^k. It is again understood that the velocity components of the increment must be divergence-free. It is well known that this method converges quadratically once U^k gets sufficiently close to U_0.

While this procedure is simple in principle, its application to the incompressible equations raises many difficulties. The first one is due to the size of the Jacobian, typically $4 \times NI \times NJ$ (where NI and NJ are the number of grid points in each spatial direction) when dealing with the two velocity components, pressure and temperature. The order of the Jacobian can thus become prohibitively large for problems in which the instability occurs at large values of the Rayleigh number and which therefore need fine mesh sizes to be adequately resolved. Second it is often difficult to provide an explicit matrix representation of the Jacobian. Lastly, even if this Jacobian is made available, its inversion is virtually impossible due to its poor conditioning.

While the direct inversion of the full steady non-linear (5) or linearized (6) equations is almost impossible, it turns out that efficient schemes have been derived to solve the unsteady Stokes problem. This operator stems from the time discretization of the unsteady nonlinear equations (1)-(4) using an implicit timestepping scheme for the viscous diffusion terms coupled to an explicit treatment of the non-linear terms. Such a first order scheme can be written as

$$\frac{U^{n+1} - U^n}{\delta t} = L.U^{n+1} - N(U^n).U^n \tag{7}$$

where it is again understood that the velocity field which is part of U^{n+1} be divergence free. Equation (7) can be cast into the following form

$$(I - \delta t\, L)\, U^{n+1} = (I - \delta t\, N(U^n))\, U^n \tag{8}$$

Efficient schemes have been developed to integrate these equations either for the velocity-pressure formulation or for the 2D stream function-vorticity formulation. They are generally based either on the direct inversion of the Uzawa algorithm or on so-called influence matrix techniques that make use of the linear relationships between the pressure and the divergence of the velocity field at the boundary of the computational domain in the velocity-pressure formulation or between the vorticity and normal derivative of the stream function at the boundary in the stream function-vorticity formulation [15],[16], [17],[18]. These algorithms are generally restricted to simple geometries, as the building of the influence matrix can become complex for irregular geometries, in addition to the fact that most of these methods use direct methods for the inversion of the Helmholtz type equations. Due to the explicit treatment of the nonlinear terms these methods suffer from conditional timestep stability criteria which may require integrating for a large number of timesteps, in particular when approaching bifurcation points.

Tuckerman [2],[3] showed how to make use of such algorithms to compute quite easily the steady state solution. Let us call $E_{\delta t}\, U^n = \frac{U^{n+1}-U^n}{\delta t}$ where n and $n+1$ represent two consecutive time steps. From equation (8), one has

$$E_{\delta t}\, U^n = \frac{(I-\delta t\, L)^{-1}(I-\delta t\, N(U^n)) - I}{\delta t} U^n \tag{9}$$

$$= (I-\delta t\, L)^{-1}(L - N(U^n))\, U^n \tag{10}$$

and $E_{\delta t}\, U = 0$ has as roots the steady solutions of equation (5), since $(I-\delta t\, L)^{-1}$ is non singular. Applying Newton's method to $E_{\delta t}\, U = 0$ is equivalent to determining increments ΔU^k to the current iterate U^k that are solutions of the linear system

$$DE_{\delta t}\, \Delta U^k = -E_{\delta t}\, U^k \tag{11}$$

where $DE_{\delta t}$ is the Jacobian of $E_{\delta t}$ evaluated on U^k. It is easy to see that

$$DE_{\delta t} = (I-\delta t\, L)^{-1}(L - DN_{U^k}) \tag{12}$$

$$= \frac{(I-\delta t\, L)^{-1}(I-\delta t\, DN_{U^k}) - I}{\delta t} \tag{13}$$

where DN_{U^k} is the Jacobian of N evaluated on U^k. Starting with an initial guess U^0, for each Newton iteration the right-hand side of equation (11) is computed by integrating over one timestep the unsteady nonlinear equations and evaluating the increment from U^n to U^{n+1}. The increment ΔU^k can then be computed using matrix free methods (we have used GMRES [19]), which only require computing the action of the operator $DE_{\delta t}$ on a given vector. One such evaluation is equivalent to integrating over one timestep the unsteady Stokes problem.

Note that the timestep δt acts as a convergence parameter and should be chosen in order to improve convergence of the global iterative scheme. Numerous experiments have shown that δt should be taken very large (10^4 or larger). In this case the unsteady term in the unsteady Stokes operator becomes negligible and one essentially deals with the Stokes operator which is strongly elliptic. This remark turns out to be essential when solving the Stokes operator with iterative methods, which are needed if one wants to address relatively complex geometries.

Numerical Implementation We have chosen to discretize the equations using the classical staggered grid arrangement in which the computational domain is covered with grid cells. The scalar variables pressure and temperature are defined at cell centers whereas the velocity components are defined at the cell boundaries. If the domain is covered with $NI \times NJ$ cells in the x and z direction respectively, the total number of unknowns is thus $(NI + 1)NJ + NI(NJ + 1) + 2 \times NI \times NJ$. The diffusive and convective fluxes are discretized with centered differences.

As already said, a very efficient Stokes solver is needed, since it is called many times. Previous implementations have generally made use of direct Stokes solvers [2],[3],[4],[5]. For complex geometries, iterative Stokes solvers are more flexible and seem therefore preferable. A further requirement is that this solver remain efficient for high spatial resolutions since the problems we want to address are characterized by thin wall boundary layers that require good spatial resolutions as the Rayleigh number increases. Amongst all iterative methods, multigrid methods are known to achieve an efficiency which does not deteriorate as the spatial resolution increases [20],[21]. They work best for strongly elliptic operators and, in view of the remark made above, they therefore seem to be ideally suited for our purpose.

We have thus implemented the multigrid procedure proposed by Vanka [8]. On a given grid the local smoother consists of solving for each cell the linear system of order 5 made of 4 momentum equations (two in each direction) coupled to the continuity equation. Whereas we have been somewhat unsuccessful in the past in applying this method to the full non-linear equations at large Rayleigh number, its application here is much more straightforward. Since we are dealing with a linear problem a straightforward correction scheme can be used instead of the full multigrid algorithm FMG FAS used in the nonlinear case where the current solution must be defined on the coarse grids. We have chosen to define the linear operators directly on the coarse grids as the natural discretization of the operators on the corresponding grids. The restriction of the residuals of the momentum and continuity equations are defined as the weighted sums of the residuals. The extension of the corrections are defined as bilinear interpolations.

This algorithm is made to work for "complex" geometries by defining a phase indicator, defined at the same location as the scalar variables, which takes value 1 for a fluid cell and 0 for a solid cell. This phase indicator is defined on the fine grid and its values on the coarser grids are defined recursively in the following manner. Since a coarse cell corresponds to the clustering of 4 fine cells, it is assumed to be a solid cell iff all 4 fine cells are solid, otherwise it is considered a fluid cell. With the help of this phase indicator all the operators and their boundary conditions can be defined automatically on all grids.

Figure 1 below illustrates the efficiency of the steady solving algorithm in the case of the differentially heated cavity of aspect ratio 4. For all solutions the steady state solution was obtained in less than 15 Newton iterations.

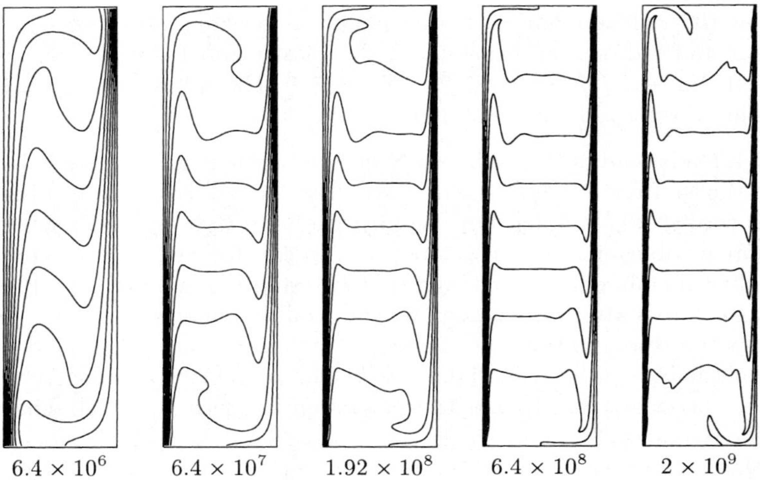

Figure 1 Temperature field for increasing values of Rayleigh number. All solutions for $Ra > 10^8$ are unstable.

4 Determination of the Jacobian Spectrum

4.1 General Considerations

The stability of any steady solution U_0 to infinitesimal perturbations is investigated by letting any solution be written as $U_0 + \tilde{U}$ where $\tilde{U} = (\tilde{u}, \tilde{w}, \tilde{p}, \tilde{\theta})$ is the perturbation vector. Inserting this expression into the unsteady equations and neglecting second order terms, yields

$$\frac{\partial \tilde{U}}{\partial t} = (L - DN_{U_0})\tilde{U} \tag{14}$$

where $(L - DN_{U_0}) = \mathcal{L}_{U_0}$ is the Jacobian of the Navier-Stokes equations evaluated at U_0. Since $\mathcal{L}_{U_0}$ is independent of time, solutions to equation (14) are of the form $\tilde{U}(x,z,t) = e^{\sigma t}\,\zeta(x,z)$. Inserting this expression into equation (14) shows that $\zeta(x,z)$ is an eigenvector of the Jacobian $\mathcal{L}_{U_0}$ corresponding to the eigenvalue σ. Since $\mathcal{L}_{U_0}$ is real, its eigenvalues are either real or complex conjugates. If all eigenvalues have negative real part the solution U_0 is linearly stable. If one eigenvalue at least is of positive real part then the solution U_0 is linearly unstable.

In bounded domains like those of interest here, the spectrum of the continuous Jacobian $\mathcal{L}_{U_0}$ consists of an infinite countable number of eigenmodes. After spatial discretization, the number of eigenmodes of the discrete Jacobian becomes finite and only part of these eigenmodes are good approximations of those of the continuous Jacobian. It is therefore useless, if possible at all, to compute all the spectrum of the discrete Jacobian since the eigenvalues of interest for stability investigations are those of maximum real part. Computing these leading eigenvalues is far from trivial. As previously stated, when the configuration at hand requires large spatial discretization direct methods are not usable, and one has to resort to iterative methods, which generally consist in applying repeatedly some linear operator on a given initial vector. It is indeed well known that applying repeatedly an operator on a given initial vector will eventually polarize the resulting vector sequence along the eigenvector corresponding to the eigenvalue of largest modulus. Thus, in order to determine the eigenvalues of maximum real part of the Jacobian, the general technique is to apply power iterate methods with a function of the Jacobian that will turn the leading eigenvalues of interest into those of maximum modulus of the function of the Jacobian (see [22]). The best candidate for this is obviously the exponential of the Jacobian but as previously stated this cannot be done in practice and one has to use an approximation of the exponential such as $(I + \delta t\,\mathcal{L}_{U_0})$ with $\delta t \ll 1$. Another possible candidate is an inverse power iterate with a shift that will make the eigenmode closest to the shift dominant [23],[24]. This method was successfully applied for real shifts [23] but still suffers from some unresolved problems for complex shifts [24].

We have used the ARPACK library developed by Sorensen [25] and Lehoucq [26] with the technique described just above to compute several eigenvalues and the corresponding eigenmodes. The ARPACK library [11] works on the basis of an Arnoldi-Krylov algorithm where an invariant subspace of specified dimension is determined through the repeated action of an operator acting on a sequence of vectors. It offers the possibility to specify the desired eigenmodes such as those of maximum real part, or those of maximum modulus. We have tried two possibilities: either using directly the Jacobian $\mathcal{L}_{U_0}$ requiring the eigenvalues of maximum real part or using $I+\delta t\,\mathcal{L}_{U_0}$, (where the timestep δt has been set to $\simeq 10^{-3}$), requiring the eigenvalues of maximum modulus.

In either case, the explicit action of the Jacobian on a given vector was performed in two steps. In the first step, the spatial part was computed explicitely, *i.e.* the discretized form of the operator $-U_0\nabla . - .\nabla U_0 + \nabla^2$ was applied on a vector consisting of $(\tilde{u}, \tilde{w}, \tilde{\theta})$. In matrix notation this reads :

$$\begin{pmatrix} -U_0\nabla - \cdot\frac{\partial \overline{u}}{\partial x} + \frac{Pr}{\sqrt{Ra}}\nabla^2 & -\cdot\frac{\partial \overline{u}}{\partial z} & 0 \\ -\cdot\frac{\partial \overline{w}}{\partial x} & -U_0\nabla - \cdot\frac{\partial \overline{w}}{\partial z} + \frac{Pr}{\sqrt{Ra}}\nabla^2 & Pr \\ -\cdot\frac{\partial \overline{\Theta}}{\partial x} & -\cdot\frac{\partial \overline{\Theta}}{\partial z} & -U_0\nabla + \frac{1}{\sqrt{Ra}}\nabla^2 \end{pmatrix} \begin{pmatrix} \tilde{u} \\ \tilde{w} \\ \tilde{\theta} \end{pmatrix}$$

where $-U_0\nabla$ stands for $-\overline{u}\frac{\partial}{\partial x} - \overline{w}\frac{\partial}{\partial z}$. This step is followed by a projection step in which the resulting velocity field is projected onto the space of divergence free vector fields. This projection $\mathcal{P}$ reads $\mathcal{P} = I - \nabla(\nabla.\nabla)^{-1}\nabla.$ and is performed through a scalar multigrid algorithm for the computation of the pressure.

Either evaluation thus reads $\mathcal{P}\,\mathcal{M}_{U_0}$ or $\mathcal{P}\,(I + \delta t\,\mathcal{M}_{U_0})$ where $\mathcal{M}_{U_0}$ stands for the matrix just above. We have found that the use of $\mathcal{P}\,(I + \delta t\,\mathcal{M}_{U_0})$ requiring the eigenvalues of maximum modulus works much better than the other alternative. Note that this is equivalent to performing an explicit timestepping of the linearized Navier-Stokes equations. The reason why we have not used a straightforward modification of the implicit-explicit timestepping scheme (7) is that the corresponding timestepping scheme would not be a polynomial of $\mathcal{L}_{U_0}$. Although the eigenvalues would yield good approximations, the corresponding eigenvectors would not be exactly those of $\mathcal{L}_{U_0}$ and this is important for later purpose.

Figure 2 shows the 100 most unstable eigenmodes of the Jacobian for the differentially heated cavity corresponding to four values of the Rayleigh number. This figure shows that the number of unstable eigenmodes increases with increasing Rayleigh number.

5 Determination of the Adjoint Jacobian Spectrum

5.1 General Considerations

Our motivation for computing the spectrum and eigenvectors of the adjoint of the Jacobian is twofold. Firstly it is known that the eigenmodes of the adjoint contain the spatial information concerning the localization of the maximum sensivity of the corresponding eigenmode of the Jacobian [12, 27]. This has obvious implications in active control, where it is important to determine the optimum localization of forcing terms. Our second motivation is to develop a method to reduce the Navier-Stokes equations to a low order differential system able to represent the dynamics of the flow at low cost [28, 29]. This also has obvious applications in active control in order to dispose of a fast predictive dynamical model that one can use in a feedback loop. To this aim, we propose to express the solution $U(x, z, t)$ of the Navier-Stokes equations at a given Rayleigh number as

$$U(x, z, t) = U_0(x, z) + \tilde{U}(x, z, t) \tag{15}$$

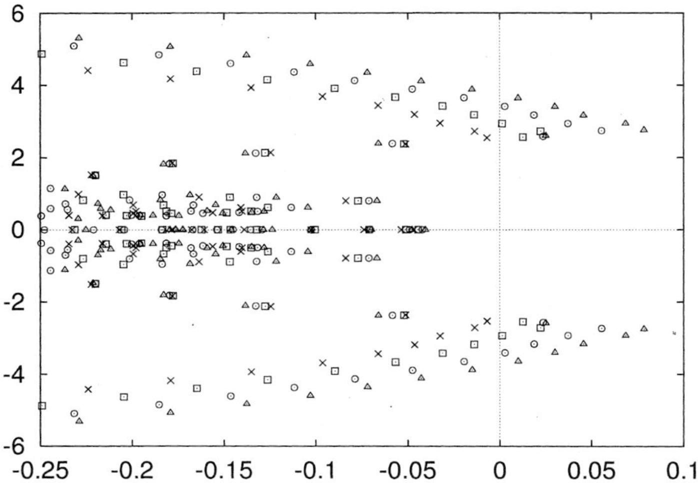

Figure 2 Jacobian spectrum for a differentially heated cavity of aspect ratio 4 (spatial resolution 128 × 256) : × : $Ra = 1.024 \times 10^8$; □ $Ra = 1.28 \times 10^8$; ∘ : $Ra = 1.6 \times 10^8$; △ $Ra = 1.92 \times 10^8$.

where $U_0(x, z)$ is the stable or unstable steady solution for the given Rayleigh number. The nonlinear Navier-Stokes equations can then be written as:

$$\frac{\partial \tilde{U}}{\partial t} = (L - DN_{U_0})\tilde{U} - \tilde{U}.\nabla\tilde{U} \qquad (16)$$

$$= \mathcal{L}_{U_0}\tilde{U} - \tilde{U}.\nabla\tilde{U} \qquad (17)$$

We decompose $\tilde{U}(x, z, t)$ into the form:

$$\tilde{U}(x, z, t) = \sum_k A_k(t)\, \zeta_k(x, z) \qquad (18)$$

where the $\zeta_k(x, z)$'s are selected eigenvectors of the Jacobian $\mathcal{L}_{U_0}$ and the $A_k(t)$'s are then the expansion coefficients. Inserting this expansion into the non-linear Navier-Stokes equations yields

$$\sum_k \frac{dA_k(t)}{dt}\zeta_k(x, z) = \sum_k \sigma_k A_k(t)\zeta_k(x, z) - \sum_{l,m} A_l A_m \zeta_l(x, z).\nabla\zeta_m(x, z) \qquad (19)$$

where σ_k is the eigenvalue corresponding to $\zeta_k(x, z)$.

In order to arrive at the differential system, one must take the scalar product of equation (19) with a set of $\xi_i(x, z)$ which is bi-orthogonal to $\zeta_k(x, z)$, ie

$$(\xi_i(x, z), \zeta_k(x, z)) = \delta_{ik} \qquad (20)$$

where $(f,g) = \int \overline{f} g$ is the hermitian scalar product. Since $\mathcal{L}_{U_0}$ is not self-adjoint, its eigenmodes are not orthogonal, and the bi-orthogonal set are the eigenmodes $\xi_k(x,z)$ of the adjoint of the Jacobian. Upon taking the scalar product of equation (19) with $\xi_k(x,z)$, one obtains the nonlinear differential system:

$$\frac{dA_k}{dt} = \sigma_k A_k - \sum_{l,m} \mu_{klm} A_l A_m \tag{21}$$

where $\mu_{klm} = (\xi_k(x,z), \zeta_l(x,z).\nabla\zeta_m(x,z))$. Note that this expression is exact and does not depend on any assumption related to the amplitude of the coefficients.

The adjoint operator is defined relative to the inner product $(f,g) = \int \overline{f} g$ such that

$$(V, \mathcal{L}U) = (\mathcal{L}^* V, U) \tag{22}$$

We recall here that $\mathcal{L}$ and $\mathcal{L}^*$ are characterized by the same set of eigenvalues. In expanded form, the definition of the adjoint $\mathcal{M}^*_{U_0}$ for the spatial part $\mathcal{M}_{U_0}$ reads ;

$$\left(\tilde{u}^* \; \tilde{w}^* \; \tilde{\theta}^* \right) (\mathcal{M}_{U_0}) \begin{pmatrix} \tilde{u} \\ \tilde{w} \\ \tilde{\theta} \end{pmatrix} = \left(\tilde{u} \; \tilde{w} \; \tilde{\theta} \right) (\mathcal{M}^*_{U_0}) \begin{pmatrix} \tilde{u}^* \\ \tilde{w}^* \\ \tilde{\theta}^* \end{pmatrix} \tag{23}$$

Using integration by part and invoking the fact that the velocity fields are divergence free, $\mathcal{M}^*_{U_0}$ can be written in continuous form as :

$$\begin{pmatrix} U_0 \nabla - \cdot \frac{\partial \overline{u}}{\partial x} + \frac{Pr}{\sqrt{Ra}} \nabla^2 & - \cdot \frac{\partial \overline{w}}{\partial z} & - \cdot \frac{\partial \overline{\Theta}}{\partial x} \\ - \cdot \frac{\partial \overline{u}}{\partial z} & U_0 \nabla - \cdot \frac{\partial \overline{w}}{\partial z} + \frac{Pr}{\sqrt{Ra}} \nabla^2 & - \cdot \frac{\partial \overline{\Theta}}{\partial z} \\ 0 & Pr & U_0 \nabla + \frac{1}{\sqrt{Ra}} \nabla^2 \end{pmatrix}$$

5.2 Numerical Implementation

It would be tempting to discretize directly the above operator, which would then be used like in the direct method, followed by a projection step to extract the divergence part of the velocity. This is not the correct way to proceed since for the adjoint eigenmodes $\xi_i(x,z)$ to be orthogonal to the $\zeta_k(x,z)$'s, the $\xi_i(x,z)$'s must be the eigenmodes of the *adjoint of the discrete Jacobian* that was used to compute the eigenmodes $\zeta_k(x,z)$'s [30]. There are thus two issues related to this difficulty. One is linked to the sequence in which these two steps are performed while the other is linked to the spatial discretization itself.

Assume for instance that the eigenmodes of the Jacobian have been computed using the operator $\mathcal{P}\,(I + \delta t\, \mathcal{M}_{U_0})$. The formal adjoint of this operator is $(I + \delta t\, \mathcal{M}^*_{U_0}))\mathcal{P}^t$ where the superscript t indicates transposition and it is the vector that comes from the action of the above operator on a previous iterate that has to be used to built the Krylov subspace.

Let us now deal with the spatial discretization. It is known that the laplacian is self-adjoint with respect to the weighted scalar product

$$(f,g)_d = \sum_i \overline{f(x_i)} g(x_i) dx_i$$

where dx_i is the mesh size, even non uniform. As for the nonlinear terms, let us briefly illustrate the computation of the discrete adjoint, for the temperature field for instance. The classical space-centered expression resulting from the finite volume integration of the convective term in conservative form $\frac{\partial u\theta}{\partial x}$ used for the Jacobian reads, with standard notations :

$$\overline{u}_{i+1/2} \frac{\tilde{\theta}_{i+1} + \tilde{\theta}_i}{2} - \overline{u}_{i-1/2} \frac{\tilde{\theta}_i + \tilde{\theta}_{i-1}}{2} \tag{24}$$

Multiplying this equation by $\tilde{\theta}_i^*$ and shifting indices to factor out $\tilde{\theta}_i$ results in the discrete adjoint form of the convective terms:

$$\frac{\overline{u}_{i+1/2}(\tilde{\theta}^*_{i+1} - \tilde{\theta}^*_i) + \overline{u}_{i-1/2}(\tilde{\theta}^*_i - \tilde{\theta}^*_{i-1})}{2} \tag{25}$$

which is merely the average value of the convective form $u\frac{\partial \theta^*}{\partial x}$ of the transport term evaluated at $x_{i+1/2}$ and $x_{i-1/2}$ [2]. Similar although more complex expressions, in particular on non-uniform meshes, can be derived for the convective terms of the momentum equations.

As for the projection $\mathcal{P}$, it is easy to see that it is self adjoint since, loosely speaking, $(\nabla)^t = (\nabla .)$. This demonstration requires appropriate functional analysis. It is guaranteed here because the use of the staggered grid arrangement yields pressure and velocity fields in compatible vector spaces.

The above procedure was tested on the differentially heated cavity. The adjoint eigenmodes were also computed using the ARPACK library. Requesting the same number of modes of largest modulus yielded the same set of eigenvalues as for the direct problem. The eigenvalues were found equal to within 10^{-10} and the two spectra can be superposed. These modes are defined to within a rotation in the complex plane (if $\xi_k(x,z)$ is an eigenmode, $\exp(i\phi)\xi_k(x,z)$ is also an eigenmode). Although the two sets computed by ARPACK are indeed bi-orthogonal, appropriate rotations were applied to the adjoint eigenmodes in order to satisfy the orthonormality condition (20). For a set of almost 400 eigenmodes for the Jacobian and as many for the adjoint problem, the scalar products are equal to 1 or 0 with a precision better than 10^{-10}.

[2] Note that this expression is not conservative. As a consequence, the celebrated skew-symmetric form of the convective term, which is half the sum the two expressions (24) and (25), and reads $\overline{u}_{i+1/2}\tilde{\theta}_{i+1} - \overline{u}_{i-1/2}\tilde{\theta}_{i-1}$, although it conserves $\tilde{\theta}^2$ at the discrete level does no longer conserve $\tilde{\theta}$!

5.3 Example of Low Order Dynamical System

The adjoint eigenmodes can be used to compute the coefficients

$$\mu_{klm} = (\xi_k(x,z), \zeta_l(x,z).\nabla\zeta_m(x,z))$$

which appear in the nonlinear ordinary differential equations. We have performed these computations for a Rayleigh number value of 6.4×10^7 in the aspect ratio 4 cavity, that is when the steady solution is still stable, using the 100 modes of largest real part as expansion basis. The differential system (21) was integrated using the same time discretization scheme (second order backwards Euler for the time derivative coupled to explicit Adams-Bashforth for the nonlinear terms) as that used for the full Navier Stokes equations. The Navier-Stokes equations were integrated in fluctuating form, starting from a random initial field of small amplitude. The initial values for the coefficients of the differential system were obtained from the projection of the random field on the expansion basis which is done by taking the scalar product of this random field with the adjoint eigenmodes.

Figure 3 shows the comparison of the fluctuating temperature at a given monitoring point between the full nonlinear equations and the synthetized temperature corresponding to the differential system during the transient leading back to steady state. Both signals are virtually undistinguishable except for very small times when the high frequency of the initial condition is not adequately represented on the eigenmode expansion basis. Work is in progress in the unstable case.

6 Conclusions

We have presented a general methodology for the investigation of the stability of natural convection flows in enclosures of complex geometry. This methodology consists of a steady state solver and of algorithms to compute the leading part of the spectrum of the Jacobian on a steady state solution and that of the adjoint of the Jacobian. We have implemented the steady state solver based on Newton's iterations with the help of a Stokes solver based on an iterative multigrid algorithm, which is capable of handling both rectangular and complex geometries, such as a cavity with internal plates. We have been able to follow the steady solution in a differentially heated cavity of vertical aspect ratio 4 for Rayleigh values more than one order of magnitude above the critical value. This has allowed us to confirm the asymptotic structure of the boundary layer regime.

Computation of the leading eigenmodes, those of largest real part of the Jacobian was performed through the use of the ARPACK library. We have shown that provided spatial resolution is used the critical Rayleigh number thus determined agrees very well with the accurate determination previously performed using a time dependent code using Chebyshev collocation. The

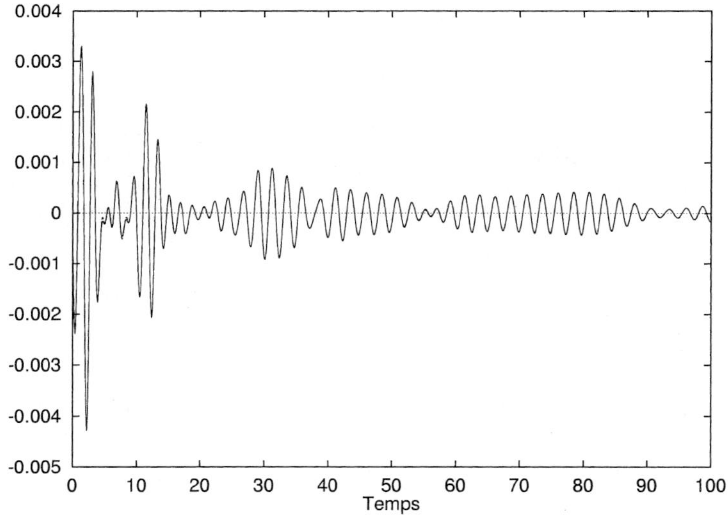

Figure 3 Time trace of fluctuating temperature for differentially heated cavity at $Ra = 6.4 \times 10^7$; nonlinear computation : full line ; dynamical model : dashed line

advantage of this new approach is the possible computation of several hundred eigenmodes. In the differentially heated cavity the oscillating eigenmodes can be classified into boundary layer modes and internal wave modes.

Care was taken to compute the eigenmodes of the adjoint of the discrete Jacobian, in order to ensure the orthogonality of the two sets to within machine accuracy. This allowed us to built a low order dynamical system by expanding the unsteady solution as the sum of the steady state solution and of a linear combination of selected eigenmodes. The comparison of the time evolution of the full nonlinear computation with the solution of the differential system was excellent in a stable case.

We are confident that this general methodology will prove useful in understanding transition to unsteadiness and chaos of internal flows as well as for controlling them.

Acknowledgements

We would like to thank Drs D.C. Sorensen and R.B. Lehoucq for providing a copy of the ARPACK library. The computations were performed on a C90 at IDRIS and on an SGI O2K located at CRIHAN. This work was carried out in the framework of the AMETH program.

References

[1] K.A. Cliffe: "Numerical calculations of two-cell and single-cell Taylor flows". J. Fluid Mech., **135**, 1983, pp. 219-233.

[2] L.S. Tuckerman: "Steady-state solving via Stokes preconditioning: Recurrence relations for elliptic operators". Lecture notes in Physics, edited by D.L. Dwoyer, M.Y. Hussaini, and R.G. Voigt, Springer, New York, 1989, pp. 573

[3] C.K. Mamum and L.S. Tuckerman: "Asymmetry and Hopf bifurcation in spherical Couette flow". Phys. Fluids, **7**, 1995, pp. 80-91

[4] S. Xin, P. Le Quéré, and O. Daube: "Natural convection in a differentially heated horizontal cylinder: Effects of Prandtl number on flow structure and instability". Phys. Fluids, **9**, 1997, pp. 1014-1033

[5] O. Daube, P. Le Quéré, N. Cousin-Rittemard and R. Jacques: "Influence of curvature on transition to unsteadiness and chaos of rotor-stator disk flows". submitted to J. Fluid Mech.

[6] A. Bergeon, D. Henry, H. BenHadid, and L. S. Tuckerman: "Marangoni convection in binary mixtures with Soret effect". J. Fluid Mech., **375**, 1998, pp. 143-177.

[7] E. Chenier: "Etude de la stabilité linéaire des écoulements thermocapillaires et thermogravitationnels en croissance cristalline". Thesis, Université Paris XI, LIMSI report 97-26, 1997

[8] S. P. Vanka: "Block-implicit multigrid solution of Navier-Stokes equations in primitive variables". J. Comp. Phys., **65**, 1986, pp. 138-158.

[9] Trefethen L.N.: "Pseudo-spectra of linear operators". SIAM Review, **39**, 1997, pp 383-406.

[10] A. Yu Gelfgat and I. Tanasawa: "Numerical analysis of oscillatory instability of buoyancy convection with the Galerkin spectral method". Numer. Heat Transfer A, **25**, 1994, pp. 627-648.

[11] R.B. Lehoucq, D.C. Sorensen and C. Yang: "ARPACK User's guide", S.I.A.M., Philadelphia, 1998

[12] D. C. Hill: "Adjoint systems and their role in the receptivity problem for boundary layers". J. Fluid Mech., **292**, 1995, pp. 183-204.

[13] P. Le Quéré: "Transition to unsteadiness, routes to chaos and simulation of turbulent flows in cavities heated from the side : a review of present status". Heat Transfer 1994: Proc. 10th Int. Heat Transfer Conf., Brighton, UK, **1**, 1994, pp. 281-296.

[14] E. Gadoin, P. Le Quéré and O. Daube: "A general methodology for investigating flow instabilities in complex geometries : application to natural convection in enclosures". Int. J. Num. Meth. Fluids, **37**, 2001, pp. 175-208.

[15] L. Kleiser and U. Schumann: "Treatment of incompressibility and boundary conditions in 3-D numerical spectral simulations of plane channel flows". Notes in Num. Fluid Mech., Hirschel Ed., Vieweg, **2**, 1980, pp. 165.

[16] P. Le Quéré and T. Alziary de Roquefort: "Computation of natural convection in 2-D cavities with Chebyshev polynomials". J. Comp. Phys., **57**, 1985, pp. 210-228.

[17] L.S. Tuckerman: "Divergence free velocity field in non-periodic geometries". J. Comp. Phys., **80**, 1989, pp. 403-441.

[18] O. Daube: "Resolution of the 2D Navier-Stokes equations in velocity-vorticity form by means of an influence matrix technique". J. Comp. Phys., **103**, 1992, pp. 402-414.

[19] Y. Saad and M.H. Schultz: "GMRES: a generalized minimal residual algorithm for solving nonsymetric linear systems". S.I.A.M. J. Sci. Stat. Comput., **7**, 1986, pp. 856-869.

[20] P. Wesseling: "An introduction to multigrid methods". Wiley, Chichester, 1992.

[21] W. Hackbusch: "Multigrid methods and applications". Springer Verlag, Berlin, 1985.

[22] K.N. Christodoulou and L.E. Scriven: "Finding leading modes of a viscous free surface flow: an asymmetric generalized eigenproblem", J. Sci. Comp., **3**, 1988, pp. 355-406.

[23] D. Barkley and L.S. Tuckerman: "Stokes preconditioning for the inverse power method", in Lecture Notes in Physics: Proc. of the Fifteenth Int. Conf. on Numerical Methods in Fluid Dynamics ed. by P. Kutler, J. Flores, and J.-J. Chattot (Springer, New York, 1997), 1997, pp. 75-76

[24] L.S. Tuckerman, F. Bertagnolio, O. Daube, P. Le Quéré and D. Barkley: "Stokes preconditioning for the inverse Arnoldi method". Notes on Numerical Fluid Mechanics, ed. by D. Henry and A. Bergeon, Vieweg, **74**, 2000, pp. 241-255.

[25] D.C. Sorensen: "Implicit Application of Polynomial Filters in a k-Step Arnoldi Method", SIAM J. Matr. Anal. App., **13**, 1992, pp. 357-385.

[26] R.B. Lehoucq: "Analysis and implementation of an implicitely restarted Arnoldi iteration", PhD Thesis, Rice University, 1995

[27] P. Luchini and A. Bottaro: "Görtler vortices: a backward-in-time approach to the receptivity problem". J. Fluid Mech., **363**, 1998, pp. 1-23.

[28] B.G. Allan: "A reduced order model of the linearized incompressible Navier-Stokes equations for the sensor/actuator placement problem", ICASE Report No. 2000-19, 2000

[29] H.M. Park and M.W. Lee: "Control of Navier-Stokes equations by means of mode reduction". Int. J. Numer. Meth. Fluids, **33**, 2000, pp. 535-557.

[30] M. B. Giles, and N. A. Pierce: "An introduction to the adjoint approach to design". ERCOFTAC Workshop on adjoint methods. Proc. to appear (ed. by Kluwer Academic Publishers), Toulouse, France, 21-23 June 1999.

Application of Unsteady Fluid-Structure Methods to Problems in Aeronautics and Space

A. DERVIEUX[1], B. KOOBUS[2], E. SCHALL[3], R.LARDAT[1], and C. FARHAT[4]

[1] INRIA, BP 93, 06902 Sophia-Antipolis (France),
Alain.Dervieux@inria.fr,
http://www-sop.inria.fr/tropics
[2] Univ Montpellier II, Dept Mathematiques, CC.051,
34095 MONTPELLIER Cedex 5 (France)
[3] IUT Genie Thermique et Energie,
1 avenue de l'Universite 64 000 PAU (France)
[4] Department of Aerospace Engineering Sciences ,
University of Colorado at Boulder ,
Campus Box 429, Boulder, CO 80309-0429 , USA

Summary

We are interested by the unsteady fluid-structure coupling with a compressible fluid model performed in a cooperation between an INRIA group in CFD and the Center for Aerospace structure of University of Boulder. A particular focus is put on the fluid part, which involves the MUSCL-FEM technology relying on a vertex centered upwind finite volume formulation for unstructured tetrahedrizations, and a mesh deformation technique combined with an Arbitrary Lagrange-Euler formulation. Three typical applications will then be presented.

1 Introduction

It is the role of laboratories to investigate which innovative methods will allow the building of affordable schemes addressing complete nonlinear algorithm for the calculation of complex flows coupled with structures. Such an investigation has been performed in a collaborative mode by CAS of university of Boulder and several teams in INRIA. We refer to [1] for a pedagogical presentation of the methods developed by the two teams.

The present paper proposes several demonstrative examples of compressible fluid-structure nonlinear analyses that have been performed with these new tools. These examples will prove, we hope so, that the algorithms under study are able to address these problems. They will also carry new problems, that we shall identify, and discuss. Some new solutions are then described and discussed.

2 Three-Field Transient Model

The problem of the motion of the fluid/structure interface that occurs in coupled aeroelastic problems is addressed by solving the fluid equations on deformable dynamic meshes. An Arbitrary Lagrangian Eulerian (ALE) [2],[3] formulation is used in order to perform the integration of the fluid equations on a moving mesh. The coupled aeroelastic problem to be solved can then be viewed as a three-field problem [1],[3]: the fluid, the structure and the dynamic mesh which is represented by a pseudo-structural system. The semi-discrete equations governing the three-way coupled problem can be written as follows:

$$\begin{aligned} \frac{\partial}{\partial t}(V(x,t)w(t)) + F^c(w(t),x,\dot{x}) &= R(w(t),x) \\ M\frac{\partial^2 q}{\partial t^2} + f^{int}(q) &= f^{ext}(w(t),x) \\ \tilde{M}\frac{\partial^2 x}{\partial t^2} + \tilde{D}\frac{\partial x}{\partial t} + \tilde{K}x &= K_c q \end{aligned} \tag{1}$$

where t designates time, x the position of a moving fluid grid point, w is the fluid state vector, V results from the finite element/volume discretization of the fluid equations, F^c is the vector of convective ALE fluxes, R is the vector of diffusive fluxes, q is the structural displacement vector, f^{int} denotes the vector of internal forces in the structure, f^{ext} the vector of external forces, M is the finite element mass matrix of the structure, $\tilde{M}$, $\tilde{D}$ and $\tilde{K}$ are fictitious mass, damping and stiffness matrices associated with the moving fluid grid and K_c is a transfer matrix that describes the action of the motion of the structural side of the fluid/structure interface on the fluid dynamic mesh.

3 Numerical Methodology

In this section, we give the main features of the numerical methods employed in this work for solving the coupled non-linear aeroelastic problem given by Eqs.(1) . For more details, the reader is invited to examine the references given in the text.

3.1 Discretization of Transient Non-Linear Aeroelastic Problems

Spatial discretization of fluid

The spatial discretization of the fluid equations is based on a finite element/finite volume formulation on unstructured meshes. It combines a Roe upwind scheme for computing the convective fluxes, and a Galerkin centered method for evaluating the viscous terms, see [4]. For addressing the problem of flow simulations on moving grids, an ALE formulation is incorporated in the flow solver. The numerical algorithms used with this ALE formulation satisfy the Geometric Conservation Laws

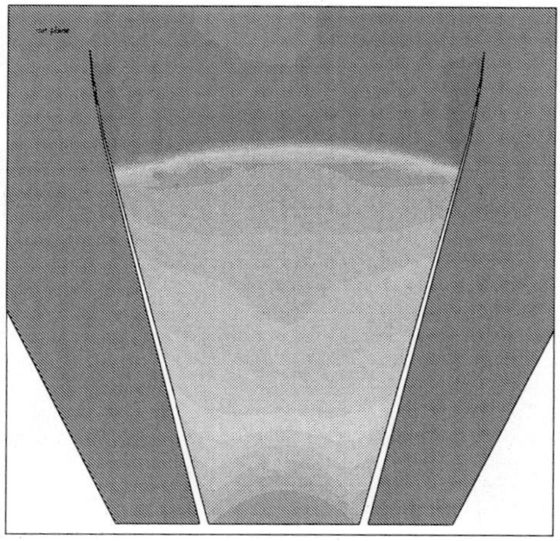

Figure 1 3D supersonic flow in a flexible nozzle; axial cut; nozzle inlet is the bottom section, the outlet and free flow field are near the top of the picture; we present Mach number contours (ranging from 1. to 4.); the strong shock inside the nozzle carries new types of instabilities.

(GCL) [5],[6] that govern flow computations on moving grids.
Second-order space accuracy is achieved through a piecewise linear interpolation method based on the MUSCL (Monotonic Upwind Scheme for Conservation Laws) procedure [7], with an upwind bias β.

The Roe splitting is the basis of the stabilisation process. In order to control the level of numerical viscosity, and following an idea of [8], a scalar coefficient γ is used to weight the stabilization term:

$$\begin{aligned} Flux(W_L, W_R, \boldsymbol{n}) &= 0.5\,(Flux(W_L, W_L, \boldsymbol{n}) + Flux(W_R, W_R, \boldsymbol{n})) \\ &- 0.5\,\gamma\,|J|(W_L - W_R) \end{aligned} \tag{2}$$

in which L stands for left, R for right, $\boldsymbol{n}$ is the normal vector to cell boundary and J the scalar product of the Jacobian of fluxes with the normal vector $\boldsymbol{n}$. Low Mach unsteady flows are taken into account thanks to adequate preconditioning of the upwind stabilization terms [9].

Time discretization of fluid

For solving accurately and efficiently the flow equations given by the first of Eqs.(1) on dynamic meshes, a second-order time-accurate implicit algorithm preserving the GCL [10] is employed. The time discretization is based on a second-order backward difference scheme. The non-linear flow equations derived from the time-discretization are solved by a defect-correction (Newton-like) method [11].

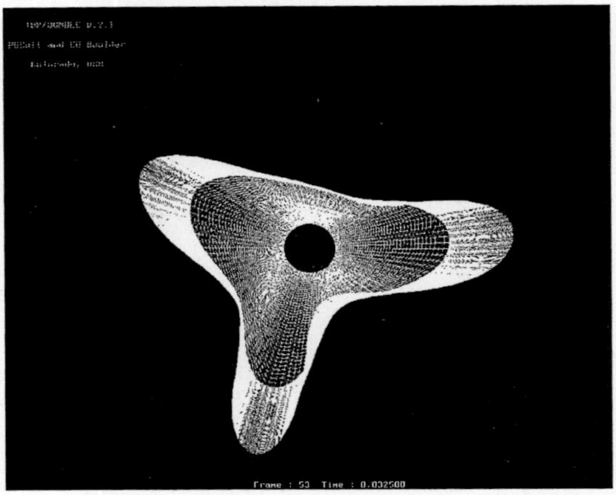

Figure 2 3D supersonic flow in a flexible nozzle, 3-lobe unstable mode; view from the bottom of the nozzle, the circle at center is the nozzle inlet, the deformed external boundary is the nozzle outlet; deformation is artificially amplified by a factor 16

Structure

The structure is represented by a finite element model, and its dynamics behavior is predicted using the true displacement, velocity and acceleration degrees of freedom. The structural equations of dynamic equilibrium given by the second of Eqs. (1) are solved with a second-order time-accurate implicit scheme where the trapezoidal method is used.

Coupled system

At selective time-steps of an aeroelastic simulation, the dynamic fluid mesh is updated to conform the most recently computed configuration of the structure. The points lying on the fluid/structure boundary are first adjusted to conform to the new position of the surface of the structure, then the remainder of the fluid grid points are repositioned accordingly. In the methodology used in this work, the new position of the interior grid points is determined from the displacement solution of a discrete pseudo-structural problem representing the unstructured dynamic fluid mesh. The pseudo-structural system is constructed by lumping a fictitious mass at each vertex of the fluid mesh and attaching fictitious lineal springs to each edge connecting two vertices [12],[3],[1]. In order to prevent possible collapse mechanisms of tetrahedra, the lineal springs can be combined with torsional springs ([13]). This discrete system is represented by the third of Eqs. (1) where $\tilde{M} = \tilde{D} = 0$.

Finally, in fluid/structure interaction problems the fluid and structure meshes have often non-matching discrete interfaces. In that case, we use the load and mo-

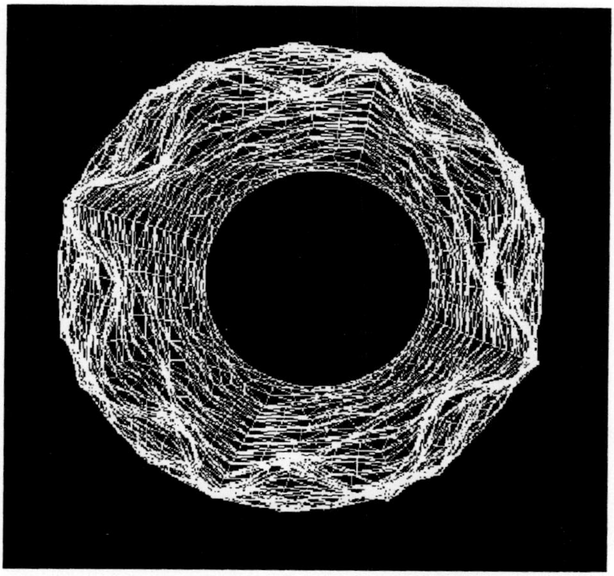

Figure 3 3D supersonic flow in a flexible nozzle, with fixed outlet: static structural deformation (amplified)

tion transfer algorithms described in [14] for evaluating properly the pressure forces on the surface of the structure, and transferring correctly the structural motion to the fluid mesh. In particular, the loads induced by the fluid on the structure are computed in a conservative way.

The solution of the coupled fluid/structure problem [1] is computed by a staggered solution procedure in the time domain [3],[15]. More precisely, we use the staggered algorithm given in [16] which satisfies the GCL as well as the continuity of both the displacement and velocity fields at the fluid/structure interface. This algorithm can be written as follows:

1. Using the mesh position $x^{n-\frac{1}{2}}$, and the mesh velocity $\dot{x}^n$ that matches the structural velocity $\dot{q}^n$ on the fluid/structure interface, update the mesh coordinates as follows:

$$x^{n+\frac{1}{2}} = x^{n-\frac{1}{2}} + \Delta t\dot{x}^n$$

2. Using $x^{n-\frac{1}{2}}$, $x^{n+\frac{1}{2}}$ and $\dot{x}^n$, update the fluid state vector $w^{n+\frac{1}{2}}$ in a manner that satisfies the GCL;
3. Using the pressure computed from $w^{n+\frac{1}{2}}$, compute q^{n+1} and $\dot{q}^{n+1}$ using the midpoint rule.

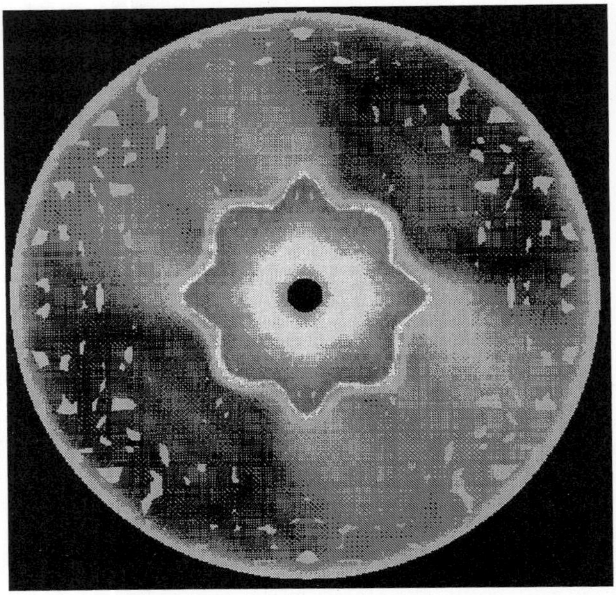

Figure 4 3D supersonic flow in a flexible nozzle, with fixed outlet: Mach distribution inside the deformed divergent

4 Applications

Although application of unsteady solvers for fluid-structure interaction is not a new topic, the progress of numerical methods is making many promises concerning the new type of outputs than can be yielded by this approach.

4.1 Deformations in a Rocket Divergent

In some crucial aeronautical applications such as wing flutter, some modes of the structure are able to pull energy from the fluid flow. We have experienced this kind of behavior with the deformations of a nozzle subject to the load of a supersonic jet ([17]). A first series of calculations showed the paramount role of the stiffening of the nozzle lip. Three-lobe clover-like deformations (Fig.2) are rapidly excited when a shock wave lies inside the nozzle (Fig.1). When the lip is assumed to be fixed, structural deformations become similar to buckling (Fig.3) with some spatial effects on the flow (Fig. 4) that also presents small temporal pulsations.

4.2 Deformation of a Launcher Body

The main quality of the CFD scheme is the accuracy of the consistent part, i.e. the scheme without stabilization. This part is very accurate on a very large family of meshes. Unfortunately, this part is not stable and the necessary stabilization for shock capturing can carry an enormous amount of numerical viscosity. The second example (presented in details in [18]) will illustrate this point.

The flow around a spatial launcher applies strong forces on the launcher body that can bend when it takes a not so large angle of attack (5 degrees; farfield Mach number is 0.9). The CFD model is Euler, the CSM model is a unique axial beam with structural properties equivalent to those of a more complex CSM model. We study the resulting static deformation (nonlinearly coupled steady flow and steady structure).

A preliminary study of C_x evaluation shows that the problem is not trivial, except, maybe, with really fine meshes. Two meshes (29000 vertices, 123000 vertices) are considered (cf. Fig.5). We also consider several values of parameter δ of the Roe solver stability term. The global numerical viscosity coefficient is the product of δ by the MUSCL upwind bias, β. We emphasize that we cannot compute with the option $\delta\beta = 0$. But we can extrapolate from more viscous computations for both meshes and we observe that fortunately both meshes give very similar extrapolated values (cf. Fig.6). The corresponding deformation is essentially a bending. We present some pictures of it in Fig.7.

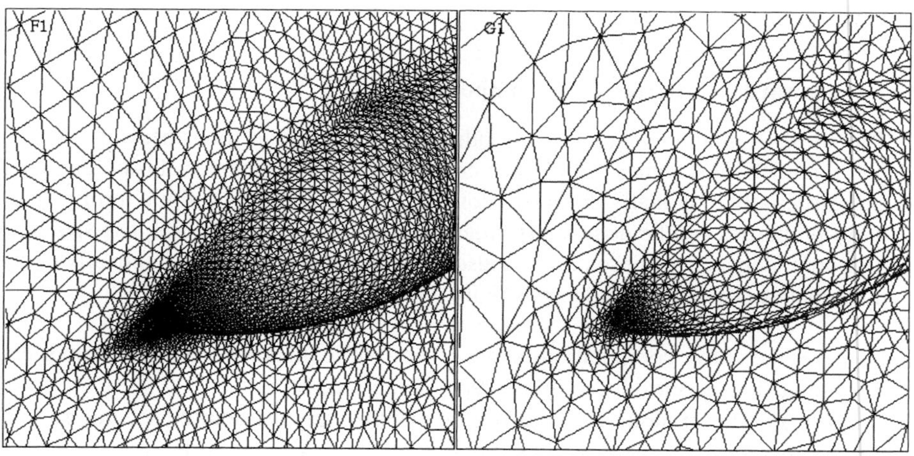

Figure 5 Two meshes for the launcher problem (skin mesh and vertical plan)

4.3 Acoustics Coupling in a Thrust Inverter

In some other examples, the coupling arises from unsteadiness that takes its origin from the fluid itself. In the 3D flow inside and out from an aircraft thrust inverter,

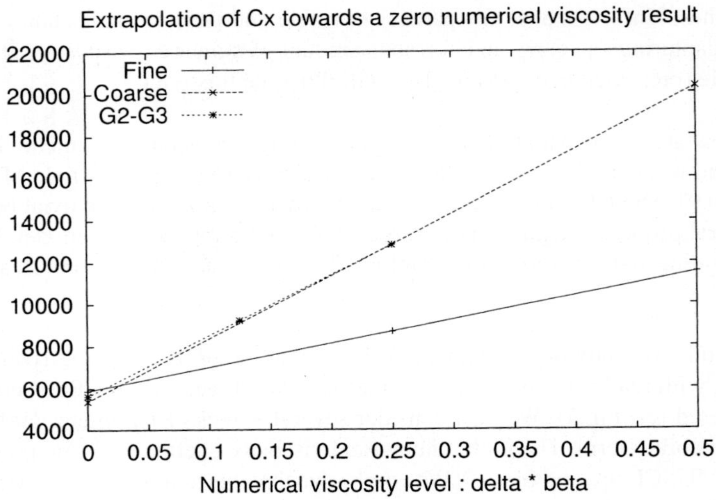

Figure 6 Launcher: computation of Cx for different meshes and numerical viscosities

see Fig.8), the sudden enlargement provokes vortex shedding that couples with low frequency acoustic waves. This can couple with the flap that deviates the flow, with some beating phenomenon (Fig.9). See also [19].

4.4 Towards a Smarter CFD Component

The above computations (and many others) show the good robustness of the proposed numerical schemes. Vertex-centered finite-volume schemes allow a compact and essentially diagonally dominant contribution to matrix by elliptic terms. Positivity properties can be deduced from angle conditions on the mesh. The spatially first-order convective approximation also contributes to essentially diagonally dominant matrix coefficients and to positivity through TVD mechanisms.

However the above computations show several difficulties related to the numerical viscosities that are introduced in this type of approximation:

- the amount of numerical viscosity to stabilize an unstructured scheme can be much larger due to the larger number of neighbors for a given node.

- TVD schemes ensuring positiveness are much more viscous than usual second-order schemes.

In practice, dissipation errors are dominant in both steady and unsteady cases. One way to measure that point is to observe how much central differenced schemes are more accurate when they can be applied more or less stably. The problem of upwind schemes is in fact much more important for the unsteady case. Improving these schemes is the motivation of many investigations in CFD laboratories.

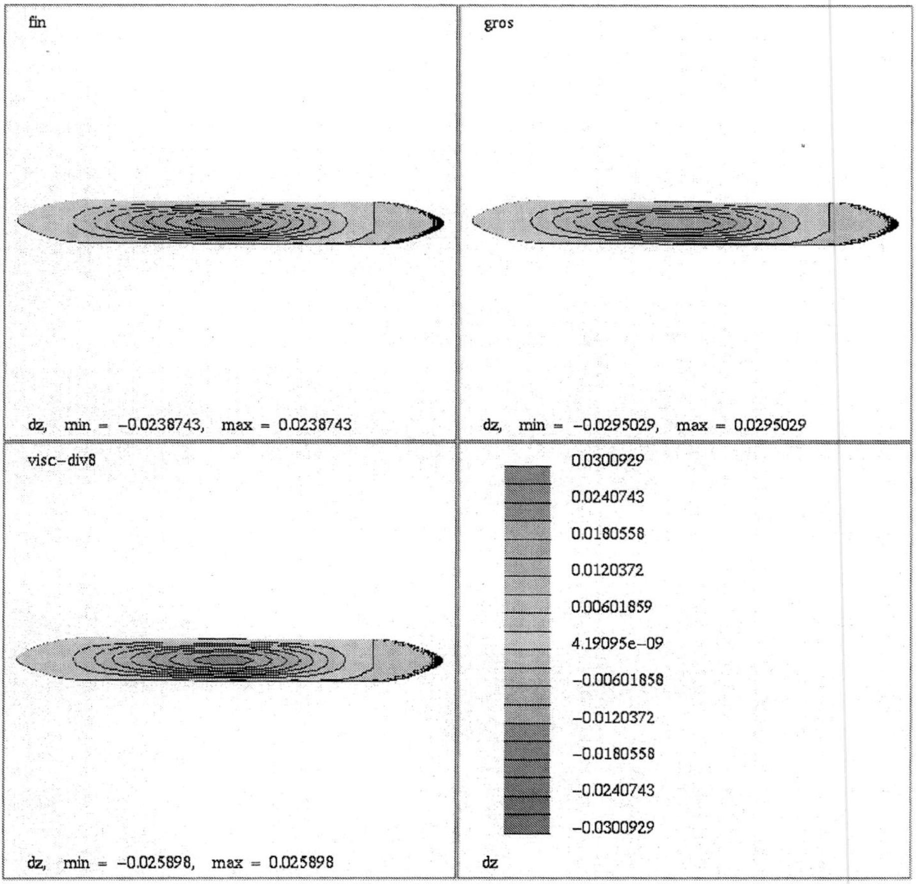

Figure 7 Spanwise deformation of the launcher for coarse grid, fine grid with mean level of numerical viscosity, fine grid with smaller level of numerical viscosity

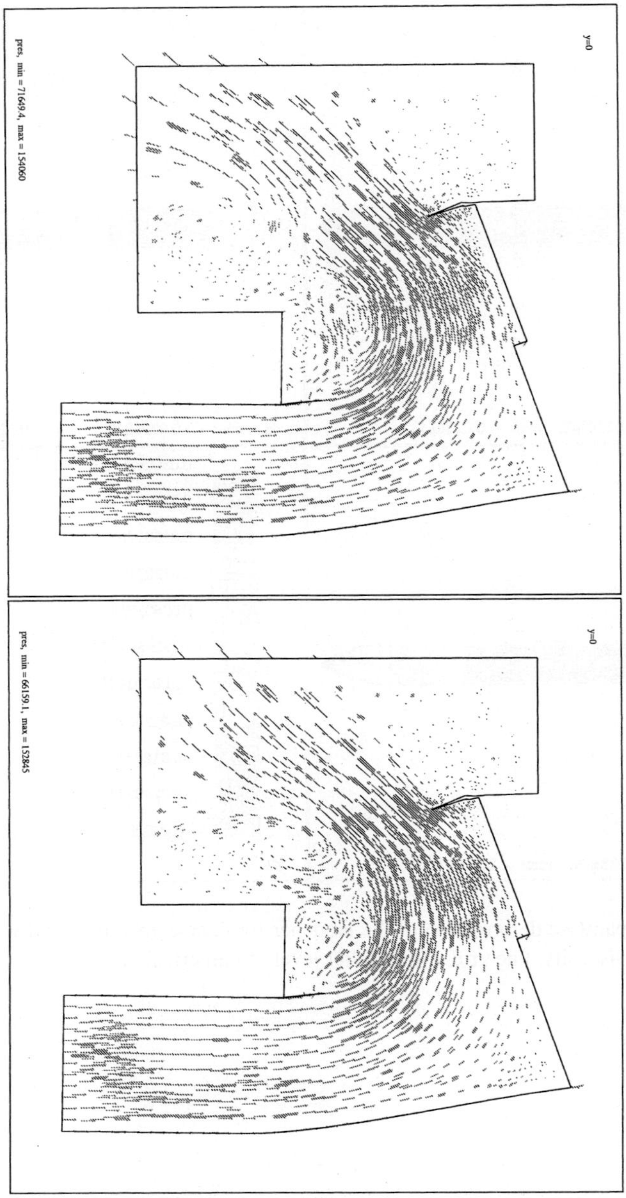

Figure 8 3D subsonic flow in a thrust inverter, section at symmetry plane $y = 0$, the velocity field is presented for two particular times of the periodic behavior (airflow from engine is getting in through vertical boundary at bottom-left, the upper part is the free flow field region; the flexible outlet flap under study is on right boundary at 2/3 of total height

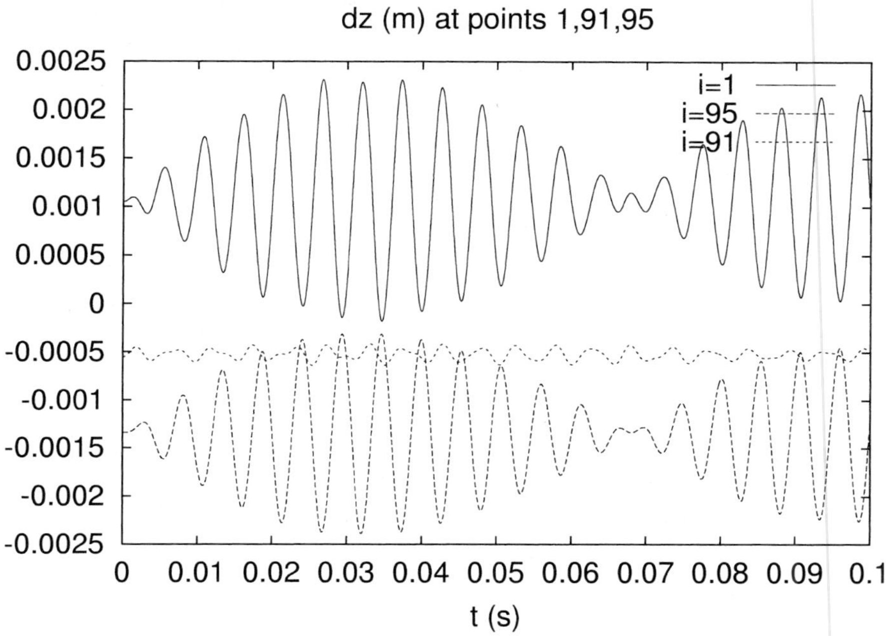

Figure 9 3D subsonic flow in a thrust inverter, motion of three particular points of the outlet flap; the mean position of each point give the static deformation due to mean flow; the point near symmetry axis does not oscillate much, the two other ones are is phase opposition; battements show that the acoustic excitation has not the same period as the structural mode that is excited

The strategy that we are now exploring consists in seeking first a better balance between the two different parts in truncation error, dissipation and dispersion, and, second, a more accurate design of a positive scheme.

Superconvergent scheme An option with a very low level of dissipation is available thanks to a new MUSCL-type formulation in which stabilization is obtained by a sixth order viscosity. The new scheme, an extension of [20], enjoys fifth-order accuracy on cartesian meshes and is much more accurate on unstructured meshes than usual second-order ones due to the very small impact of the new numerical dissipation on truncation error. This scheme is being applied to both acoustics ([21])and to LES ([22]).

Positiveness A new edge-based limiter inspired by [23] enjoys a strict density positiveness property for Euler computations. It is fully compatible with the superconvergence scheme and is today used for supersonic LES ([24])

5 Concluding Remarks

This paper has presented some recent applications of a now well established method for the nonlinear unsteady analysis of the interaction between a compressible fluid and a structure. The above method has been also applied to wing flutters, to aeroacoustic coupling in a rocket booster, see [25], and fast transients in closed buildings. We have also discussed the current evolution of the CFD component in order to answer to necessary improvements or to new needs. In particular, the need of better stabilisation with more accurate results is at the center of our concern. Answers are investigated with higher-order upwinding and more sophisticated TVD methods.

At least two future extensions of the presented method are very challenging for the numerical CFD component. Firstly, in a rather classical direction, we wish to compute in a near future unsteady Reynolds Averaged Navier Stokes flows with accurate capture of boundary layers. This demands a high quality in both approximation and in solution algorithm. Secondly, there is a need of investigation of unsteady flows where unsteadiness is directly produced by turbulence. One way to attack this difficulty is to perform Large Eddy Simulations, which also require improved approximations.

Acknowledgements

This work was in part performed inside the INRIA programme "Aerostructure" and supported by CNES Aérospatiale, Hispano-Suiza and SEP.

References

[1] C. Farhat , "High Performance Simulation of Coupled non-linear Transient Aeroelastic Problems", AGARD Report R-807, Special Course on Parallel Computing in CFD (l'A rodynamique numérique et le calcul en parallèle, North Atlantic Treaty Organization (NATO), October 1995.

[2] Donea J., "An Arbitrary Lagrangian-Eulerian Finite Element Method for Transient Fluid-Structure Interactions", Comput. Meths. Appl. Mech. Engrg. 1982, **33**, p. 689-723

[3] C. Farhat , M. Lesoinne and N. Maman, "Mixed Explicit/Implicit Time Integration of Coupled Aeroelastic Problems: Three-Field Formulation, Geometric Conservation and Distributed Solution", Internat. J. Numer. Meths. Fluids **21** (1995), p. 807-835

[4] A. Dervieux, "Steady Euler Simulations Using Unstructured Meshes", Von Karman Institute Lecture Series, 1985.

[5] M. Lesoinne and C. Farhat, "Geometric Conservation Laws for Aeroelastic Computations Using Unstructured Dynamic Meshes", AIAA Paper 95-1709, 12th AIAA Computational Fluid Dynamics Conference, June 19-22, 1995, San Diego, CA.

[6] B. Nkonga and H. Guillard, "Godunov Type Methods on Non-Structured Meshes for Three-Dimensional Moving Boundary Problems", Comp. Methods Appl. Mech. Engrg. **113** (1994) p. 183-204

[7] B. Van Leer, "Towards the Ultimate Conservative Difference Scheme V: a Second-Order Sequel to Godunov's Method", J. Comp. Phys. 1979, **32**, p. 361-370

[8] R. Carpentier, Approximation d'écoulements instationnaires. Application à des instationnarités tourbillonnaires. Thesis, Nice, 1995

[9] E. Schall, C. Viozat, B.Koobus and A. Dervieux, "Low Mach thermical flows with implicit upwind methods", INRIA report, to appear, 2002

[10] B. Koobus and C. Farhat, "Second-Order Time-Accurate and Geometrically Conservative Implicit Schemes for Flow Computations on Unstructured Dynamic Meshes", Computer Methods in Applied Mechanics and Engineering, **170**, p. 103-130 (1999).

[11] R. Martin and H. Guillard, "A Second-Order Defect Correction Scheme for Unsteady Problems", Comput. and Fluids 1996, **25**, p.9-27

[12] J.T. Batina, "Unsteady Euler Airfoil Solutions Using Unstructured Dynamic Meshes", AIAA paper 89-0115, AIAA 27th Aerospace Sciences Meeting, Reno, Nevada, January 9-12, 1989.

[13] C. Farhat, C. Degand, B. Koobus and M. Lesoinne, Torsional Springs for Two-Dimensional Dynamic Unstructured Fluid Meshes, Computer Methods in Applied Mechanics and Engineering, **163**, pp. 231-245, 1998.

[14] C. Farhat, M. Lesoinne and P. Le Tallec, "Load and Motion Transfer Algorithms for Fluid/Structure Interaction problems with Non-Matching Discrete Interfaces: Momentum and Energy Conservation, Optimal Discretization and Application to Aeroelasticity", Computer Methods in Applied Mechanics and Engineering, **157**, pp. 95-114, 1998.

[15] C. Farhat, M. Lesoinne, P. Stern and S. Lanteri, "High Performance Solution Procedure of Three-Dimensional Non-Linear Aeroelastic Problems via Parallel Partitioned Algorithms: Methodology and Preliminary Results", Advances in Engineering Software,1997, **28**, p. 43-61

[16] Farhat C. and Lesoinne M., "On the Accuracy, Stability and Performance of the Solution of Three-Dimensional Non-Linear Transient Aeroelastic Problems by Partitioned Procedures", AIAA Paper 96-1388, 37th AIAA/ASME/ASCE/AHS /ASC Structures, Structural Dynamics and Materials Conference, April 18-19, 1996, Salt Lake City, UT.

[17] E. Schall, R. Lardat, A. Dervieux, B. Koobus and C.Farhat," Aeroelastic Coupling between a Thin Divergent and high pressure jets", Revue Europeenne des Elements Finis, **9**,6-7 , 835-851,2000

[18] R. Lardat, B. Koobus, F. Ruffino, C. Farhat and A. Dervieux, "Premières investigations du couplage fluide-structure autour d'un lanceur spatial générique", INRIA report 4314, nov. 2001 (in french)

[19] R. Lardat, B. Koobus, E.Schall, A. Dervieux and C. Farhat, "Analysis of a Possible Coupling in a Thrust Inverter", Revue Europeenne des Elements Finis, **9**,6-7 ,819-834,2000

[20] C. Debiez and A. Dervieux, " Mixed Element Volume MUSCL methods with weak viscosity for steady and unsteady flow calculation". Computer and Fluids, 1999,**29**, 89-118

[21] I. Abalakin, T. Kozubskaya, A. Dervieux A vertex centered high order muscl scheme applying to linearised Euler acoustics, INRIA report, 2002

[22] S. Camarri, B. Koobus, M. Salvetti and A. Dervieux, "Numerical diffusion based on high-order derivatives in MUSCL schemes for LES on unstructured grids", in the proceedings of DLES-4, Direct and Large-Eddy Simulation-IV, July 18-20,2001, Twente (Holland), and INRIA report, 2002

[23] P.-H. Cournede, Ch. Debiez and A. Dervieux "A positive MUSCL scheme for triangulations" INRIA report 3465, 1998

[24] Y. Hoarau, S. Bourdet, M. Braza, B. Koobus, A. Dervieux, LES for supersonic flows with unstructured tetrahedrizations, INRIA report (2002) and IUTAM Symposium on unsteady separated flows, april 8-12, 2002, Toulouse, France

[25] R. Lardat, R. Carpentier, B. Koobus, E. Schall, A. Dervieux, C. Farhat, J.-F Guery. and P. Della Pieta, "Interaction Between a pulsating flow and a perforated membrane", Revue Europeenne des Elements Finis, **9**,6-7 ,805-817,2000

Numerical Simulation in Aeroelasticity For Complex Configurations in a Transonic Flow

J.-P. GRISVAL
ONERA, DDSS/MS, BP 72, 92322 Châtillon Cedex, France

Summary

This paper presents numerical approaches for computing fluid/structure interactions in a transonic flow using Euler/Navier-Stokes solvers. The first method concerns the flutter simulation using the linearized Euler equations in the frequency domain. The second method concerns the direct fluid structure coupling in the time domain.

The Euler/Navier-Stokes equations are solved on a multi-block structured mesh and are based on an ALE formulation. Specific tools are developed for the mesh deformations. A Finite Element Formulation is used to model the structural part of the aeroelastic problem. The use of a modal reduction for the structure and a dual timestep technique to simulate the unsteady aerodynamics reduces considerably the CPU times without detracting from the accuracy and the robustness of the simulations. These numerical methods are applied for static aeroelasticity and flutter prediction for complex civil and military configurations in transonic flow. The tools are tested for a Limit Cycle Oscillations (LCO) problem in 2D.

Keywords:
Aeroelasticity, Aircraft, Fluid-Structure interaction, Numerical Methods, Transonic Flow.

1 Introduction

A large number of parameters are required to predict the aeroelastic stability of aircraft and missiles in order to cover the complete flight envelope and all the structural and weight configurations of a project. The volume of computations needed is such that the linear numerical aerodynamic methods are widely used to predict aeroelastic stability, calculate loads and determine the response to atmospheric turbulence. These methods are developed to take flight controls into account, including their non-linearities (saturations, delays), as well as the structural nonlinearities encountered on fixed-wing aircraft and missiles (play and friction). Such methods are generally based on solving the linearized velocity or acceleration potential equations by integral methods. The steady and unsteady flows are independent and steady flow calculations are not needed to perform the aeroelastic stability prediction. These methods require meshing of the aerodynamic surfaces alone and allow us to evaluate the unsteady generalized aerodynamic forces, generally in the frequency domain, for an acceptable implementation and CPU cost. The use of such methods is amply justified for subsonic and supersonic flows where viscous effects are negligible. In this approach and in the case where the structure itself exhibits linear behavior, the coupled aerodynamic-structure system is solved in the frequency domain [9].

The nonlinear aspect of the aerodynamics in transonic flow and the dependence of the steady and unsteady flows restrict the use of these methods. However these methods are used with empirical corrections. For aeroelasticity in transonic flows, numerical methods solving the transonic small perturbation equation [1, 5, 15, 19] were widely developed and used in the 1980's (XTRAN3S, CAP-TSD, PPT3D). The 3D Viscous-Inviscid Interaction approach was developed for transonic separated flow [23]. Unsteady full potential solvers including boundary layer correction have also been developed [13]. Unsteady 3D Euler and Navier-Stokes solvers were developed using structured or unstructured grids to solve the aeroelasticity problems [12, 16, 17, 18, 20, 22, 25, 26, 34].

The Structural Dynamics and Coupled Systems Department (DDSS) of ONERA develops two approaches, mainly for aircraft transonic aeroelasticity problems. The first is based on solving the linearized Euler equations in the frequency domain, possibly taking steady viscous effects into account [27, 28, 30]. The second concerns direct static and dynamic fluid-structure coupling in the time domain using either a modal reduction for the structure or a complete structural model [13, 29]. With this approach, use of the dual timestep, as described herein, to simulate the unsteady fluid flow represents a major progress in aeroelastic simulation. Below, we briefly review the aerodynamic model and the mechanical model used and describe the fluid-structure coupling strategies developed. An application of the tools to a wing model for which we have an experimental database is presented. Then, the tools are applied for complex configurations of civil and military aircraft and for the Limit Cycle Oscillations prediction.

2 Aerodynamic Model

2.1 Unsteady aerodynamics

The CANARI code [7, 35] and elsA code developed by the Computational Fluid Dynamics and Aeroacoustics Department (DSNA) of ONERA solves the Euler or Navier-Stokes equations on multiblock structured meshes using the cell-centered Jameson Lerat scheme with implicit residual smoothing and a Jameson-like 2nd and 4th order artificial viscosity [24]. Time integration is based on a 4-step Runge-Kutta scheme. Local timestepping is used for steady state computations. A multigrid method of the embedded mesh type was introduced [6]. An ALE (Arbitrary Lagrangian Equation) formulation [11] of the Euler or Navier-Stokes equations with several turbulence models was developed for the case of meshes deformable over time. The unsteady problems can be solved either with the basic scheme or with a dual timestep method [31, 32]. When run in Euler mode, viscous effects can be taken into account for steady state solutions through a boundary layer computation or through a Navier-Stokes computation. Steady viscous effects are introduced by an injection velocity. In unsteady Euler mode, the ALE formulation or blowing condition could be used to take into account the structural deformations.

2.2 Linearized unsteady aerodynamics

The REELC code [28] developed by the Structural Dynamics and Coupled Systems Department (DDSS) of ONERA solves the linearized Euler equations for harmonic excitations of the structure. In this approach, the field of the Euler variables is divided into a steady part and a first-order fluctuation due to the motion of the structure. These fluctuations are harmonic too and they are the solutions of a steady problem written in complex variables. These solutions can be obtained by a pseudo-non stationary method. The same scheme solving the steady case is used to integrate the equations. The boundary conditions are linearized too. In earlier developments [27], a blowing condition was used to simulate the structural motions. This boundary condition proved inaccurate in the case of several configurations (fuselages, bodies, tanks). The use of the ALE formulation of Euler equations and their discretizations in a moving grid improve the method. An FAS (Full Approximation Storage) type multigrid method was introduced and reduces substantially the CPU time [30].

This approach has the following advantages:

- the flow discontinuities (shock and wake) are taken into account by the steady Euler field including static deformations,
- the stationary viscous effects could be introduced by an injection velocity,
- the CPU cost is not dependent on the excitation frequency and the computation can be conducted at zero frequency and at low frequency,
- the use of local timestep results in to very low CPU time compared to unsteady approach,
- the stationary solution of the unsteady problem is sought and the convergence of the process can be verified.

This approach makes possible the traditional aeroelastic analyses with and without control system.

2.3 Mesh deformation

Numerical techniques have been developed at the DDSS Department of ONERA to build mesh deformations. They are based on structural analogy, with discrete spring networks or elastic continuum analogy. A finite element formulation is used and special features allow the reduction of the size of the problem, especially in the Navier-Stokes meshes case.

For the spring techniques [4, 14], two mesh deformation techniques have been developed using spring networks. The first one is Batina's spring method. The other one uses a wider formulation which couples the 3 components of the displacement vector. The numerical discretisation is a finite element one and the linear relation between the local forces vector and the local displacement vector is solved.

The elastic medium analogy uses a hexahedral finite element with 8 nodes. The local stiffness matrix is computed using a numerical Gauss integration procedure with 8 Gauss points. The approximate integration using one Gauss point is possible with a specific treatment of Hour-Glass modes. The linear system, connecting the imposed motion and the induced displacement, is solved using a pre-conditioned conjugate gradient method.

These techniques have been implemented in the case of multi-block structured grids [10]. The complete mesh deformation is defined as a sequence of individual block deformations. In order to reduce CPU time and the size of the linear problem, a macro-mesh is used. This mesh is built from the original by packing cells in each direction. The macro-mesh nodes are moved using the previously described techniques, whereas inner node displacements are interpolated in the macro-cells.

3 Structural Model

Structural models can be taken into account in two ways for static and dynamic aeroelasticity. The first consists of using the complete finite element model of the structure, whereas the second uses a reduction of the mechanical model on a modal base. Since dynamic aeroelasticity problems concern the low frequency domain, using the modal base for the structure is natural and has many advantages: fewer unknowns, simplicity of representation of structural displacements, lower CPU cost of mesh movement and a better controlled analysis of the results.

4 Fluid-Structure Coupling

4.1 Fluid-solid interface

The interface region between the fluid domain and the solid domain is defined by:

- a list of nodes and element faces on the fluid side,
- a list of nodes and elements faces on the solid side.

Since nodes on both sides of the interface do not need to match, a search algorithm is used to identify the solid face that contains each fluid interface node. Once this mapping is obtained, local fluid pressure forces are computed at each fluid interface node. They are then interpolated at the solid interface nodes [21]. The resulting pressure load is used as a boundary condition to solve the structural problem. One can note that this approach leads to a total fluid pressure force being transferred to the solid side of the interface. Since the fluid mesh is usually finer than the solid mesh, this interpolation strategy would appear to be more accurate than interpolating the fluid pressure directly onto the solid interface nodes and then computing the local pressure forces on that surface using the solid discretization.

4.2 Static Coupling

The static solution of the coupled fluid-structure problem is obtained by balancing the steady aerodynamic forces and the elastic forces. Two methods can be used. The first consists of using a blowing boundary condition for the fluid to simulate the aerodynamic surface deformations in a fixed mesh. The second method is based on modification of the mesh under the effect of the structural displacements. Spring methods or structural analogy methods are used to deform the mesh. At the outcome of each multigrid cycle, the aerodynamic forces and structure are coupled in the mechanical solver.

Remarks:

- With the modal approach, the structural modes can be obtained either from experimental data or from a Finite Element Model.
- We can use blowing condition or mesh deformation, resulting of the superposition of the modal mesh deformation.
- The use of the first structural modes may cause an error on the static deformation, which a priori justifies the use of unreduced structural models.

4.3 Dynamic Coupling

Dynamic coupling is achieved by using the modal reduction of the structure and the ALE formulation of the averaged Euler or Navier-Stokes equations. For the Euler equations, a fixed mesh blowing condition can also be used. The undesirable effects of modal truncation can be compensated for by static corrections. This method allows the steady point to be preserved for unsteady simulations. For this purpose, the calculation of the displacement field is modified taking in account the static correction.

For dynamic coupling, whether the local timestep method is used or not is decisive for the choice of the time discretization of the structural equations. If the dual timestep method is not used, the structural equations are converted to a first-order differential system. The structural equations are integrated simultaneously with the fluid in the explicit phase of the Runge-Kutta scheme with 4 or 5 steps. In this case, no sub-iterations are required to balance the fluid and the structure. If the dual timestep method is used, the structural equations are discretized by the Newmark scheme [3]. The fluid and structure must be strongly coupled in order to balance the aerodynamic forces with the displacements generated. Subiterations are therefore necessary each fluid timestep. A stop criterion based on a limited number of sub-iterations and the error between two iterations of the aeroelastic forces is applied.

Remark:

- An equal distribution of the elastic energy gives the initial conditions for the dynamic coupling.

5 Numerical Applications

5.1 AMP wing

A large experimental database has been obtained on a scaled model of a transonic civil aircraft in order to validate the numerical tools [36]. It includes steady and unsteady pressures, static and dynamic deformation measurements and for flutter cases, measurements of damping and frequencies for different configurations (clean wing, wing + one or two nacelles) and for a large Mach number and incidence variations. A simple structural finite element model (fish model) was developed and updated with ground vibration test results. A mono-block structured mesh with 720,000 (209x53x65) cells ($y^+ \cong 1$) was made for the jig shape of the wing. A static fluid-structure coupling using elsA (versus Navier-Stokes, k-l turbulence model) has been performed for several aerodynamics conditions. The numerical steady pressure coefficient is shown (Figure 1) and compared with experimental data for a Mach number of 0.86. The static displacements (at the wing leading edge) and the twist are presented (Figure 2).

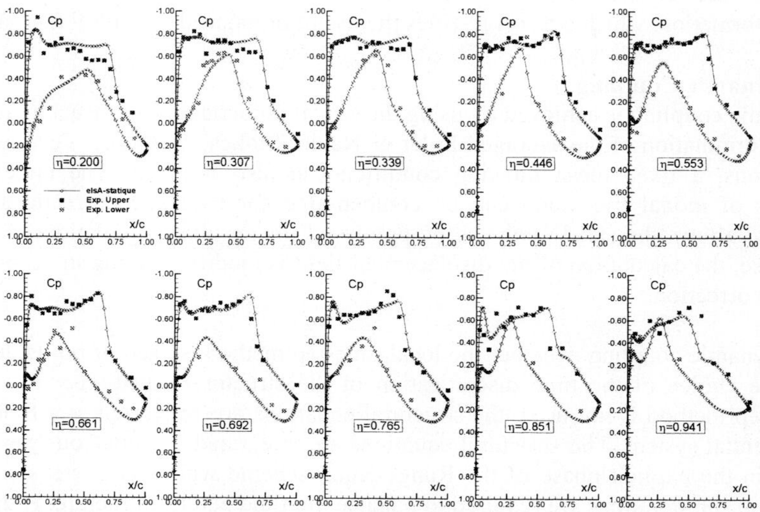

Figure 1 AMP wing, Pressure coefficient, Mach=0.862, α=1.60°, Re=3.64 10^6 Fluid structure coupling with Navier-Stokes (k-l Smith turbulence model)

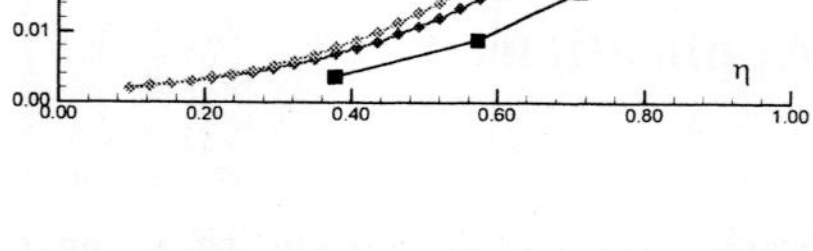

Figure 2a Static vertical deflections

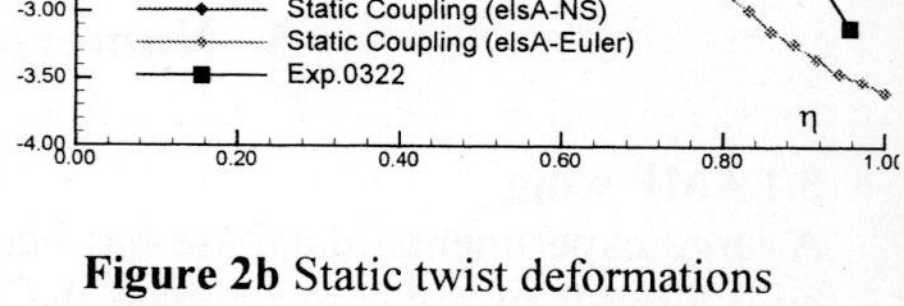

Figure 2b Static twist deformations

The unsteady aerodynamics loads were calculated with the REELC code in the frequency domain using the steady boundary layer given by the Navier-Stokes code and the static deformations as input. The flutter calculation has been performed using the "p,k" method [9]. The flutter diagram (Figure 3) presents the frequencies and damping evolution (numerical and experimental) versus the stagnation pressure in the wind tunnel. It appears a good agreement for the static aeroelasticity (steady pressures, static deformations) as well as for the dynamic aeroelasticity (frequencies and aerodynamic damping) according to the pressure in the wind tunnel and this, into high transonic flow.

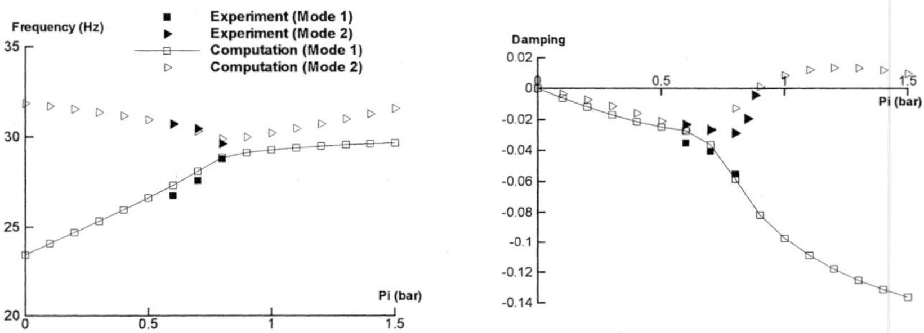

Figure 3 AMP wing, Flutter diagram, Mach=0.82, α=0.60°

5.2 Civil aircraft

A numerical application was made on a civil transport aircraft configuration. This model consisted of a transonic flexible wing at the wall, two pylons and two nacelles. This wing is equipped with a winglet at the tip. A structural model was constructed. It defines the structural deformations caused by the aerodynamic loads corresponding to the flight configuration. The Euler mesh realized for the flight included 114 structured domains and 852 boundaries. It had a total of 1,114,185 cells for Euler calculations and 1,650,000 cells for Navier-Stokes calculations. More than 170 boundaries exhibited degenerations.

5.2.1 Mesh deformation

In a first step, the aerodynamic mesh of the ground form was obtained from this mesh using the 3D mesh deformation technique. Most of the domains were deformed using the Batina spring techniques because it used less computational time. The elastic structural analogy was used for the other domains because of its robustness. A macro-block technique was used to reduce the size of the problem. The structured multi-domain mesh deformation problem was solved separately in each domain. The domains were processed sequentially and motion continuity was ensured at matching boundaries. The resulting mesh was used for a static coupling computation for several aerodynamic conditions (Figure 4).

Figure 4 Civil aircraft model, deformation of the aerodynamic mesh

5.2.2 *Steady calculation*

The steady case was computed for a Mach number of 0.82 and an incidence of 2.7 degrees corresponding to a lift coefficient Cn of 0.5. The steady case was obtained with 1500 iterations and needs 3000 s. time computing of one processor NEC SX5 (Figure 5).

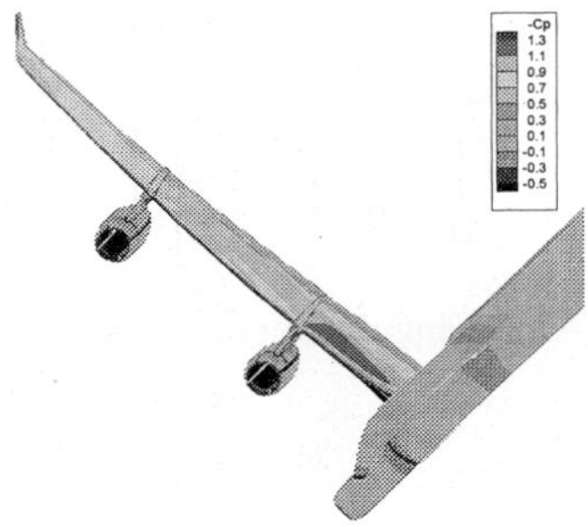

Figure 5 Civil aircraft model, pressure distribution Mach=0.82 Cn=0.5

5.2.3 *Flutter calculation*

From the converged steady field, the generalized unsteady aerodynamic forces were calculated by the REELC code (code solving the linearized Euler equations in ALE formulation for complex excitations) for the ten first symetric modes and for 13 reduced frequencies. For each reduced frequency and each mode, sufficient convergence was achieved after 500 iterations and needs 2800 s. on the NEC SX5. The complete simulation for a configuration consumes 100 hours CPU. The stability diagrams obtained by solving the flutter equation using the conventional "*p,k*" method are given (Figure 6) and shows no flutter problem in the flight domain.

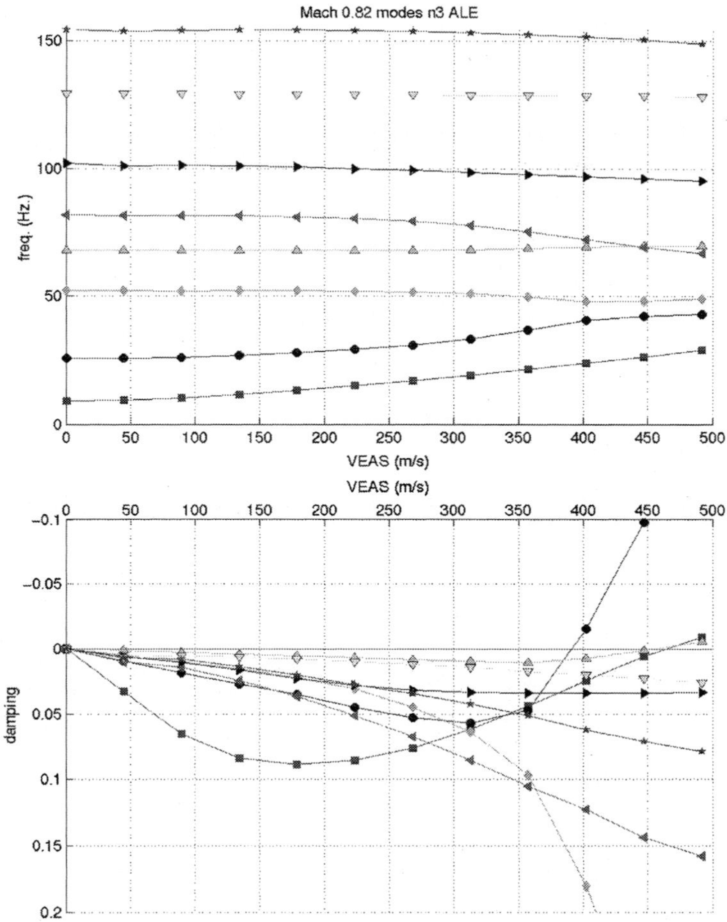

Figure 6 Civil aircraft model, Flutter diagram, Mach=0.82, Cn=0,5

5.3 Fighter model

A numerical application was made on a generic numerical model defined by Dassault-Aviation. This numerical model consisted of a wing at the wall, two pylons and two tanks and was equipped with a wing tip missile. For the study, two weight configurations were defined: one stable and the other unstable in the flight envelope. Conventional subsonic aeroelasticity tools (doublet method and ELFINI code) were used for this purpose. The model was also designed to present an aerodynamic shock between payloads at transonic speed. A multiblock structural mesh for an Euler application to this configuration was generated. It had 320,000 cells. A specific structural model, taking into account the internal fluid in the tanks and the sloshing modes, was developed.

5.3.1 Static Coupling

Static fluid-structure coupling was computed for a Mach number of 0.85, an angle of attack of 1 degree and an altitude of 3000 meters using the modal base. The steady solution resulting from balancing between the elastic and aerodynamic forces was obtained in 500 iterations. The steady field shown (Figure 7) exhibits a strong shock between tanks.

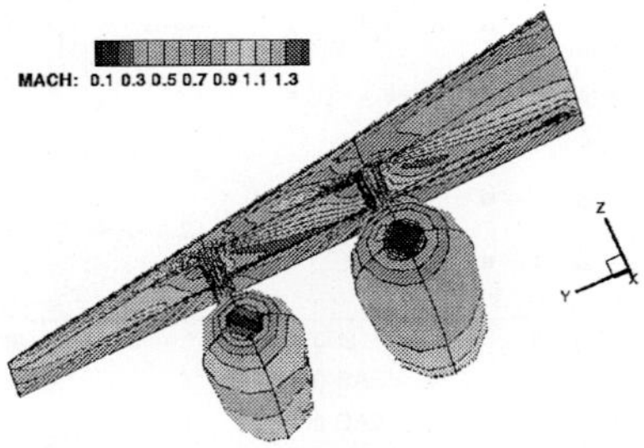

Figure 7 Fighter model, Isomach distribution, Mach=0,85, incidence=1

5.3.2 Fluid-Structure Frequency Coupling

From the converged steady field, the generalized unsteady aerodynamic forces were calculated by the REELC code for the first seven modes of each configuration and for 12 reduced frequencies. For each reduced frequency and each mode, sufficient convergence was achieved after 200 iterations. The stability diagrams obtained by solving the flutter equation by the conventional "*p,k*" method are given in Figure 8 (stable and unstable configurations).

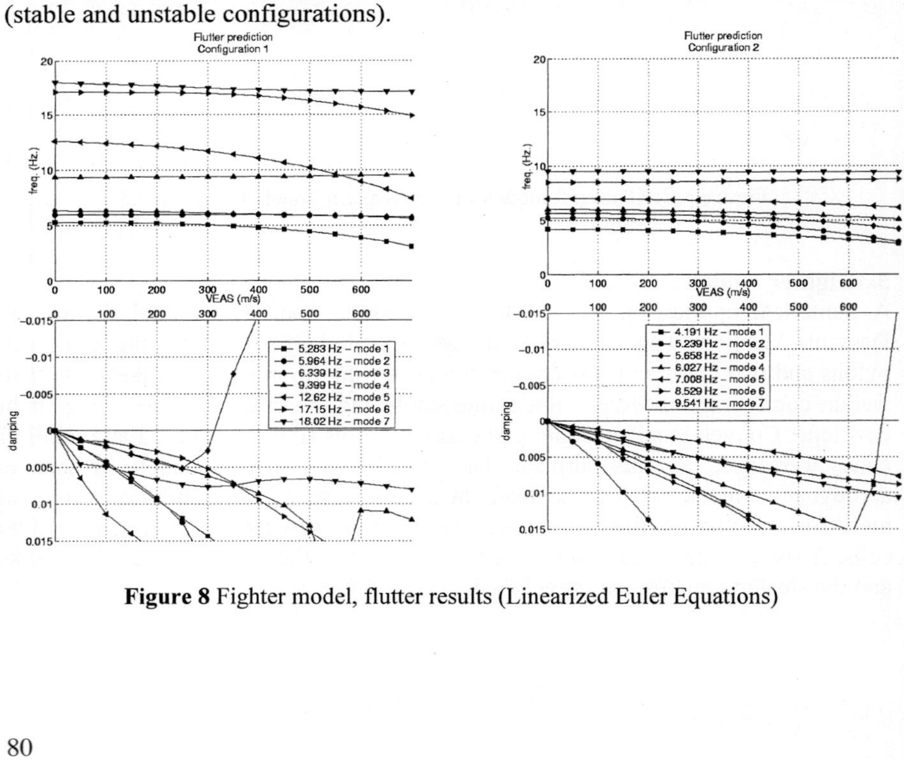

Figure 8 Fighter model, flutter results (Linearized Euler Equations)

5.3.3 Fluid-Structure Time Coupling

From the point of static equilibrium, a velocity impulse was imparted to the structure. Unsteady simulation was carried out using the dual timestep method. Convergence for a timestep was achieved when the residues between two sub-iterations were reduced by two orders of magnitude. In practice, 50 iterations were sufficient. Convergence between fluid and structure was generally achieved in 3 iterations for each timestep. With the conventional method (uniform timestep), 60,000 timesteps would have been necessary to describe a 20 Hz period, considering the stability criteria of the aerodynamic code. When the local timestep method is used, there are 16 timesteps for a 20 Hz period. The calculation is stopped after a physical time of 2 seconds. The generalized coordinates are analyzed. A good agreement was obtained between Prony's analysis [8] of these signals and the results obtained for flutter (Table 1). The Fourier transform of these signals are presented in Figure 9. Analysis of the signals also yields the aeroelastic damping values [2].

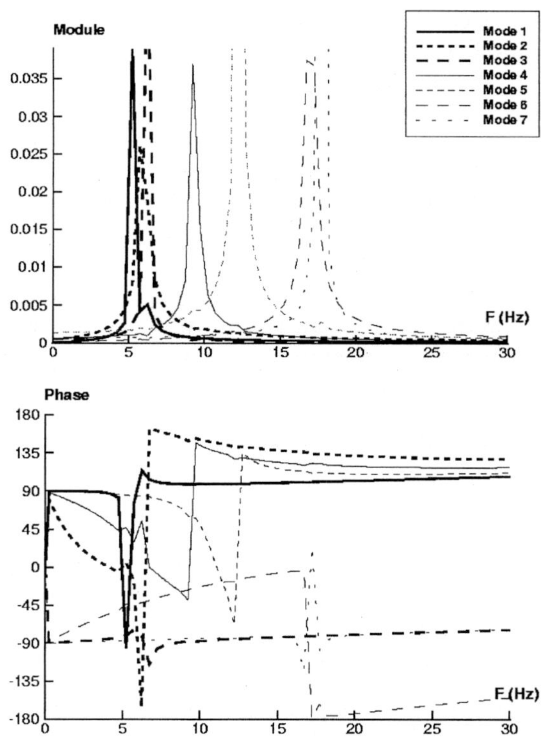

Figure 9 Fighter model, time results analysis (time coupling)

Table 1 Frequencies and aeroelastic damping comparisons

Mode	Time coupling		Frequency coupling	
	Frequency	Damping	Frequency	Damping
1	5.27	0.00033	5.16	0.0091
2	5.92	0.012	6.00	0.012
3	6.28	0.0019	6.20	0.0020
4	9.37	0.0014	9.41	0.0043
5	12.3	0.0028	12.1	0.00620
6	17.0	0.00057	17.1	0.0015
7	17.7	0.00016	17.7	0.00039

5.4 Limit Cycle Oscillation

Experimental investigations of LCO phenomena have been performed at DLR using a two-dimensional NLR7301 model [33]. Forced and self-sustained oscillation regimes have been recorded. Steady and unsteady results have been presented in the subsonic and transonic cases. At ONERA, numerical investigations will focus on the case occurring at an uncorrected Mach value of 0.82 and angle of attack 0.27 degrees. This case is supposed to exhibit limit cycle oscillations phenomena due to the unsteady interaction between shock and boundary layer. The Reynolds number for this case is about 2.5 million. A linear structural model using two degrees of freedom (i.e. heave and pitch) represents the structural model. Unsteady fluid structure coupled computations were performed using dual timestepping and turbulence model of Spalart-Allmaras at CFL 4. The computation duration was 0.5 s. of physical time. 64 physical iterations per period of 1/34 s were computed. Each of them counted two mesh displacement iterations of 200 dual iterations. The steady case iso-Mach map is shown (Figure 10) and the history of the computed vertical position and angle of attack of the model is presented in Figure 11. The computation gives a very low value for the aerodynamic damping, in accordance to experimental observations. New computations including wall wind-tunnel effect will be performed and compared to experimental data.

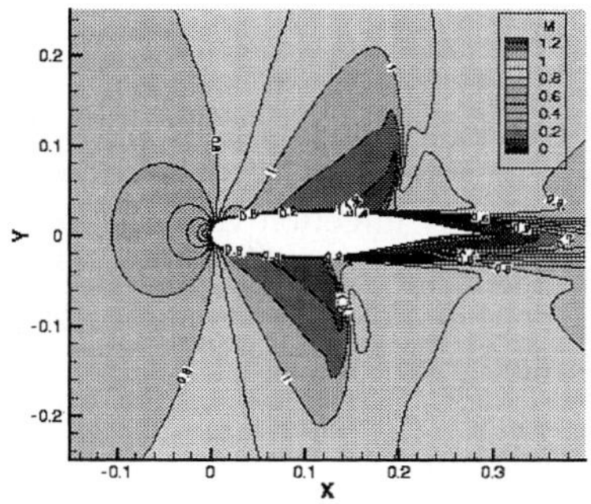

Figure 10 NLR7301 profile, Mach=0.82, α=0.27°
Steady iso-Mach values

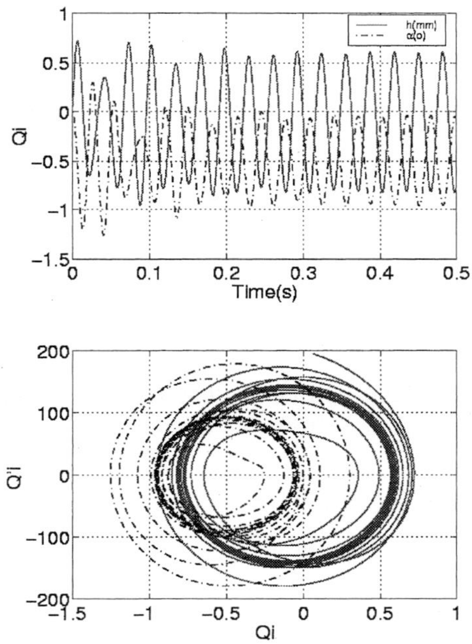

Figure 11 NLR7301 profile, History of generalized coordinates

6 Conclusion

Two approaches for computing the fluid structure interaction in a transonic flow using Euler/Navier-Stokes solvers are developed. The first concerns the Euler linearized approach including viscous effect by injection velocity. In this approach, the aerodynamic calculation and the fluid-structure coupling could be conducted separately. This approach gives flutter prediction with acceptable computing and it makes possible the aero-servo-elastic calculation.

The second way concerns the direct fluid-structure coupling in the time domain. The numerical fluid-structure coupling method using the dual timestep to simulate the unsteady fluid flow has been validated. For dynamic aeroelastic simulations, which theoretically concern the low frequencies and for which the use of a general timestep requires very small timesteps to describe unsteady phenomena, the introduction of the dual timestep has considerably decreased the CPU time without detracting from the accuracy or robustness of the predictions.

The static coupling has been made and validated for an Euler and a Navier-Stokes solvers. The dynamic simulations were made for an Euler fluid model. Taking viscous effects into account will be the subject of future work, together with the development of linearized Navier-Stokes approaches.

Acknowledgments

The author thanks the Direction Générale de l'Aviation Civile and the Service des Programmes Aéronautiques for its financial support. He would like to thank the Dassault-Aviation teams for their collaboration on the Fighter Model. The author would like to thank Alain Dugeai, Philippe Girodroux, Guy-Daniel Mortchelewicz and Marc Rapin (ONERA/DDSS) for their contribution throughout this studies.

References

[1] W.F. Ballhaus and P.M. Goorjian: "Implicit Finite-Difference Computations of Unsteady Transonic Flows about Airfoil", AIAA Journal Vol.15, No. 12,77, pp. 1728-1735.

[2] E. Balmès: "Experimental and Analytical Structural Dynamic Toolbox", Matlab User's Guide, January 1997.

[3] K.J. Bathe: "Finite element procedures in engineering analysis'', Prentice Hall, Englewood Cliffs, New Jersey 07632, 1982, pp. 512.

[4] J.T. Batina: "Unsteady Euler airfoil solution using unstructured dynamic meshes", AIAA Paper 89-0115.

[5] R.M. Bennet, J.T. Batina and H.J. Cunningham: "Wing Flutter Calculations with the CAP-TSD Unsteady Transonic Small-Disturbance Program", Journal of Aircraft, Vol. 26, No. 9, Sept. 1989, pp. 876-882.

[6] R. Collercandy: "Multigrid Strategy for Euler and Navier Stokes computations for flows around complex aerospace configurations", Proceedings of the ECCOMAS 98 Conference, Athens, 1998.
[7] V. Couaillier: "Numerical simulation of separated turbulent flows based on the solution of RANS/low Reynolds two-equation model", AIAA Paper 99-0154, 37th AIAA Aerospace Sciences Meeting and Exhibit, Reno, Nevada, USA, January 11-14, 1999.
[8] B. Darras: "Identification de systèmes vibrants par la méthode du décrément aléatoire", Revue française de mécanique, numéro 58-59, pp. 21-27, 1976.
[9] R. Dat and J.L. Meurzec: "Sur les calculs de flottement par la méthode dite du 'balayage en fréquence réduite'", La Recherche Aérospatiale No. 133, Nov.-Dec. 1969.
[10] A. Dugeai, A. Madec and A.S. Sens: "Numerical unsteady aerodynamics for turbomachinery aeroelasticity", 9th International Symposium of Unsteady Aerodynamics, Aeroacoustics and Aeroelasticity of Turbomachines, Lyon, France, September 4-7, 2000.
[11] J. Donea: "Arbitrary Lagrangian-Eulerian Finite Element Methods", Computational Methods for Transient Analysis. Edited by T. Belytschko and T.J.R. Hughes. Elsevier Sciences Publisher B.V. (1983) pp. 473-516.
[12] C. Farhat and M. Lesoinne: "Fast staggered algorithms for the solution of three-dimensional nonlinear aeroelastic problem", 85th AGARD SMP Meeting, Aalborg, Denmark, November 13-16, 1997.
[13] P. Girodroux-Lavigne, J.P. Grisval, S. Guillemot, M. Henshaw, A. Karlsson, V. Selmin, J. Smith, E. Teupootahiti and B. Winzell: "Comparative of advanced fluid-structure interaction methods in the case of a highly flexible wing (Results from the UNSI Program)", International Forum on Aeroelasticity and Structural Dynamics, Madrid, Spain, June, 2001.
[14] P.A. Gnoffo: "A finite volume adaptive grid algorithm applied to planetary entry flowfields", AIAA Journal, Vol. 21, number 9, pp 1249-1254 (1983).
[15] J.P. Grisval and J.L. Meurzec: "Couplage direct fluide structure en aéroélasticité transsonique", 73rd Meeting of the AGARD Structures and Materials Panel, San Diego, CA, USA, October 7-11, 1991, AGARD Proceedings 507, ONERA TP 1991-205.
[16] J.P. Grisval, C. Sauvignet and Z. Johan: "Aeroelasticity calculation using a finite element method", International Forum on Aeroelasticity and Structural Dynamics, Rome, Italy, June 17-20, 1997.
[17] J.P. Grisval, C. Liauzun and Z. Johan: "Aeroelasticity simulations in turbulent flows", International Forum on Aeroelasticity and Structural Dynamics, Williamsburg, USA, June 22-25, 1999.
[18] J.P. Grisval and C. Liauzun: "Application of the finite element method to aeroelasticity", Revue européenne des éléments finis. Vol. 8, no 5-6, pages 553-579, septembre 1999.
[19] G.P. Guruswamy, P.M. Goorjian and E.L. Tu: "Transonic Aeroelasticity of Wings with Tip Stores", Journal of Aircraft, Vol. 24, No. 10, Oct. 1987, pp. 688-695 and AIAA Paper 86-1007.
[20] G.P. Guruswamy and C. Byun: "Fluid structural interactions using Navier Stokes flow equations coupled with shell finite element structures", AIAA 24th Fluid Dynamics Conference, July 6-9, 1993, Orlando, FL, USA, AIAA-93-3087.
[21] R.L. Harder and R.N. Desmarais: "Interpolation using surface splines", Journal of Aircraft, February 1972.
[22] E.M. Lee-Rausch and J.T. Batina: "Wing flutter boundary prediction using unsteady Euler aerodynamic method", AIAA-93-1422-CP.

[23] J.C. Le Balleur and P. Girodroux-Lavigne : "Calculation of fully three-dimensional separated flows with an unsteady viscous-inviscid interaction method", Fifth Symposium on Numerical and Physical Aspects of Aerodynamic Flows, California States University, Long-Beach, USA January 13-15, 1992, ONERA TP 1992-01.

[24] N. Liamis: "Méthodes implicites de résolution des équations d'Euler pour des écoulements tridimensionnels instationnaires dans des turbomachines", Thèse, Université de Paris VI, 4 Octobre 1993.

[25] J.L. Meurzec, A.S. Sens and J.P. Grisval: "Flutter prediction using an aerodynamic finite element method", International Forum on Aeroelasticity and Structural Dynamics 1995, Manchester, U.K., June 26-28, 1995, ONERA TP 1995-64.

[26] G.D. Mortchelewicz and A.S. Sens: "Solution of 3-D Euler Equations with Unstructured Meshes for Aeroelasticity Problems.", International Symposium on Aeroelasticity and Structural Dynamics, Aachen, Germany, June 3-6, 1991, and TP Onera No. 1991-62.

[27] G.D. Mortchéléwicz: "Application des équations d'Euler linéarisées à la prévision du flottement", AGARD report 822, 85th Meeting of the AGARD Structures and Material Panel, Aalborg, Denmark, October 14-15, 1997.

[28] G.D. Mortchéléwicz: "Flutter simulation using the linearized Euler equations", 39th Israel Annual Conference on Aerospace Sciences, Tel Aviv, Haifa, Israel, February 17-18, 1999.

[29] G.D. Mortchéléwicz: "Flutter simulations", Aerospace Science & Technology No. 4 (2000), pp. 33-40.

[30] G.D. Mortchéléwicz: "Flutter simulation using the linearized Euler equations", 41st Israel Annual Conference on Aerospace Sciences, Tel Aviv, Israel, February 21-22, 2001.

[31] O. Rouzaud and A. Plop: "2D unsteady Navier Stokes computations at ONERA for prediction of dynamic stall", 24th European Rotorcraft Forum, Marseille, France, September 15-17, 1998.

[32] O. Rouzaud, S. Plot and V. Couaillier : "Numerical simulation of buffeting over airfoil using dual time stepping method", European Congress on Computational Methods in Applied Sciences and Engineering, ECCOMAS 2000, Barcelona, September 11-14, 2000.

[33] G. Schewe and H. Deyhle: "Experiments on transonic flutter of a two-dimensional supercritical wing with emphasis on the non-linear effects", Proceedings of the Royal Aeronautical Society conference on "Unsteady Aerodynamics", London, United Kingdom, July 17-18, 1996.

[34] D.M. Schuster, J. Vadyak and E. Atta: "Static Aeroelastic Analysis of Fighter Aircraft Using a Three Dimensional Navier-Stokes Algorithm ", Journal of Aircraft, Vol. 27, No. 9, Sep. 1990, pp. 820-825.

[35] A.M. Vuillot, V. Couaillier and N. Liamis: "3D turbomachinery Euler and Navier-Stokes calculation with multidomain cell-centered approach", AIAA Paper 93-2576, Monterey, California, June 28-July 1, 1993.

[36] H. Zingel et al.: "Measurement of steady and unsteady airloads on a stiffness scaled model of a modern transport aircraft wing", International Symposium on Aeroelasticity and Structural Dynamics, Aachen, Germany, June 3-6, 1991, and DGLR-Bericht 91-06, 1991, pp. 120-131.

Numerical Prediction of Flow Instabilities and Aeroelastic Effects

KERRY HOURIGAN[1], IVAN MCBEAN[1], MARK THOMPSON[1] and FENG LIU[2]

[1] Fluids Laboratory for Aeronautical and Industrial Research (FLAIR), Department of Mechanical Engineering, Monash University, Clayton 3800, Australia
[2] Department of Mechanical and Aerospace Engineering, University of California, Irvine, California 92697-3975, USA

Summary

Two lines of investigation being undertaken at Monash University and of relevance to the aeronautics industry will be presented. First, the prediction of vortex breakdown, a fluid structure that bedevils high angle of attack aircraft but may assist the breakup of tip vortices, has been undertaken using an in-house parallelized spectral element code. Second, aeroelasticity has been identified as one of the most important problems presently facing the designers of turbomachinery blades as well as aircraft control surfaces. A 3-dimensional solver implementing the compressible Navier-Stokes equations coupled with the k-ω turbulence model has been developed to solve the unsteady flow through oscillating annular or linear turbine cascades. Simulations have been performed in parallel in a time accurate manner using a Jameson-type dual time Runge-Kutta scheme. This is then used to investigate the effects of 3-dimensionality and secondary flows on the unsteady aerodynamics and flutter characteristics of turbine cascades.

1 Introduction

The control of the transition from laminar to turbulent flow is of crucial importance in various flows, including those found in swirling flows and the flows around bluff bodies. Swirling flow is a basic ingredient of many important industrial and aerospace flows. Modern high-performance aircraft generate considerable aerodynamic lift forces at high angles of attack. However, the resulting highly swirling vortex structures that result are prone to rapid dilatation called Vortex Breakdown, leading to restriction of the angle of attacks and the operational flight envelope. Similarly, turbomachinery can produce flows of sufficiently high swirl in which vortex breakdown and loss of efficiency are observed. On the other hand, vortex breakdown can be utilised as flame holders in combustion chambers and to break u p the tip vortices of large aircraft. In the case of bluff body flows, the transition to turbulence in the wake leads to significant drag changes on the body and the character of the wake that interferes with downstream bodies. Significant improvements in processes in the aerospace, mineral processing, wind engineering and manufacturing industries can result from the control of the transitions of flow in high swirl and wakes.

As designers in the turbomachinery industry strive to design machines that are lighter, more powerful and more efficient, blade flutter has become one of the most important limiting factors on the design process. The aeroelastic response is a complex phenomenon that is not well modeled or predicted by current design techniques. Codes that implement 2-dimensional models can simulate this behaviour in a meridional plane; however the flow structures found in blade passages are generally 3-dimensional and such models provide a qualitative rather than a quantitative analysis. Furthermore, important flow phenomena are not modeled including hub and casing vortices and tip effects.

A structured 3-dimensional Navier-Stokes code is developed to solve the unsteady governing equations. These are solved using an explicit Runge-Kutta scheme, implementing residual averaging and multigrid. The problem is then solved in a time accurate manner through a fully implicit scheme as proposed by Jameson [10]. This scheme has already been used in a 2-dimensional model of aeroelasticity in turbomachinery [19, 12]. The development of the present code is an extension of the previous two-dimensional method to three dimensions. Similar algorithms have been successfully implemented in a 3-dimensional Navier-Stokes external solver that models flow over a flexible wing [24, 14].

In this paper, the numerical methods being employed to study the transition to turbulence in swirling flows are first described. Then, the numerical methods used to study the aeroelastic response of aerofoils in turbulent flow are reported.

2 Time-Dependent Simulations of Vortex Breakdown

Two- and three-dimensional time-dependent spectral-element simulations of the swirling flow in a torsionally driven cylinder were undertaken. The aim was not to impose unphysical symmetries on the solution fields; in particular, there was no enforcement of either non-axisymmetry or time-independence.

Numerical Method

The time-dependent simulations employed a spectral/spectral-element method for axisymmetric geometries. A spectral-element discretisation was used in the r–z planes and, for the three-dimensional simulations, Galerkin-Fourier expansion in the θ direction. Within each spectral-element the solution variables ($\mathbf{u}$, p) are represented by a tensor-product of Lagrangian interpolants. The internal node points are chosen to correspond to the Gauss-Legendre-Lobatto quadrature points. This leads to considerable efficiency gains when evaluating the integrals from the application of the Galerkin weighted residual method used to obtain the discretised equations. The order of the interpolation can be selected during run-time. This is a significant advantage since the spatial convergence of the simulations can be easily validated. When run in this mode, for *smooth* problems the method achieves spatial exponential (or spectral) convergence rates usually associated with global spectral methods.

The spatially-discrete equations are then independently discretised in time using a classical three-step splitting scheme described in [13]. This involves splitting each time-step into substeps accounting separately for convection, pressure gradient/continuity and diffusion. The convection substep employs the third-order Adams-Bashforth extrapolation and the diffusion substep uses the second-order Crank-Nicolson method (with the θ correction to improve high-frequency stability (see [4]). Overall second-order accuracy is achieved by using the higher-order pressure boundary condition described in [13].

The implementation has been validated on numerous problems and the expected convergence behaviour confirmed. For example, it has been used to predict the two-shedding modes occurring in a circular cylinder wake [22], the flow transitions in the wake of a sphere [23], and the feedback stability mechanism governing lock-on states for the flow past long rectangular plates ([8, 21]). More details of the implementation can be found in these papers and references therein.

A combination of temporal and spatial resolution studies were undertaken for the three-dimensional simulations described in the following section.

Parallelisation Issues

Three-dimensional simulations are inherently expensive and put considerable strain on computational resources. Fortunately, the spectral/spectral-element method described here can be parallelised efficiently allowing accurate simulations to be performed on clusters using the MPI (message passing interface) routines.

The application of the time-splitting scheme to the Navier-Stokes equations results in the following set of equations

$$\frac{\hat{\mathbf{u}} - \mathbf{u}^n}{\Delta t} = \mathrm{N}(\mathbf{u}), \tag{1}$$

$$\frac{\hat{\hat{\mathbf{u}}} - \hat{\mathbf{u}}}{\Delta t} = -\nabla P/\rho, \tag{2}$$

$$\frac{\mathbf{u}^{n+1} - \hat{\hat{\mathbf{u}}}}{\Delta t} = \mathrm{L}(\mathbf{u}). \tag{3}$$

Here, $\mathrm{N}(\mathbf{u})$ and $\mathrm{L}(\mathbf{u})$ represent the nonlinear (convection) and linear (diffusion) terms in the Navier-Stokes equations, P is the pressure, ρ the fluid density, $\mathbf{u} = (u_z, u_r, u_\theta)$ is the velocity vector and, at the completion of the three steps, the time has been advanced from time t^n to t^{n+1}. Further details are given in ([13])

For a Cartesian coordinate system, if a (Galerkin) Fourier expansion is used to describe the dependence of a variable on the azimuthal coordinate, the second and third substeps decouple into a set of equations for each Fourier plane. This means these substeps can be done efficiently in Fourier space. In addition, the equations for each Fourier mode (consisting of two planes) can be solved on a separate processor and the associated memory requirements shared between the processors. The introduction of axisymmetric coordinates introduces some complications. In this case, the equations for u_r and u_θ from the diffusion step are coupled; however, decoupling can be achieved by introducing the complex variables $u_1 = u_r + iu_\theta$ and $u_2 = u_r - iu_\theta$. Once this is substitution is performed, the pressure and diffusion substeps can still be solved in Fourier space with each Fourier mode evaluated on a different processor.

The convection substep in usually performed in physical space (because it is non-linear) using an explicit scheme such as Adams-Bashforth. The transformation to physical space means that significant interprocess communication is required. Fortunately, optimised MPI routines exist to make this process reasonably efficient. The physical space domain is split into a number of approximately equal parts corresponding to the number of processors used. The different Fourier mode components are collected on each of these parts and a fast Fourier transform used to convert to physical space velocity components. The velocity components are then updated. To take account of the convection substep of the time-splitting algorithm, a Fourier transform is applied to transform back to Fourier space and the Fourier components are again distributed amongst the processors so that each processor once again has one (or more) Fourier mode(s). Overall, the interprocess communication between the processors is considerable, and hence the method requires high-speed intercommunication. Tests on a Compaq Alpha cluster built from ES40 processor units with a low-latency high-bandwidth Quadrics interconnect indicate typical parallel efficiencies of 80% or greater are achieved on 32 nodes. In addition, memory requirements are shared between the processors and hence order-of-magnitude larger simulations are possible using multi-processor machines. The code has been validated for axisymmetric and non-axisymmetric problems (*e.g.*, [23].

Three-Dimensional Simulation Results

Two cases are examined in this section. The first is vortex breakdown in a confined circular cylinder where rotation is induced by a spinning lid. The second case is vortex breakdown in a swirling pipe flow. Both cases lead to many different flow behaviours depending on the governing parameters.

Case A: Spinning-Lid Rig Three-dimensional simulations of the confined flow in a circular cylinder at $Re \geq 1850$ were undertaken to establish the final asymptotic three-dimensional time-dependent state. This study was motivated by a similar numerical investigation of the flow using a non-axisymmetric mesh which showed that the flow evolved to an asymptotic state with four-fold rotational symmetry ([20]). Other studies based on stability analyses, (e.g., [6]) have suggested that the flow should remain axisymmetric to much higher Reynolds numbers.

The mesh used was the same as for a two-dimensional resolution study extended to three dimensions using a Fourier expansion in the azimuthal direction. Typically, 32 Fourier planes were used. Simulation undertaken evolved the flow from an initially-stationary state in a confined cylinder. In order to accelerate the development of any three-dimensionality, a white noise perturbation of amplitude 10^{-3} was added to each velocity component at each node after the first few time-steps. The flow field was subsequently evolved for a time corresponding to approximately 100 lid revolutions. All Fourier components (except the zeroth mode) decayed quickly leaving the flow in the same axisymmetric state as calculated with the axisymmetric code. In addition, an independent stability analysis indicated the stability of the flow to non-axisymmetric perturbations verifying the results of ([6]).

Case B: Swirling Flow in a Pipe A three-dimensional simulation of the swirling flow in a pipe is shown in Figure 1. In this case $Re = 1280$ with the pipe geometry and inlet velocity profile based on the experiments of Faler and Lebovich ([5]).

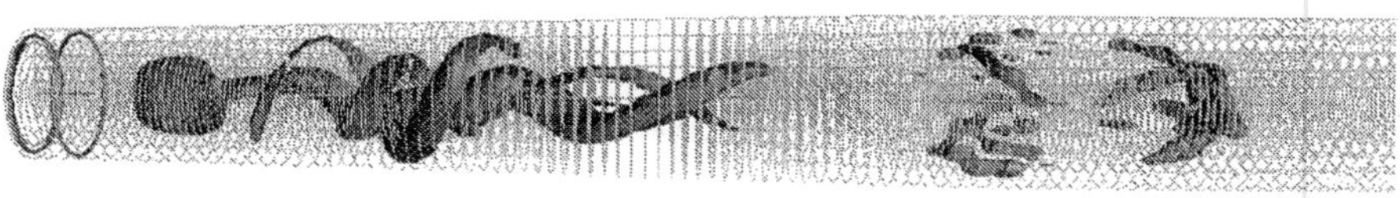

Figure 1 Transition state of swirling flow in a pipe showing vortex breakdown bubble and downstream spiral structure. Vortical flow structures are highlighted using the method of Jeong and Hussain. ([11])

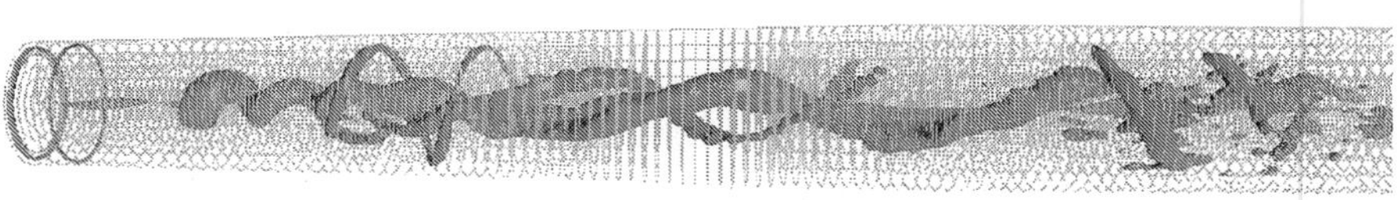

Figure 2 Final spiral breakdown state.

In the first part of the figure, the transition stage is shown with a vortex breakdown bubble apparent and spirals developing downstream. In the second figure, the bubble has disappeared and has been replaced by a spiral breakdown. Note that the isosurface structures near the pipe entrance are due to the developing boundary layer and initial adjustment of the flow. An interesting feature is the outer spirals close to the pipe outer wall. These may be a transient feature indicating a convective instability although they appear both during the initial development and after the flow has evolved to its asymptotic state.

3 Fluid Model for the Aeroelastic Problem

The present three-dimensional multiblock and parallel code has been developed from a proven steady solver designed to model turbomachinery cascade flow [15, 18, 16]. The governing equations for the unsteady fluid problem in a Eulerian reference frame with a moving mesh.

$$\frac{\partial}{\partial t}\iint_{\Omega} \mathbf{w}\, d\Omega + \oint \mathbf{f}\, dS_x + \mathbf{g}\, dS_y + \mathbf{h}\, dS_z = 0 \tag{4}$$

where

$$\mathbf{w} = \begin{pmatrix} \rho \\ \rho u \\ \rho v \\ \rho w \\ \rho E \end{pmatrix} \tag{5}$$

$$\mathbf{f} = \begin{pmatrix} \rho\bar{u} \\ \rho u\bar{u} + p \\ \rho v\bar{u} \\ \rho w\bar{u} \\ \rho E\bar{u} + pu \end{pmatrix}, \quad \mathbf{g} = \begin{pmatrix} \rho\bar{v} \\ \rho u\bar{v} \\ \rho v\bar{v} + p \\ \rho w\bar{v} \\ \rho E\bar{v} + pv \end{pmatrix}, \quad \mathbf{h} = \begin{pmatrix} \rho\bar{w} \\ \rho u\bar{w} \\ \rho v\bar{w} \\ \rho w\bar{w} + p \\ \rho E\bar{w} + pw \end{pmatrix} \tag{6}$$

The time dependent and semi-discrete form of the governing equations may be written as

$$\frac{d\mathbf{w}}{dt} + R(\mathbf{w}) = 0 \tag{7}$$

A dual time stepping scheme [10] is used to calculate the unsteady flow problem. A second order accurate, fully implicit scheme is used to integrate Equation (7) to evolve the unsteady problem in a time accurate manner.

The discrete form of (7) is

$$\frac{3w^{n+1} - 4w^{n} + w^{n-1}}{2\Delta t} + R(w^{n+1}) = 0 \tag{8}$$

This equation may be recast into

$$\frac{d\mathbf{w}}{dt^{*}} + R^{*}(\mathbf{w}) = 0 \tag{9}$$

where

$$R^*(\mathbf{w}) = \frac{3w}{2\Delta t} + R(w) - \frac{2}{\Delta t}w^n + \frac{1}{2\Delta t}w^{n-1} \tag{10}$$

The steady state solution w in equation (9) is then equivalent to the time accurate solution w^{n+1} of equation (8). Any efficient algorithm may be used to obtain the steady-state solution to (9). In this paper, the above mentioned Runge-Kutta type scheme with multigrid is used. Minimum modification of the steady solver if required to make it time accurate in the above manner.

4 Multiblock and Parallel Implementation

A method using multiple blocks of structured grids is used to maximize the use of computational resources and to allow the generation of grids for complex geometries. While each block consists of a structured grid, the blocks can be connected to each other in an unstructured manner provided the mesh geometry is matched at the block interfaces.

It was decided early in the development of the code to make use of MPI and some of the high level programming constructs available in Fortran90. A number of different objects were created in the code data structure to facilitate the parallel calculation of the fluid problem. Each fluid block is treated as a single object or entity. A schematic of the multiblock data structure is shown in Figure 3. A processor may be allocated more than one fine grid block and each fine grid block will have associated a number of coarser, multiblock grids. The machine calculates for each multigrid level simultaneously, then copies the solution or interpolates the residual to the next multigrid level.

A subface is defined as another object. This is used for the interface between the present block and another block, or a region to which a single boundary condition is to be applied. These objects are cycled through each time the boundary of the blocks are to be updated, upon which communication is effected or a boundary condition is applied. So that separate boundary conditions are not required for each coordinate direction, subface objects are transformed into a single coordinate system. This increases code complexity marginally. However it maintains that the code that is relatively compact.

The structured cell numbering within each block is unimportant as transformations are used to reorient the face so that the numbering matches with the neighboring face. Consider the example of a C-grid where a single block is wrapped around the blade. At the interface downstream of the trailing edge, the cell numbering will be different on the upper and lower blocks, thus requiring reorientation of the 2-dimensional arrays.

In keeping with the use of high levels of Fortran90 code, the communication module also exploits some of the more sophisticated MPI routines. The use of MPI derived

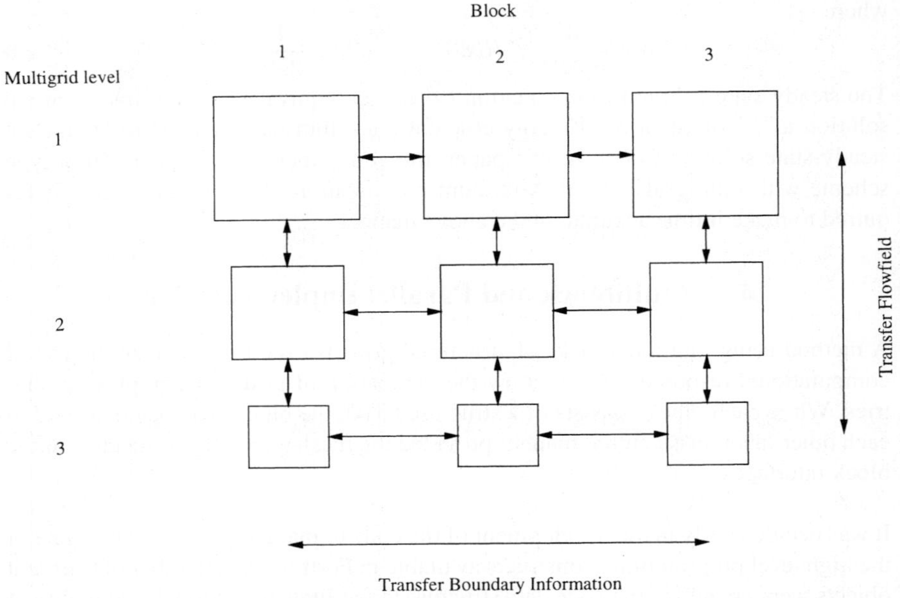

Figure 3 Schematic of multigrid and multiblock communication.

types allows the direct access of memory for the transfer of data, reducing the number of copies required during the communication of ghost cells. Due to the repetitive nature of the CFD computation, the "pipe-lining" of message passing calls is also implemented.

5 The Moving Grid

The movement of the fluid boundary requires the fluid grid to be regenerated over the entire flow domain. Thus in the multiple block code, given that the grid for each structured block is regenerated independently, the position of the corner points of each block must be somehow defined. This is effected by using a spring network analogy as proposed by Batina [2] to maintain grid regularity, which is particularly important where Navier-Stoke calculations are performed.

The network is formulated by connecting each block corner with hypothetical springs and corner positions are determined by a solution of the static equations. This simple and efficient calculation is performed on a single processor. Initially an unstructured grid network is constructed on the root processor. This contains nodal locations for each block corner and the connecting node information. New nodal positions are determined for free nodes through a predictor-corrector scheme. These are distributed to the respective processors, where transfinite interpolation (TFI) is performed to interpolate the local grid at the previous time step to the new position. The details of this method may be found in Wong [24].

6 Model Validation and Results

Forced Airfoil Oscillation

To demonstrate the validity of the moving mesh, multiblock and unsteady implementation, the NACA64A010 case is presented. Computational results are compared in Figure 4 for different configurations and with experiment. An unsteady Euler calculation is performed in the flow solver. In the first case, a single block O-grid is used in combination with TFI to deform the grid to the oscillating airfoil. The far field boundary remains rigid. The second case involves a mesh that is not deformed, but rotates rigidly with the displacement of the airfoil surface. For the third calculation, the same grid as used for the single block cases is divided into 32 equal blocks, with 4 blocks in the radial direction and 8 in the circumferential direction. In this case the block corners were located using the spring analogy. The results for inviscid flow compare similarly with results presented elsewhere [1, 17] and there is little difference between the results for the different configurations.

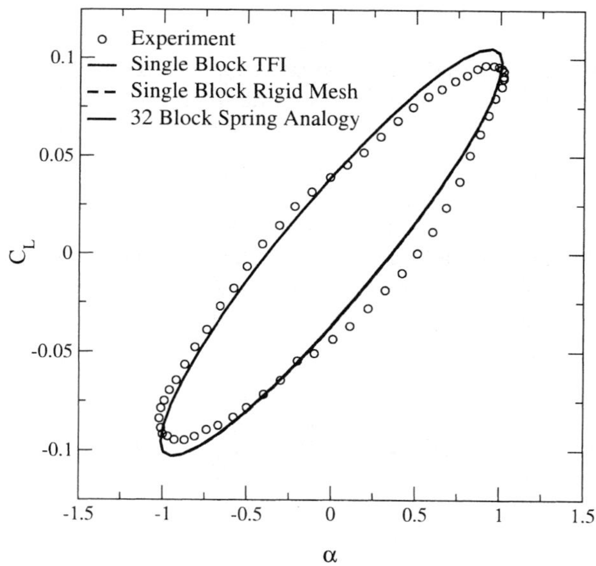

Figure 4 Comparison of unsteady NACA64A010 results with experiment.

7 Coupled Airfoil Model

With the integration of the structural solver, it was necessary to validate the implementation of the coupled model. The Isogai wing model [9] is a simple case that exhibits unsteady fluid-structure interaction and has proved useful in testing numerical models. It has been used previously by other researchers where the mesh moved in a rigid fashion [1]. The structural parameters for the case were chosen to simulate the vibrational characteristics of a swept back wing that are often used in military fighter aircraft.

The model is shown in Figure 5. Note the springs attached to both the plunging and pitching axes and the axis of rotation is actually well forward of the airfoil leading edge. Initially the airfoil is forced to oscillate for one period in pitch and then released. Once released the structural equations are applied to ascertain the new airfoil location at every time step. The aerodynamics force the response of the structural system. Inner iterations are used whereby the structural equations are updated within each real fluid time step. The initial oscillation was necessary to perturb the model from rest, as some disturbance is required to move the model from a stable configuration.
The important parameters for this simulation are the free stream Mach number Ma and the flutter velocity V_f. The flutter velocity is defined as

$$V_f = \frac{U_\infty}{b\omega_\alpha\sqrt{\mu}}. \tag{11}$$

This is used to determine the effect of freestream velocity U_∞ on the stability and involves the ratio of fluid momentum to structural inertial terms,

$$\mu = \frac{m}{\pi\rho b^2},$$

the airfoil chord b, the fluid density ρ and the structural natural frequency ω_α.
Predicted results for the flutter boundary are shown in Figure 6 and compare well with those of Alonso [1]. The flutter boundary is defined as where the amplitude of oscillation neither increases nor decreases with time. Simulations were carried

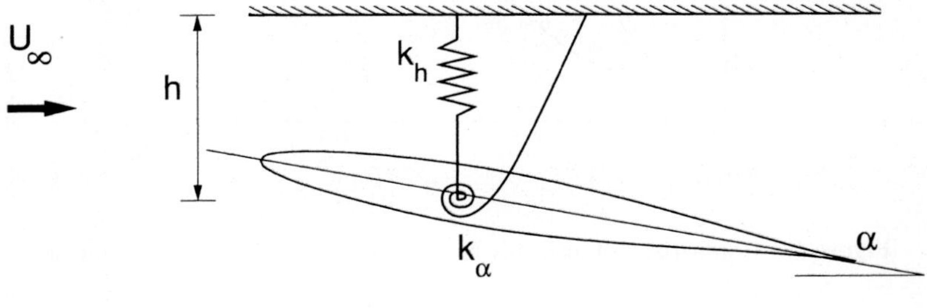

Figure 5 Isogai wing model.

out at fixed far field Mach number while the flutter velocity was varied. Each point required approximately five simulations to locate the flutter boundary. The line of best fit indicates the flutter boundary for the model.

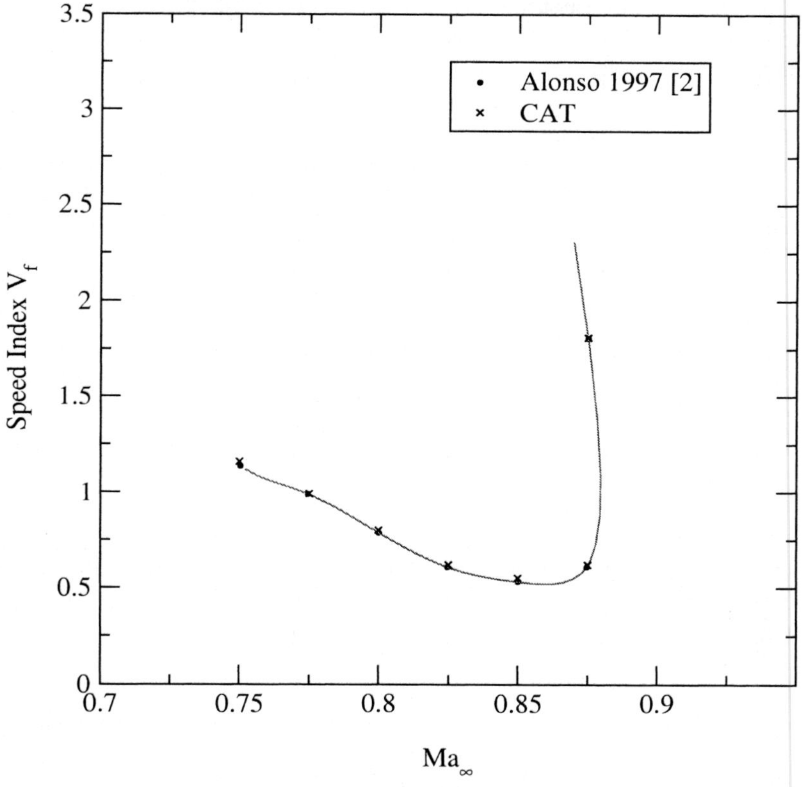

Figure 6 Flutter boundary for Isogai wing model

8 Turbomachinery Cascade

Few 3-dimensional experimental measurements exist in the field of unsteady aerodynamics in turbomachinery. A large effort has been made to compile measurements for oscillating cascades that are typical of those found in industry through the Workshop on Aeroelasticity in Turbomachines [3], however measurements in these cases are made at mid span and blade motion is symmetric in the radial plane to minimise the 3-dimensional effects on results.

The Standard Test Case 4 is described as a highly loaded turbine rotor, involving typical sections of modern free standing turbine blades [3]. The flow is high subsonic and the blade normally exhibits flutter in the first bending mode. To simulate the unsteady flow, the blade is forced to oscillate in this bending mode by translation at an angle to the axial axis in the radial plane. Viscous effects in two-dimensional models have been simulated numerically for this case by other authors using the Navier-Stokes equations with the algebraic Baldwin Lomax Model [7] and the k-ω turbulence model [12].

Recently, complete conditions at the cascade inlet and outlet plane have been made available for Standard Test Case 4. This allows the authors to validate the 3- dimensional implementation and investigate the difference between a number of different cascade models. Blade stability will be calculated by way of the energy method.

Inviscid and Navier-Stokes simulations were performed for both 2-dimensional and 3-dimensional configurations of Standard Test Case 4. Within Test Case 4 there are a number of measurements; results here are compared with Test 627. In this case, the passage flow was in the high subsonic regime with a reduced frequency of $k_c = 0.1187$ and involving a bending amplitude of $b_c = 3.8 \times 10^{-3}$. At the inlet $Ma_{in} = 0.18$ and at the outlet $Ma_{out} = 0.9$. It was found that using the present method approximately 4 oscillations were required for a converged unsteady solution as described in previous work [12].

The unsteady results were investigated along the span of the blade, as shown in Figure 7. The effects of the passage vortices on the unsteady pressure coefficient appear minimal and the coefficient on the suction side collapses onto each other towards the trailing edge. However at 10 percent span for the inviscid result, the suction side pressure deviates from the viscous results. Unsteady pressure coefficients in this figure are referenced to the static pressures at the mid-span of the passage inlet.

9 Concluding Remarks

Two lines of research, involving different numerical techniques for parallelised processing, are looking at the problem of transition to turbulence and of aeroelastic response. The spectral element method allows high order prediction of flow instabilities; in particular, it has been applied to the problem of highly swirling flows and vortex breakdown. A novel multiblock and parallel, integrated structural and fluid solver has been used for the investigation of aeroelastic response in compressible flows. The implementation is general and is not limited to particular geometries and thus is flexible in that it may be applied to a broad range of problems. The moving mesh and structural model allow for the coupled solution of aeroelastic problems. A number of different cases have been presented that compare computed results with experiment or other numerical results.

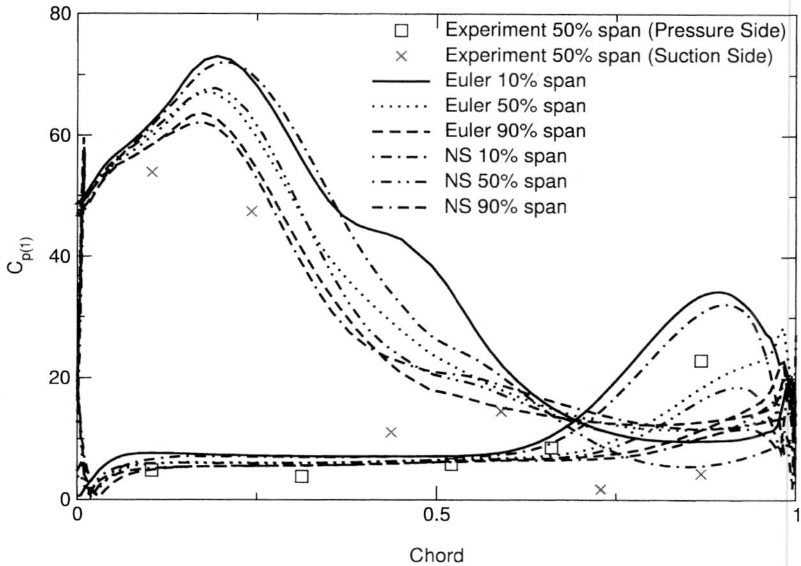

Figure 7 Comparison of unsteady pressure coefficient at different span-wise locations for 3D model for IBPA=180.

Acknowledgments

Ivan McBean acknowledges the support of an Australian Postgraduate Award. Computing support from VPAC and APAC are gratefully acknowledged. Michael Jones is thanked for running the spectral element simulations.

References

[1] Alonso, J. & Jameson, A. (1994) Fully-implicit time-marching aeroelastic solutions. In *AIAA 32nd Aerospace Sciences Meeting*. AIAA, January. 94-0056.

[2] Batina, J. T. (1990) Unsteady Euler airfoil solutions using unstructured dynamic meshes. *AIAA Journal*, **28**(8), 1381–1388.

[3] Bölcs, A. & Fransson, T. H. (1986) *Aeroelasticity in turbomachines: Comparison of theoretical and experimental cascade results*. Lausanne: EPFL.

[4] Canuto, C., Hussaini, M.Y., Quarteroni, A. & Zang, T. (1987) *Spectral methods in fluid dynamics*. Springer-Verlag.

[5] Faler, J. H. & Leibovich, S. (1978) An experimental map of the internal structure of a vortex breakdown. *J. Fluid Mech.* **86**, 313-335.

[6] Gelfgat, A.Yu., Bar-Yoseph, P. Z. & Solan, A. (2001) Three-dimensional instability of axisymmetric flow in a rotating lid-cylinder enclosure *Journal of Fluid Mechanics*, **438**, 363-377.

[7] Grueber, B. & Carstens, V. (1998) Computation of the unsteady transonic flow in harmonically oscillating turbine cascades taking into account viscous effects. *Journal of Turbomachinery*, **120**, 104-111.

[8] Hourigan, K, Thompson, M. C. & Tan, B. T. (2001) Self-Sustained Oscillations in Flows around Long Flat Plates. *J. Fluids and Structures*, **15**, 387-398.

[9] Isogai, K. (1979) On the transonic-dip mechanism of flutter of a sweptback wing. *AIAA Journal*, **17**(7), 793–795.

[10] Jameson, A. (1991) Time dependent calculations using multigrid, with applications to unsteady flows past airfoils and wings. In *AIAA 10th Computational Fluid Dynamics Conference*. AIAA, June.

[11] Jeong, J. & Hussain, F. (1995) On the identification of a vortex. *J. Fluid Mech.*, **285**, 69-94.

[12] Ji, S. & Liu, F. (1999) Flutter computation of turbomachinery cascades using a parallel unsteady Navier-Stokes code. *AIAA Journal*, **37**(3), 320–327.

[13] Karniadakis, G. E., Israeli, M. & Orszag, S. A. (1991) High-Order splitting methods for incompressible Navier-Stokes equations. *J. Comp. Phys.*, **97**, 441.

[14] Liu, F., Cai,J., Zhu, Y., Tsai, H. M. & Wong, A. S. F. (2000) Calculation of wing flutter by a coupled CFD-CSD method. In *AIAA 38th Aerospace Sciences Meeting & Exhibit*. AIAA, January. AIAA 2000-0907.

[15] Liu, F. (1991) *Numerical Calculation of Turbomachinery Cascade Flows*. PhD thesis, Princeton University, May.

[16] Liu, F., Ian K. Jennions, I. K. & Jameson, A. (1998) Computation of turbomachinery flow by a convective-upwind-split- pressure (CUSP) scheme. In *36th Aerospace Sciences Meeting and Exhibit*. AIAA, January.

[17] Liu, F. & Ji, S. (1996) Unsteady flow calculations with a multigrid navier-stokes method. *AIAA Journal*, **34**(10),2047–2053.

[18] Liu, F. & Zheng, X. (1994) Staggered finite volume scheme for solving cascade flow with a k-ω turbulence model. *AIAA Journal*, **32**(8),1589–1597.

[19] Sadeghi, M. and Liu, F. (2000) Computation of mistuning effects on cascade flutter. In *AIAA 38th Aerospace Sciences Meeting & Exhibit*. AIAA, January. AIAA 2000-0230.

[20] Sotiropoulos, F. & Ventikos, Y. (2001) The three-dimensional structure of confined swirling flows with vortex breakdown. *J. Fluid Mech.*, **426**, 155-175.

[21] Tan, B. T., Thompson, M.C. & Hourigan, K. (1998) Flow Around Long Rectangular Plates under Cross Flow Perturbations. *Int. J. of Fluid Dynamics (WWW)*, **2**, Article 1.

[22] Thompson, M. C., Hourigan, K. & Sheridan, J. (1996) Three-Dimensional Instabilities in the Wake of a Circular Cylinder. *International Journal of Experimental Heat Transfer, Thermodynamics, and Fluid Mechanics*, **12**, 190-196.

[23] Thompson, M. C., Leweke, Th. & Provansal, M. (2001) Kinematics and dynamics of sphere wake transition. *Journal of Fluids and Structures*, **15**, 575-586.

[24] Wong,A. S. F., Tsai, H. M., Cai, J., Zhu, Y. & Liu, F. (2000) Unsteady flow calculations with a multi-block moving mesh algorithm. In *38th Aerospace Sciences Meeting and Exhibit*. AIAA, January. AIAA-2000-1002.

Explicit Finite Element Modelling of Impact Events on Composite Aerospace Structures

M.L. SCOTT[1], M.Q. NGUYEN[1], D.J. ELDER[1], J. BAYANDOR[1], S.P. RAJBHANDARI[1] and R.S. THOMSON[2]

[1] Sir Lawrence Wackett Centre for Aerospace Design Technology
Department of Aerospace Engineering, Royal Melbourne Institute of Technology
GPO Box 2476V, Melbourne, Victoria, 3001, Australia

[2] Cooperative Research Centre for Advanced Composite Structures
506 Lorimer Street, Fishermans Bend, Victoria, 3207, Australia

Summary

Advanced fibre composite materials are now widely used in aerospace structures due to their superior performance characteristics such as high specific strength and stiffness. However, their susceptibility to non-visible or barely-visible impact damage is a significant constraint in achieving optimum structural performance. Recent developments in explicit finite element codes such as MSC.Dytran, LS-Dyna, Pam-Shock and Radioss indicate that there is now considerable potential for the accurate prediction of transient dynamic behaviour and onset of critical impact damage in composite structures. A series of test cases has been developed to investigate the critical parameters associated with the modelling of composite stiffened panels used in large civil transport aircraft components.

1 Introduction

The finite element (FE) software codes, LS-Dyna (Version 950e), MSC.Dytran (Version 2000), Pam-Shock (Version 2000) and Radioss (Version 4.1), are becoming common commercial tools employed within various engineering industries. Both the aerospace and automotive industries have accepted simulation as part of the design process to minimise costs and to create more efficient structures. Prototyping and testing are always performed to verify the design, but simulation has become standard practice throughout the design process.

As explicit FE codes improve and advanced material models become available, such tools will find more widespread application within the aerospace sector, as 'what-if' simulations become manageable with increasing computing power and greater modelling realism. This study aimed to determine the readiness of existing explicit FE analysis tools to predict barely visible impact damage (BVID) within stiffened composite panels, using a single layer 2-D shell model. This is achieved by evaluating each of the four aforementioned programs in their ability to construct an FE model of the stiffened panel, solve for BVID and retrieve results which are compared to experimental data. The ability of each code to model the damage that arises from the impact loads provides a gauge to the suitability of the codes for composite design and analysis.

As the aim of this study was to evaluate the explicit, dynamic, analytical capabilities of various software codes specific to the requirements of BVID analysis, it was necessary to establish a standardised approach in which to undertake the evaluation process. Some features that were highlighted early in the study as important software features were failure models, material data entries, computation time and overall solution accuracy. From these criteria, a benchmark model was developed and analysed within LS-Dyna, MSC.Dytran, Pam-Shock and Radioss.

When considering structural analysis applications, implicit FE methods can be used in static, dynamic time and frequency domain analyses, where linear or moderate nonlinear effects are investigated. The implicit method formulates groups of matrices that allow the structural problem to be characterised by mathematical representation of key qualities of the structure, such as mass and stiffness. With the use of a computer, solutions of the matrix relationships can be obtained to allow a solution to the boundary value differential equation sets. This requires the inversion of the stiffness matrix, which is a computationally expensive task. This is compared to the explicit FE method where only dynamic time domain analysis is possible. This approach employs a simpler approach by iteratively solving the basic equation (force = mass x acceleration). The solution is progressed from a known boundary condition and incremented with very small time steps to produce a sequential solution to the initial value boundary problem.

The advantage with explicit FE codes over implicit methods is that the nature of the computational method (i.e. extremely small time steps coupled with an iterative solution method) produces an unequalled ability to solve time domain dynamic problems with extreme nonlinearity from material and geometrical effects. In this study of BVID resulting from low speed impacts, the geometric nonlinear ability of the explicit codes is used in the form of impactor / composite contact and the large deflections resulting from impact. The nonlinear material damage analysis capabilities of Pam-Shock are also used, while the other codes considered (MSC.Dytran, LS-Dyna and Radioss) only consider elastic damage.

2 The Benchmark Problem

The use of fibre reinforced composite materials, such as carbon, glass or aramid, has been steadily increasing in nearly all areas of the transport industry. Within the aerospace sector, these materials are finding widespread application in secondary structures like rudders, flaps, fairings and wing covers [1,2]. As a result of their low density and high stiffness properties, coupled with new cost-effective manufacturing technologies, interest has been developing in the application of such materials as primary structures for civil transport aircraft.

Unfortunately composite materials suffer from a major drawback, which is the difficulty in detecting damage arising from impact or defects. A risk associated with the maintenance and application of advanced composite materials is that catastrophic failure of the plies embedded within the composite laminate can occur after being subjected to an impact load, even though on the surface, the material may outwardly appear to be structurally intact. This may lead to the false impression that the component still retains its mechanical properties, when this may not necessarily be the case, thereby giving rise to a host of potential safety issues [3].

Over the past decade, test programs coupled with explicit finite element (FE) modelling codes have been employed to develop an understanding of the effects of impacts and the damage caused by such events on the mechanical performance of advanced composites [4,5]. However, there has been a growing trend for researchers over recent times to favour approaches using FE models, in comparison to the time consuming and costly practices associated with experimental testing. This paper therefore, concentrates on the impact analysis and damage response of composite stiffened panels. It details the modelling techniques used to generate results and compares the FE solution with experimental test data. The following points were considered in choosing the benchmark problem for evaluation of the codes:

(i) Impactor Parameters: The impactor condition (energy, mass, velocity and geometry) has an important effect on damage state. BVID typically results from low-velocity impact incidents such as dropped tools ($v_0 \leq 5$ ms, $m_i \leq 5$ kg, $E_{kinetic} \leq 60$ J). For low velocity impacts, the delamination and matrix damage area is proportional to the impact energy. At higher velocity, fibre fracture becomes dominant and delamination area decreases. The low-velocity state of the impactor was used in the benchmark problem and is shown in Figure 1.

(ii) Geometry: Research with coupon specimens [6,7] has shown that impact damage can lead to reductions as high as 50% in compressive properties. Other testing work conducted on stringer-stiffened panels [8] has shown that the reduction in compressive strength due to low energy impact can be up to 20%.

This indicates that the structural geometry, which dictates the elastic response under impact, has a strong influence on the form and severity of impact damage [8,9,10]. A stiff structure will develop a very high contact force resulting in a large shear stress along the mid-plane which can result in a large internal shear delamination. A flexible structure, in contrast, develops high bending strains, resulting in a large amount of surface failure due to compression or tension.

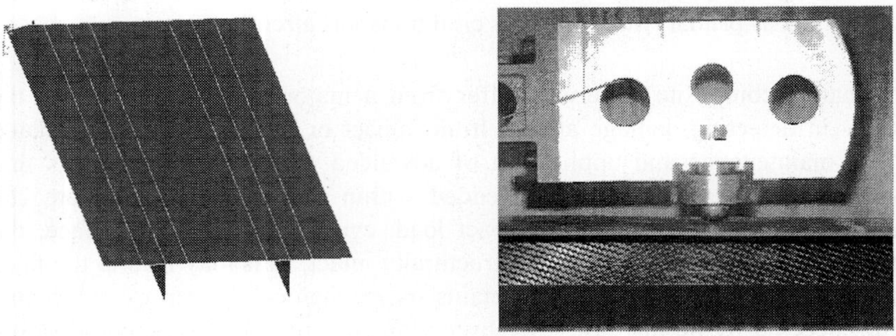

(a) (b)

Figure 1 (a) Geometry of the stiffened panel and (b) impactor contact on the panel

3 Composite Impact Degradation Models

The damage development in laminated composites subjected to impact is quite complex. This is due to the fact that there are several interacting failure modes present during impact. To predict damage behaviour, it is required that impact forces and induced stresses are fully determined and an appropriate failure criterion for initial failure identified. Stiffness is a dominant parameter and controls the mode of fracture. At low velocities, more flexible structures mainly respond by bending, which produces high tensile stresses in the lowest ply deflected and at the lowest interface to form a delamination. This delamination, in turn, is deflected by the matrix cracks of the upper layer, whereas at intermediate velocities, damage also occurs due to high contact stresses on the impact surface. Both types of damage modes are present in the case considered in this work. For stiffer composite structures however, although the damage pattern is similar, they appear in a reverse order. For low velocity impacts, damage is initiated by contact stresses. Bending-induced damage is observed for the medium level impact event. In high velocity impacts, different failure modes can be recorded through microscopic examination of the specimens. The two different types of damage development in composite components are shown in Figure 2.

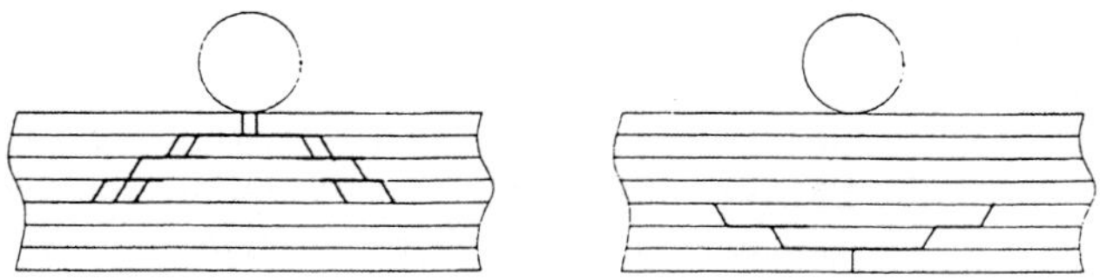

(a) (b)

Figure 2 Damage development in (a) rigid and (b) flexible structures [11]

3.1 The MSC.Dytran Degradation Model

An issue of critical importance is modelling the behaviour of the composite lamina and laminate following initial failure. The first aspect of degradation modelling is defining which properties are degraded depending on the type of failure that has occurred. For instance, Table 1 shows the default settings in MSC.Dytran for the modes of failure and the associated properties that are degraded. It is possible to select different failure theories for each failure mode.

In order to avoid numerical instabilities or shock waves from developing in the analysis following failure, the properties affected by failure cannot be reduced to zero in one time interval. The properties must be decayed over a number of time steps, or over a period of time, to avoid such instabilities. This decay is also required from a physical point of view since in the analysis, the structure is discretised into finite elements. Failure, that in reality occurs at a point then propagates, is numerically the failure of an entire element. Hence, numerical failure is directly related to the element size, and therefore parameters that control the rate of decay must also relate to the element size. Most codes have parameters in which the user can define the number of steps or the time period over which the degradation occurs. However, few codes have the capability to include a term representing the residual strength.

Table 1 Default degradation rules in MSC.Dytran [12]

Material Constant	**Failure Mode**				
	Fibre Tension	Fibre Compression	Matrix Tension	Matrix Compression	Shear
E_{11}	X	X			
E_{22}	X	X	X	X	
G_{12}	X	X	X	X	
ν_{12}	X		X		X

3.2 LS-Dyna Degradation Model

LS-Dyna basically has three standard degradation laws, which are for brittle, plastic and an 'evolutionary' type of failure. The brittle failure condition is applied when the composite ply is deemed to have reached a failure condition, the ply is immediately degraded to have zero stiffness and strength. The overall stiffness and stress distribution of the laminate is adjusted accordingly. This is associated with material models 22, 54 and 55 [13]. For plastic failure, when the composite ply has reached its failure condition, the ply stiffness is immediately degraded to zero, but still has the same force resistance contribution to the overall laminate. This is associated with material models 54 and 55.

For the evolution law failure condition, the stiffness of each ply can be represented by a continuous curve that has a linear elastic portion, followed by a small plastic section and finally a softening behaviour. The curve is smooth and continuous and represents a typical nonlinear material that may be associated with many types of real materials. This is associated with material model 58a and 58b where special treatments are provided for both fabric and uni-directional composite materials.

3.3 Pam-Shock Bi-Phase Composite Failure and Degradation Model

Two models are available in Pam-Shock to model laminated composite materials. The first is the bi-phase model and the second is termed the homogeneous model, which provides an overall description of the ply and its responses to dynamic loading. The bi-phase cumulative damage model distinguishes between the fibre and the matrix behaviour, failure and degradation and then superimposes the effects of the two phases. The bi-phase model projects a model of a heterogeneous material adapted to reinforced unidirectional composite materials with continuous fibres. This model was initially introduced to study the response of a full composite road vehicle during crash events.

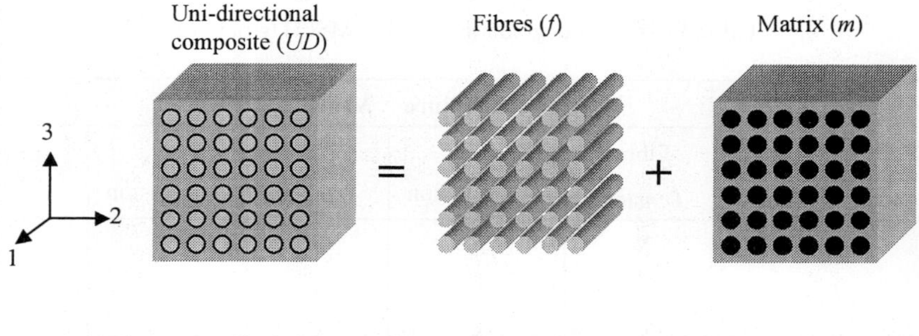

Figure 3 Bi-phase composite model available in Pam-Shock [14]

In the bi-phase model, the behaviour of a unidirectional ply is described by defining two phases, as shown in Figure 3. In this model, fibres have an uni-directional behaviour that is "brittle elastic with damage", while the matrix is "brittle orthotropic elastic", or "elastic damaging". The tensile and compressive behaviour of each phase is considered. The total material stiffness is determined through superposition of the two phases, whereas the stresses are calculated separately and, together with damage, can be propagated independently according to the criteria selected in the pre-processing stage.

3.4 Radioss Degradation Model

Radioss has a plastic degradation rule: when the composite ply has reached failure in accordance with the Tsai-Wu failure criterion, the ply stiffness is immediately degraded to zero, while retaining the force resistance contribution to the overall laminate. Like that of most other codes, Radioss also has the capability for the implementation of user defined material and failure laws.

4 Finite Element Flat Shells in Explicit Codes

Unlike the implicit codes where element choice is usually a function of optimising the quality of the output required, explicit codes have been designed for a different purpose and this is reflected in the elements provided. In normal explicit crashworthy analysis (for the car industry say), computational efficiency of the finite elements is of the utmost importance. As a general comment, the crashworthiness analysis requires ever increasing model sizes to keep pace with the current requirements of the transport safety industry. Because of an ongoing increase in model size over the past 20 years, the model run times for the car industry have been increasing in duration despite a continuous increase in computer speeds [15].

In responding to this need for fast computation times, much effort has gone into the development of elements that require the minimum mathematical instructions per time step. These elements are known as under integrated (UI) elements. They are formulated using the simplest of numerical constructs but will still provide robust predictions under large strain regimes. Additionally these explicit flat elements (unlike the implicit four-node shell formulations) are capable of significant warping - up to 20° without unduly affecting the element accuracy. All codes reviewed in this work have a selection of UI shell elements that satisfy the fast computation needs for large models.

Two of the codes reviewed, LS-Dyna and Pam-Shock, currently provide, in addition to the UI elements, a range of fully integrated, four-node, flat shell elements. These elements are known as fully integrated (FI) or selectively integrated (SRI) elements and are only available in the four-node flat shell. No curved FI or SRI elements are currently available in any of the codes reviewed, although MSC.Software is developing SRI elements for MSC.Dytran. Although these integrated elements provide better accuracy and guaranteed convergence, they are computationally about 4 to 20 times slower than the fastest UI element [15].

The analysis requirements for general crashworthiness analysis differ from those for the study of impact damage in composite structures. Where crashworthiness analysis deals with large models subject to extremely large deformations, the study of low speed impact generally requires much smaller models undergoing small deformations near the elastic limit of the material. The significance of this is that low speed impact analysis requires stress/strain accuracy predictions, not computational speed, and as yet, some of the explicit codes reviewed do not provide the most accurate 2-D shell element formulations that may be required.

4.1 Available 2-D Flat Shell Element Types

The total number of three- and four-node flat shell elements capable of bending, shear and membrane actions for each code is listed in Table 2, with detail of the four-node elements provided in Table 3.

Table 2 Available 2-D element types

Code	Avaliable 2-D Shell Types
MSC.Dytran	1 Tri and 3 Quad elements (Total of 4 flat 2-D elements)
LS-Dyna	3 Tri and 10 Quad elements (Total of 13 flat 2-D elements)
Pam-Shock	1 Tri and 3 Quad elements (Total of 4 flat 2-D elements)
Radioss	1 Tri and 2 Quad elements (Total of 3 flat 2-D elements)

All codes use one integration point per ply, with the exception of LS-Dyna, which can use more than one per ply. This is of benefit when a small number of thick plies are used and material degradation is specified (i.e. Elasto-Plastic stress distribution within a ply). All of the codes seem to provide comparable through-thickness integration abilities. For the other codes, the plies should be sub-divided to provide further integration points. Furthermore, all of the codes use the Mindlin - Reissner theory for determining transverse shear deformations. For isotropic plates, the factor '*k*' is used to convert the parabolic shear distribution for a rectangular section into an equivalent constant stress block for shear stress deformation [16]. All FE models in the benchmark study used $k = 5/6$, although this is not strictly correct for the unidirectional plies used in the skin.

MSC.Dytran, Pam-Shock and Radioss calculate average shear stress over the composite depth, meaning that all plies have the same transverse shear stress. LS-Dyna has the option of determining the ply shear stress in accordance with either the average value or with conventional theory to produce a more accurate estimate of the transverse shear stress that is unique to each ply for any given element.

Table 3 Typical four-node elements available (not all LS-Dyna elements shown)

Code	Element	Type	Plan Integration Points	Through Thickness Integration Points	Shear Distribution	Shell Bending Accuracy
MSC. Dytran	Key-Hoff	UI	1	1 per ply	Average	See note i
	Belytschko-Tsay	UI	1	1 per ply	Average	See note i
	Hughes-Lui	UI	1	1 per ply	Average	See note i
LS-Dyna	Hughes-Lui	UI	1	1 per ply	Avge or Calculated	See note i
	Belytschko-Tsay	UI	1	1 per ply	Avge or Calculated	See note i
	Belytschko-Wong-Chiang	UI	1	1 per ply	Avge or Calculated	See note i
	S/R Hughes-Lui	SRI	2x2	1 per ply	Avge or Calculated	See note ii
	Fully Integrated	FI	2x2	1 per ply	Avge or Calculated	See note iii
Pam-Shock	Belytschko-Tsay	UI	1	1 per ply	Average	See note i
	Belytschko-Wong-Chiang	UI	1	1 per ply	Average	See note i
	Hughes-Tezduyar	SRI	2x2	1 per ply	Average	See note ii
Radioss	Belytschko-Tsay	UI	1	1 per ply	Average	See note i
	Belytschko-Leviathan	UI	1	1 per ply	Average	See note i

Notes:

i) This element cannot guarantee convergence for out-of-plane bending.

ii) This element can guarantee convergence for out-of-plane bending.

iii) This element can guarantee convergence for out-of-plane bending, however it may suffer from an overly stiff solution due to shear lock.

5 Simulation Results and Discussion

The results from simulations using MSC.Dytran, LS-Dyna and Pam-Shock are presented below; results from Radioss are not presented due to its limited composite functionality. The FE model of the 1000 x 330 mm composite, blade-stiffened panel consisted of approximately 1000 shell elements, as shown in Figure 4(a). The 25 mm diameter impactor tup was idealised as a rigid shell, as shown in Figure 4(b). The panel mesh was refined considerably under the impactor to more accurately represent the expected local deformation, strains and damage development.

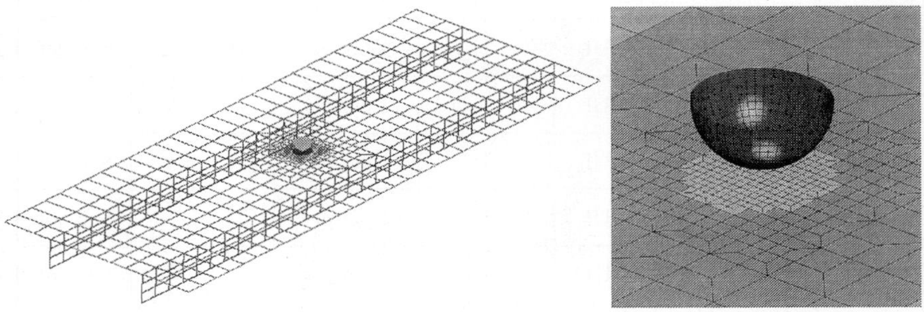

(a) (b)

Figure 4 (a) Typical FE mesh used in the impact analyses and (b) detail of the impactor

The effect of introducing damage and degradation into the MSC.Dytran FE simulation is shown in Figure 5 for 30 J and 40 J impacts. In this case, the Chang-Chang failure criterion coupled with brittle degradation was used. In both cases, the initiation of damage and degradation is associated with a sharp drop in contact force. This first drop occurred at a contact force of 4 kN for both energy levels. The introduction of damage and degradation resulted in a reduction in the magnitude of the peaks in the force-time history and also extended the contact duration. There were also more high frequency variations in the contact force due to the sudden degradation of the lamina stiffness properties as failure was predicted to occur.

Comparisons between the experimental and predicted force-time histories are presented in Figure 6 for the 30 J and 40 J impact cases. All FE simulation results for these cases include damage development and degradation. The Chang-Chang failure criterion with brittle degradation was used for both LS-Dyna and MSC.Dytran, while the bi-phase model was used for Pam-Shock. For both impact energies, the general shape of the curves and the peak forces were predicted reasonably well. However, all simulations predicted a trough between the two peaks that was much less significant in the test results. The most significant difference was that the predicted contact duration was shorter in the analyses than occurred during testing. The initial and final slopes of the force-time histories were predicted to be much steeper than the test results.

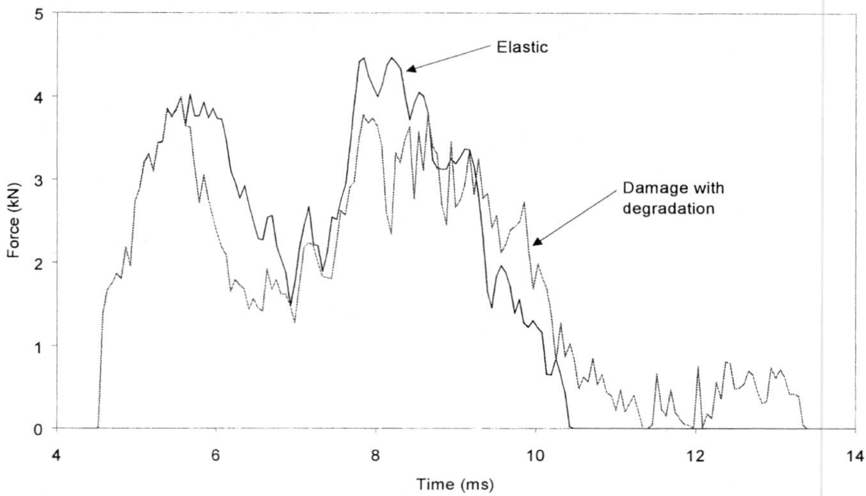

(a)

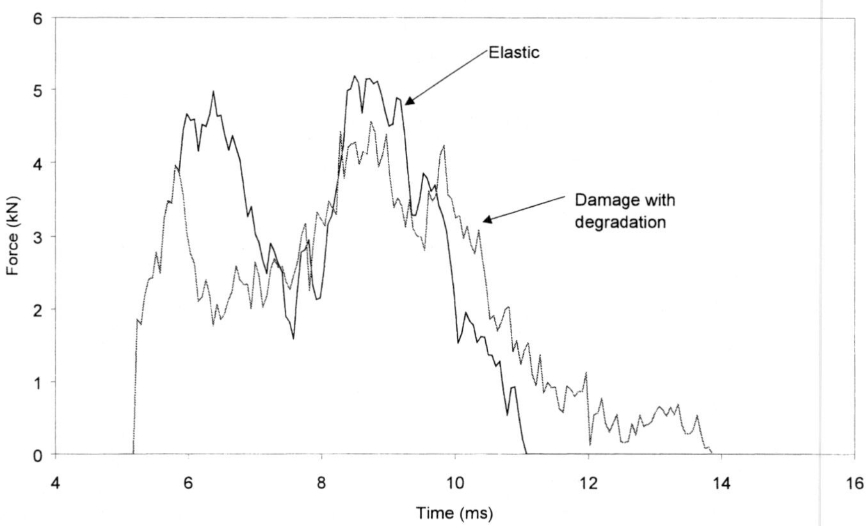

(b)

Figure 5 Effect of damage and degradation on the force-time histories for (a) 30 J and (b) 40 J impacts using MSC.Dytran

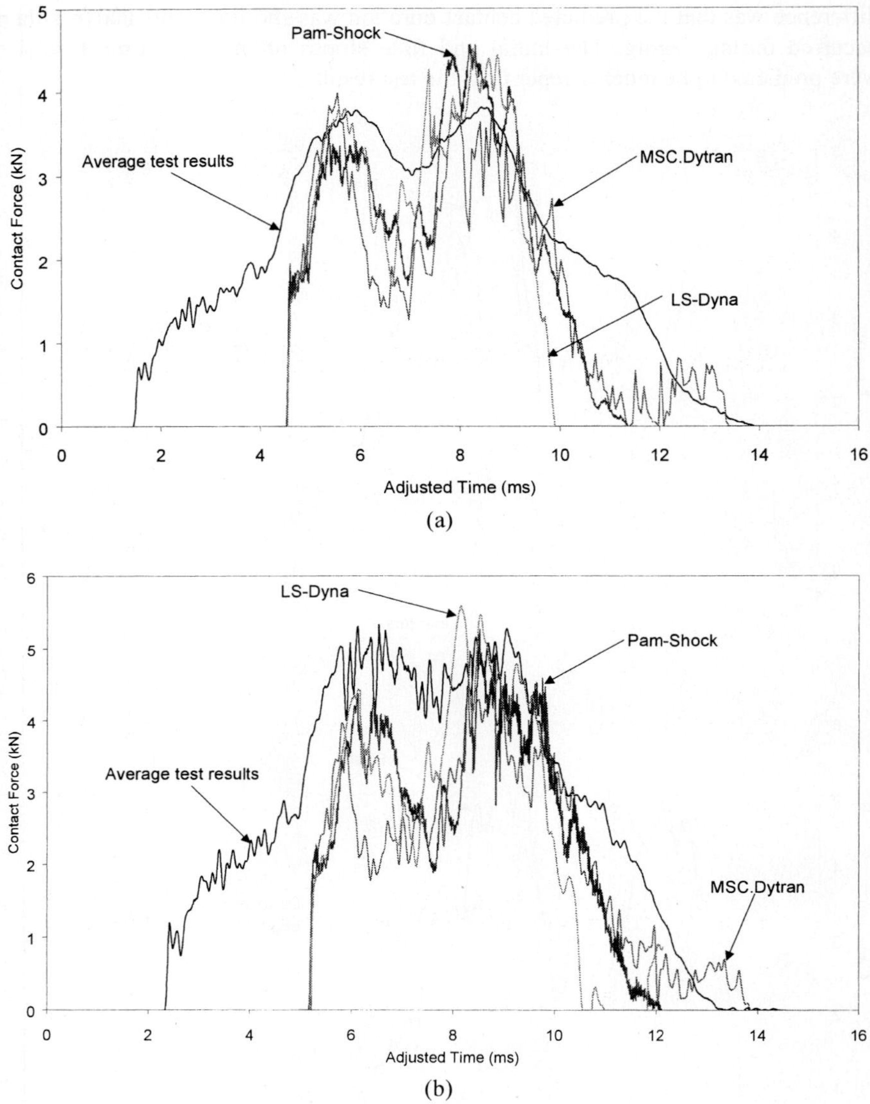

Figure 6 Test and analysis force-time histories for (a) 30 J and (b) 40 J impacts

The major cause for these discrepancies is believed to be the quality of the end boundary conditions used during testing. The clamping method, which consisted of four G-clamps onto wooden blocks, would have allowed significant movement and added a degree of damping to the system, thus affecting the dynamic response of the panel. A less significant source of error would be 3-D effects not accounted for in modelling the panel and impactor. In particular, modelling the impactor as a rigid projectile was a simplification of the actual pendulum impactor, which travels through an arc during the impact event, and has its own elastic dynamic response.

6 Conclusions

An ongoing study is being conducted into the readiness and suitability of four explicit FE analysis codes, LS-Dyna, MSC.Dytran, Pam-Shock and Radioss, for modelling impact events on laminated composite structures. It was found that each code offers unique analytical strengths in particular areas, but is limited in others. The results to date indicate that the force-time history of low velocity impact events, especially the peak forces, can be predicted with adequate accuracy. It was also found that the different composite failure and degradation models available had a small but significant influence on the force-time histories. In this study, limitations with the experimental boundary conditions adversely affected the contact duration comparison, which was under predicted in all simulations. Research is continuing into the prediction of damage in composite panels resulting from low velocity impact events.

Acknowledgements

The authors would like to acknowledge the assistance that they have received from Mr J.W.H. Yap and Mr S. McKay of the Sir Lawrence Wackett Centre for Aerospace Design Technology, and Dr D. Hachenberg of Airbus Deutschland. Also acknowledged is the support given to this project by Mr G. Diegelmann, Mr S. Kent and Mr E. de Vries of MSC.Software Australia, Mr D. McGuckin and Mr A. Chhor of Pacific ESI and Mr T. Nuygen of Advea Engineering.

References

[1] B.D. Agarwal and L.J. Broutman. "Analysis and Performance of Fiber Composites", 2nd Ed., John Wiley and Sons Inc, New York, 1990.

[2] B. Milson, E. Greenhalgh, R. Thompson and B. Driffle. "Investigation of the impact response of a CFRP wingbox under load for impact energies above the damage threshold", DERA/SMC/SM3/TR980152/1.0, February 1998.

[3] L.J. Hart-Smith. "Should fibrous composite failure modes be interacted or superimposed?" Composites, 24 (1), 1993, pp. 53-55

[4] S. Abrate. "Impact on laminated composite materials", Applied Mechanics Review, 44, 1991, pp. 155-189.

[5] S. Abrate. "Modeling of impacts on composite structures", Composite Structures, 51, 2001, pp. 129-138.

[6] D. Hull and B.Y. Shi. "Damage mechanism characterisation in composite damage tolerance investigations". Composite Structures, 23, 1993, pp. 99-120.

[7] W. Cantwell and J. Morton. "An assessment of residual strength of an impact damaged carbon-fibre reinforced epoxy". Composite Structures, 14, 1990, pp. 303-317.

[8] E. Greenhalgh, D. Hughes, S. Singh and D. Roberts. "Impact damage growth and failure of carbon-fibre reinforced plastic skin-stringer panels", Defence Research Agency, Hampshire, UK. DRA/SMC/CR961118/1.0, 1996.

[9] G.A.O. Davies and X. Zhang. "Impact damage predictions in carbon composite structures". International Journal of Impact Engineering, 16 (1), 1995, pp. 149-170.

[10] X. Zhang. "Impact damage in composite aircraft structures - experimental testing and numerical simulation". Journal of Aerospace Engineering, 212 (4), 1998, pp. 245-259.

[11] J. Bayandor, R.S. Thomson, M.L. Scott, M.Q. Nguyen and D. Elder. "Investigation of impact and damage tolerance in advanced aerospace composite structures", I-Crash 2002, Melbourne, Australia, 25-27 February, 2002 (to be published in the International Journal of Crashworthiness).

[12] MSC.Software Corporation. "MSC.Dytran Version 4.7 Users Manual", Los Angeles, 1999.

[13] LS-Dyna. "Key Word Users Manual - Volume II", Livermore Software Technology Corporation, Version 960, 2001.

[14] D. Coutellier and P. Rozycki. "Multi-layered, multi-material finite element for crashworthiness studies", Composites Part A, Vol. 31, 2000, pp. 841-851.

[15] P.A. Du Bois. "Crashworthiness Engineering with LS-Dyna", Notes from course at RMIT by ADVEA Engineering, 1999.

[16] J.M. Whitney and N.J. Pagano. "Shear deformation in heterogeneous anisotropic plates", Journal of Applied Mechanics, 1970, pp. 1031-1036.

Dynamics and Control of Smart Structures Using Piezoelectric Actuators and Sensors

LIYONG TONG and DONGCHANG SUN
School of Aerospace, Mechanical and Mechatronic Engineering,
The University of Sydney, NSW 2006, Australia

Summary

In this paper, three piezoelectric modal sensor/actuator design methods for vibration control of smart plates are introduced. In the first method, the modal actuator and modal sensor are designed by modulating the thickness of the piezoelectric actuator and sensor layers. In the second method, the distributed piezoelectric element (DPE) method, the modal actuator and sensor are designed based on segmentation of the piezoelectric actuator layer fully covered on the host plate. In the third method, the quasi-modal sensors and quasi-modal actuators are developed using several piezoelectric patches, and optimization criteria are given for finding the optimal locations and sizes of these patches in order to minimize the observation and control spillovers.

1 Introduction

Smart structures with distributed piezoelectric sensors and actuators have been attracting significant studies since the middle of 1980s. Revolutionary changes have been made in vibration control of flexible structures due to the replacement of the traditional point sensor and actuators by the distributed ones. The independent modal space control (IMSC), which is based on modal sensors and modal actuators, is very effective because all the control energy can be fully concentrated on the target modes. However, the exact modal sensor and actuators cannot be realized unless using the distributed piezoelectric layers. Lee and Moon ([1],[2]) present a method to design the modal sensors/actuators by shaping the electric pattern of the piezoelectric layers, and based on which, critical active damping in vibration control of a beam is performed successfully. Later on, this idea is generalized to the modal control of the smart plates and shell structures ([3]-[8]). However, it is difficult to apply this method to the two-dimensional structures, such as plates and shells, except some special cases.

In this paper, three modal sensor/actuator designing methods for vibration control of smart plates are introduced. In the first method, the modal actuator is designed to excite one or more selected modes simultaneously by shaping the thickness of one piezoelectric actuator layer, while the piezoelectric sensor layer can also be made to sense one or several designated modes in a similar way. In the second method, the distributed piezoelectric element (DPE) method [9],[10], the modal actuator and sensor are created based on the segmentation of the piezoelectric actuator layer fully covered on the host plate. In the third method, the quasi-modal sensors and quasi-modal actuators [11] are developed by using the discretely distributed piezoelectric patches, and two practical criteria are given for finding the optimal locations and sizes of both piezoelectric sensor and actuator elements. Simple examples for the three methods are also given.

2 Basic Equations for the Piezoelectric Smart Plates

Consider a thin plate, on both surfaces of which two piezoelectric layers are bonded as the distributed sensor and actuator, as shown in Figure 1. Assume that the piezoelectric layers are much thinner than the host plate and they are perfectly bonded, and that the bonding layers are so thin that their effects on the whole plate can be neglected.

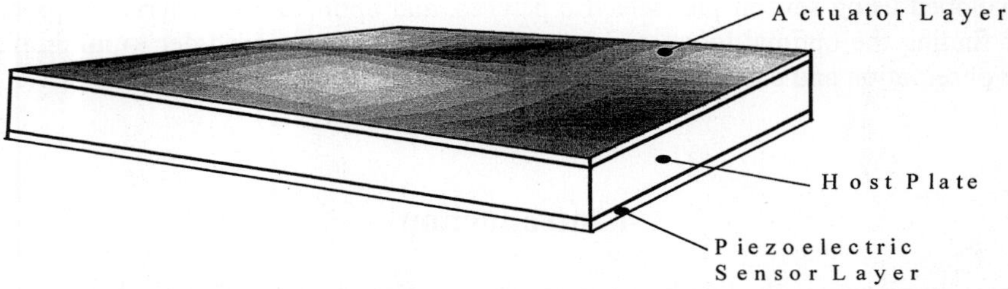

Figure 1 The plate with piezoelectric layers

The charge and current output of the sensor layer can be derived as

$$q(t) = -\iint_S F_s(x,y)\nabla^2 w\mathrm{d}x\mathrm{d}y, \quad I(t) = -\iint_S F_s(x,y)\nabla^2 w_{,t}\mathrm{d}x\mathrm{d}y \tag{1}$$

where $q(t)$ and $I(t)$ are the charge and current outputs generated by the piezoelectric sensor layer respectively, $w(x,y,t)$ the transverse displacement of the smart plate, S the area covered by the sensor layer, $F_s(x,y) = e_{31}^s(x,y)r_s(x,y)$ the spatial distribution function of the sensor layer, $e_{31}^s(x,y)$ the piezoelectric stress coefficient, $r_s(x,y)$ the z-coordinate of the mid-plane of the sensor layer from the neutral plane of the smart plate.

The differential equation of motion of the smart plate can be derived as

$$\rho h w_{,tt} + \nabla^2(D\nabla^2 w) = -\nabla^2[F_a(x,y)V(x,y,t)] \tag{2}$$

where $\rho h(x,y)$ is the equivalent mass density in unit area of the plate, $D(x,y)$ the equivalent bending stiffness of the plate, $F_a(x,y) = e_{31}^a(x,y) r_a(x,y)$ the spatial distribution function of the actuator layer. The superscript or subscript s and a represent the sensor and actuator respectively.

3 Modal Actuator/Sensor by Modulating Thickness of Piezoelectric Layers

3.1 Modal Actuator Design

The modal actuator can actuate the selected one or more modes without affecting the remaining modes. The modal actuator can be designed by employing the orthogonality of the modal functions. Consider the case that the voltage applied on the actuator layer is distributed uniformly in space, i.e. $V(x,y,t) = V(t)$.

The transverse displacement $w(x,y,t)$ can be expressed as

$$w(x,y,t) = \sum_{i=1}^{\infty}\sum_{j=1}^{\infty} \eta_{ij}(t) W_{ij}(x,y) \tag{3}$$

where $\eta_{ij}(t)$ and $W_{ij}(x,y)$ are the ij th modal coordinates and modal shape functions. Substituting equation (3) into equation (2), we have

$$\ddot{\eta}_{ij}(t) + \omega_{ij}^2 \eta_{ij}(t) = -V(t) \iint_S \nabla^2 [F_a(x,y)] W_{ij}(x,y) dxdy \quad i,j = 1,2,\cdots,\infty \tag{4}$$

If the spatial distribution function of the actuator layer is designed as

$$F_a(x,y) = A_a D \sum_{k=1}^{K}\sum_{l=1}^{L} A_{kl}^a \nabla^2 W_{kl}(x,y) + C_a \tag{5}$$

where A_{kl}^a, A_a and C_a are constants which can be adjusted to form an easily realized distribution function, substituting equation (5) into equation (4) will yield

$$\ddot{\eta}_{ij}(t) + \omega_{ij}^2 \eta_{ij}(t) = \begin{cases} -A_a A_{ij}^a \omega_{ij}^2 V(t) & i \le K, \ j \le L \\ 0 & \text{otherwise} \end{cases} \tag{6}$$

which shows that the actuator layer can produce the modal forces for actuating the selected modes without exciting the other modes.

A feasible method to realize the modal actuator is to modulate the z-coordinate of the mid-plane of the actuator layer by changing its thickness and keeping the piezoelectric stress coefficient e_{31}^a as a constant. In this case, the z-coordinate of the mid-plane of the actuator layer can be determined as

$$r_a(x,y)=\frac{1}{e_{31}^a}F_a(x,y)=\frac{1}{e_{31}^a}\left[A_aD\sum_{k=1}^{K}\sum_{l=1}^{L}A_{kl}^a\nabla^2W_{kl}(x,y)+C_a\right] \tag{7}$$

The relationship between the z-coordinate of the mid-plane of the actuator layer and its thickness can be expressed as

$$r_a(x,y)=[z_2+(z_2+h_a)]/2 \tag{8}$$

where z_2 is the z-coordinate value of the interfaces between the piezoelectric layers and the host plate. The thickness of the actuator, h_a, can be obtained as

$$h_a(x,y)=\frac{2}{e_{31}^a}\left[A_aD\sum_{k=1}^{K}\sum_{l=1}^{L}A_{kl}^a\nabla^2W_{kl}(x,y)+C_a\right]-2z_2(x,y) \tag{9}$$

To ensure positive thickness of the actuator layer everywhere so that its polarity does not need to be changed, the constants A_a and C_a in equation (5) should be selected properly.

3.2 Modal Sensor Design

Only the information of the selected modes is included in the output signal of a modal sensor, whereas information of the remaining modes is not mixed. Employing Green's formula for equation (1), we have

$$q(t)=-\iint_S w\nabla^2F_s dxdy-\oint_\Gamma w\frac{\partial F_s}{\partial n}d\Gamma+\oint_\Gamma F_s\frac{\partial w}{\partial n}d\Gamma \tag{10}$$

Substituting equation (3) into equation (10), yields

$$q(t)=-\sum_{i=1}^{\infty}\sum_{j=1}^{\infty}\eta_{ij}\left[\iint_S W_{ij}\nabla^2F_s dxdy+\oint_\Gamma W_{ij}\frac{\partial F_s}{\partial n}d\Gamma-\oint_\Gamma F_s\frac{\partial W_{ij}}{\partial n}d\Gamma\right] \tag{11}$$

If the spatial distribution function of the sensor layer satisfies

$$\begin{aligned}&F_s(x,y)=A_sD\sum_{k=1}^{K}\sum_{l=1}^{L}A_{kl}^s\nabla^2W_{kl}(x,y)+C_s && x,y\in S\\ &F_s=0,\ \frac{\partial F_s}{\partial n}=0 && x,y\in\Gamma\end{aligned} \tag{12}$$

then employing the orthogonality of the modal functions, the following equation will be derived

$$q(t)=-A_s\sum_{k=1}^{K}\sum_{l=1}^{L}A_{kl}^s\omega_{kl}^2\eta_{kl}(t),\ I(t)=-A_s\sum_{k=1}^{K}\sum_{l=1}^{L}A_{kl}^s\omega_{kl}^2\dot{\eta}_{kl}(t) \tag{13}$$

where A_s and C_s are constants. Equation (13) shows that the charge output of the sense layer with the selected distribution function is proportional to the designated modal coordinate, and therefore, such a piezoelectric sensor layer becomes a modal sensor. The spatial distribution of the sensor layer can be solved from equation (12), and then its thickness function can be determined. In fact, if the plate is clamped along its four sides, the boundary condition in equation (12) can be satisfied automatically. For a simply supported plate, the second boundary condition can be satisfied too, and the first boundary condition can be satisfied by choosing $C_s = 0$. In this case, to keep the thickness positive, the polarity direction of the parts with negative thickness should be reversed.

3.3 Simulation Example

Consider a 1m×0.7m×1mm simply supported rectangular plate with fully covered actuator and sensor layers, which are made of the same PZT material whose piezoelectric stress constant is 23.31 N/Vm. The Young's modulus of the host plate is 210 GPa, and its mass density is 8000 kg/m^3. The first five (11, 21, 12, 31, 22) modes are selected to be controlled by using the modal sensor and the modal actuator. By choosing $A_{11}^a = 1.0$, $A_{21}^a = 0.8$, $A_{12}^a = 0.7$, $A_{31}^a = 0.6$, $A_{22}^a = 0.5$, $A_a = 2.5 \times 10^{-5}$ and $C_a = 0.013$ in equation (5) respectively, the modal actuator

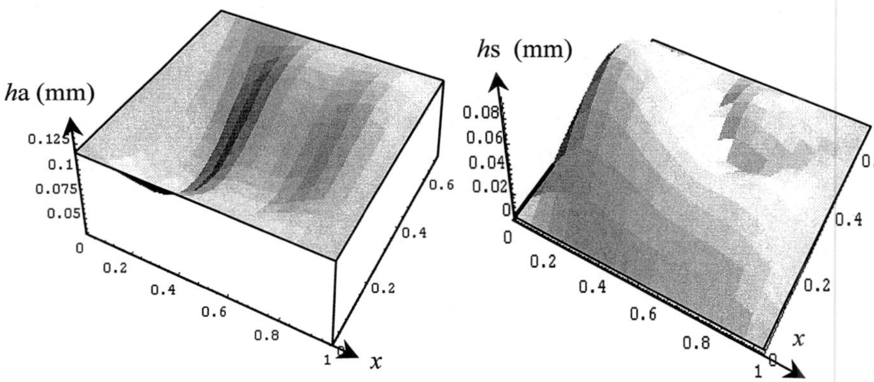

(a) Modal actuator (b) modal sensor

Figure 2 Modal actuator/sensor designed for a simply supported plate

is designed and shown in Figure 2a, whose maximum thickness is less than 0.12mm. The modal sensor is designed with the parameters $A_{11}^s = 1.0$, $A_{21}^s = 0.7$, $A_{12}^s = 0.6$, $A_{31}^s = 0.5$, $A_{22}^s = 0.4$, $A_s = 3 \times 10^{-5}$ and $C_s = 0$ so that its maximum thickness is less than 0.1mm, as shown in Figure 2b. The drawback of this method is that it may not be suitable for the changes of the target modes and parameters of the structures by redesigning the control software.

4 Distributed Piezoelectric Element Method

In order to overcome the disadvantage mentioned in previous section, the distributed piezoelectric element method can be employed. In the following sections, the thickness of each piezoelectric layer or patch is assumed to be constant.

4.1 Modal Sensor Design

To extract the modal coordinates from the voltage output of the sensor lamina, cut the sensor layer into $M \times N$ rectangular elements, and their charge outputs can be obtained as follow.

$$q_{mn}(t) = -e_{31}^{s} \iint_{S_{mn}} r_s \nabla^2 w dxdy \qquad m = 1,2,\cdots,M;\ n = 1,2,\cdots,N \tag{14}$$

Using modal expansion and only $M \times N$ modes being considered, the sensor outputs can be further expressed as

$$q_{mn}(t) = -e_{31}^{s} \sum_{i=1}^{M} \sum_{j=1}^{N} \eta_{ij}^{*}(t) \iint_{S_{mn}} r_s \nabla^2 W_{ij} dxdy \qquad m = 1,2,\cdots,M;\ n = 1,2,\cdots,N \tag{15}$$

where $\eta_{ij}^{*}(t)$ is the ij th observed mode coordinate of the plate. Equation (15) can be rewritten in matrix form

$$\{q(t)\} = e_{31}^{s}\, [\mathrm{E}]\, \{\eta^{*}(t)\} \tag{16}$$

where $\{q(t)\}$ is the vector comprised of $M \times N$ outputs of the sensor elements, $\{\eta^{*}(t)\}$ the vector of observed modal coordinates of the plate, and [E] a $MN \times MN$ matrix whose entries are given by

$$\mathrm{E}_{ijmn} = -\iint_{S_{mn}} r_s \nabla^2 W_{ij}(x,y) dxdy \tag{17}$$

The matrix [E] can be made non-singular by properly superposing or removing output charge of elements. Without loss of generality, the matrix is assumed to be non-singular, and the $M \times N$ mode coordinates of the plate can be solved from equation (16)

$$\{\eta^{*}(t)\} = (1/e_{31}^{s})\, [\mathrm{E}]^{-1} \{q(t)\} \tag{18}$$

Equation (18) shows that the observed values of former $M \times N$ mode coordinates can be obtained from output charge of the sensor elements.

4.2 Modal Actuator Design

The modal actuator can be designed by adjusting the spatial distribution of voltage of the actuator lamina.

Transform the actuator equation (2) into modal equations

$$\ddot{\eta}_{ij}(t)+\omega_{ij}^2\eta_{ij}(t)=-e_{31}^a r_a\iint_S \nabla^2 V(x,y,t)W_{ij}(x,y)dxdy \quad i,j=1,2,\cdots,\infty \tag{19}$$

When the voltage is designed as

$$V(x,y,t)=\sum_{k=1}^{K}\sum_{l=1}^{L}p_{kl}D[-\nabla^2 W_{kl}(x,y)] \tag{20}$$

where K and L are the numbers of order of required controlled modes, $K \le M$, $L \le N$, and $p_{kl}(t)$ is the designated modal force, the modal equation of motion becomes

$$\ddot{\eta}_{kl}(t)+\omega_{kl}^2\eta_{kl}(t)=e_{31}^a r_a\omega_{kl}^2 p_{kl}(t) \quad k=1,2,\cdots,M;l=1,2,\cdots,N \tag{21}$$

Equation (21) shows that the former $M\times N$ modes can be actuated, whereas the other modes are not actuated. Therefore, the piezoelectric modal actuators can be realized perfectly when the continuous distribution of the voltage in equation (20) is applied on the actuator lamina. To realize the distribution of the voltage in equation (20) practically, the piezoelectric actuator lamina is separated into QR elements, and the voltage applied on the ijth actuator element should be

$$V_{ij}(t)=-(1/A_{ij})\sum_{k=1}^{M}\sum_{l=1}^{L}p_{kl}(t)\iint_{A_{ij}}D\nabla^2 W_{kl}dxdy \quad i=1,2,...,Q;\ j=1,2,..,R \tag{22}$$

In this way, the required voltage on each piezoelectric actuator element can be determined from equation (22) and the modal control can be approximately realized by applying these voltages to the actuator elements respectively.. An increase in the number of elements in the piezoelectric layers makes the modal sensor/actuator more accurate.

4.3 Numerical simulations

As an example, consider a 0.4m×0.3m×1mm simply supported rectangular plate with two PVDF layers as sensor and actuator layer respectively, whose parameters are given in section 3.3 and 4.3. The vibration of the plate is excited by an impulse of 0.03Ns at the point (25cm,13cm) of the plate. To perform vibration control of this plate using this method, cut the sensor and actuator layers into 4×3 elements respectively. Design the modal control forces in equation (21) by[10]

$$p_{kl}(t)=-h_{kl}\dot{\eta}_{kl}^*(t),\ k=1,3;\ l=1,2$$

to control the first six modes. For the values h_{11}=1, h_{12}=0.4, h_{21}=0.5, h_{22}=0.4, h_{31}=0.3 and h_{32}=0.25, the time history of the transverse displacement at the center of the plate and the distribution of control voltage among all actuator elements are shown in Figure 3 and 4 respectively. The results show that the selected six modes of the plate are effectively controlled within 0.2 seconds (only the uncontrolled higher order modes remain) and the maximum control voltage is less than 400V.

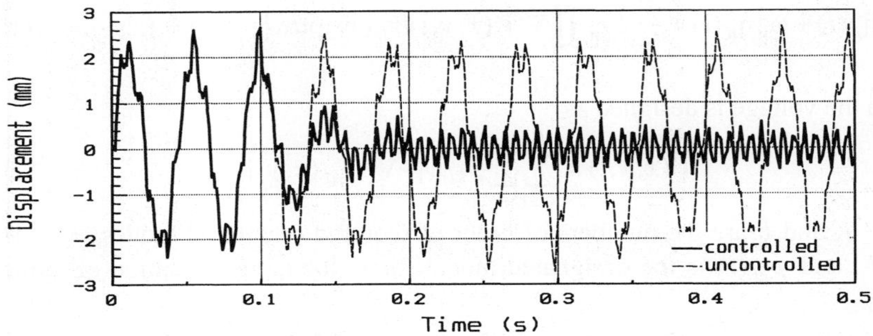

Figure 3 Center displacement of the plate [10]

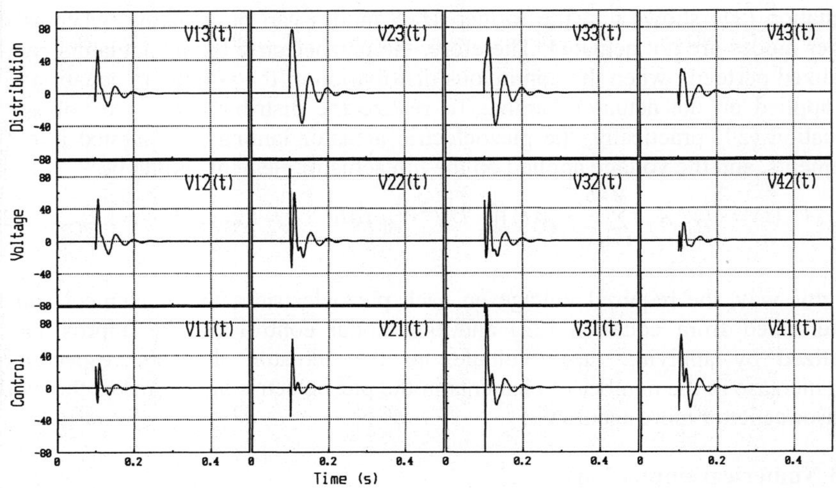

Figure 4 Control voltage distribution on actuator elements [10]

5 Piezoelectric Quasi-Modal Actuators/Sensors

For large flexible structures, fully covered piezoelectric layer and a large number of the piezoelectric elements may be less cost-effective and impractical. The quasi-modal sensors and actuator are developed in this section.

5.1 Quasi-Modal Sensor Design and Observation Spillover Estimate

The output charge of the N_s discretely distributed sensor patches can be expressed as

$$q_n(t) = -r_n e_{31n} \sum_{i=1}^{\infty} \sum_{j=1}^{\infty} \eta_{ij}(t) \iint_{S_{Sn}} \nabla^2 W_{ij} dxdy\,, \quad n = 1,2,\cdots,N_S \tag{23}$$

For convenience, the two modal indices i and j are replaced by one index i in order of increasing magnitude of frequency, then equation (25) becomes

$$q_n = \sum_{i=1}^{\infty} \alpha_{ni} \eta_i(t)\,, \quad n = 1,2,\cdots,N_S \tag{24}$$

where

$$\alpha_{ni} = -r_n e_{31n} \iint_{S_{Sn}} \nabla^2 W_i dxdy\,, \quad i = 1,2,\cdots \tag{25}$$

are the coefficients related to the locations and sizes of the sensor elements and the modal functions of the plate. Making truncation with only the first M (M>Ns) modes of the plate being considered, equation (24) becomes

$$\{q(t)\} = [A]\{\eta(t)\} \tag{26}$$

where $\{q(t)\} \in R^{N_S}$ is a vector of output charges generated by the sensor elements, $\{\eta(t)\} \in R^M$ a vector containing the first M modal coordinates of the plate, and $[A] \in R^{N_S \times M}$ a matrix related to the output charges of the sensor elements and the modal coordinates. Rewrite equation (26) as

$$\{q(t)\} = [A_1]\{\eta_1(\mathrm{t})\} + [\mathrm{A}_2]\{\eta_2(\mathrm{t})\} \tag{27}$$

where $[A_1] \in R^{N_S \times N_S}$ and $[A_2] \in R^{N_S \times (M - N_S)}$ are the matrices composed of the first N_S columns and the residual entries of $[A]$ respectively, and $\{\eta_1(t)\} \in R^{N_S}$ and $\{\eta_2(t)\} \in R^{M - N_S}$ the vectors composed of the first N_S modal coordinates and the residual M-N_S modal coordinates respectively.

Matrix $[A_1]$ can be made nonsingular as long as the piezoelectric elements are placed properly, and then we have

$$\{\eta_1(t)\} = [A_1]^{-1}\{q(t)\} - [A_1]^{-1}[A_2]\{\eta_2(t)\} \tag{28}$$

Neglecting the second term in the right terms, the first N_S modal coordinates can be estimated approximately by

$$\{\eta_1^*(t)\} = [A_1]^{-1}\{q(t)\} \tag{29}$$

where $\{\eta_1^*(t)\}$ is the vector containing the N_S observed modal coordinates. The error between the observed and the real modal coordinates can be expressed as

$$\{\Delta(t)\} = \{\eta_1^*(t)\} - \{\eta_1(t)\} = [A_1]^{-1}[A_2]\{\eta_2(t)\} \tag{30}$$

Equation (30) indicates that the observed modal coordinates obtained from equation (29) are the real ones mixed with the residual modal coordinates with higher frequencies. Therefore, equation (29) defines quasi-modal sensors. The norm of the observation error is

$$\|\Delta(t)\|^2 = \{\Delta(t)\}^T\{\Delta(t)\} = \{\eta_2(t)\}^T[A_2]^T([A_1]^{-1})^T[A_1]^{-1}[A_2]\{\eta_2(t)\} \tag{31}$$

where $[A_2]^T([A_1]^{-1})^T[A_1]^{-1}[A_2] \in R^{(M-N_S)\times(M-N_S)}$ is a symmetric positive definite matrix. It can be proved that

$$\|\Delta(t)\|^2 \le \lambda_{\max}([A_2]^T([A_1]^{-1})^T[A_1]^{-1}[A_2])\{\eta_2(t)\}^T\{\eta_2(t)\} \tag{32}$$

where $\lambda_{\max}([A_2]^T([A_1]^{-1})^T[A_1]^{-1}[A_2])$ represents the maximum eigenvalue of the matrix $[A_2]^T([A_1]^{-1})^T[A_1]^{-1}[A_2]$. Consequently, the criterion

$$J_S = \min_{S_{Sn}\,(n=1,2,\cdots N_S)} \lambda_{\max}\{[\mathrm{A}_2]^T([\mathrm{A}_1]^{-1})^T[\mathrm{A}_1]^{-1}[\mathrm{A}_2]\} \tag{33}$$

can be used to find the optimal locations and sizes of the sensor elements.

5.2 Quasi-Modal Actuator Design and Control Spillover Estimate

When N piezoelectric actuator patches are bonded discretely on the host plate, the modal equations of motion for a smart plate can be written as

$$\ddot{\eta}_i(t) + \omega_i^2\eta_i(t) = \sum_{n=1}^{N}\beta_{in}V_n(t)\,, \quad i = 1,2,\cdots \tag{34}$$

where β_{in} is the modal influence coefficients, which indicates the influence of the nth actuator elements on the ith mode. The modal influence coefficients depend on not only the modal shape functions but also the locations and sizes of the actuator elements.

Similarly, consider only the first M modes, and rewrite equation (34) in the following matrix form

$$\begin{aligned}\{\ddot{\eta}_c(t)\} + [\Omega_c^2]\{\eta_c(t)\} &= [B_c]\{V(t)\}\\ \{\ddot{\eta}_r(t)\} + [\Omega_r^2]\{\eta_r(t)\} &= [B_r]\{V(t)\}\end{aligned} \tag{35}$$

where $[B_c]$ and $[B_r]$ are the matrices related to the modal influence coefficients. The modal actuator can actuate the designated modes without affecting the rest modes. If the modal forces for the N controlled modes are designated to be $\{f_c(t)\} \in R^N$, then the voltages that can generate such N modal forces satisfy the following equations

$$[B_c]\{V(t)\} = \{f_c(t)\} \tag{36}$$

If the actuators are distributed properly, the matrix $[B_c]$ is non-singular, and therefore the voltages applied on the actuators can be obtained as

$$\{V(t)\} = [B_c]^{-1}\{f_c(t)\} \tag{37}$$

Equations (37) give the voltages needed to generate the designated modal forces according to some selected control laws. However, the obtained voltage may affect the residual modes. The norm of the modal forces $\{f_r(t)\}$ excited on the residual modes can be expressed as

$$\|f_r(t)\|^2 = \{f_r(t)\}^T\{f_r(t)\} = \{f_c(t)\}^T([B_c]^{-1})^T[B_r]^T[B_r][B_c]^{-1}\{f_c(t)\} \tag{38}$$

and

$$\|f_r(t)\|^2 \le \lambda_{\max}\{[B_c]^{-1})^T[B_r]^T[B_r][B_c]^{-1}\}\{f_c(t)\}^T\{f_c(t)\} \tag{39}$$

which indicates that a smaller maximum eigenvalue of the matrix $([B_c]^{-1})^T[B_r]^T[B_r][B_c]^{-1}$ will lead to smaller control spillover.

The relationship between the modal force $\{f_c(t)\}$ and the control energy can be obtained from equation (36) as

$$\{f_c(t)\}^T\{f_c(t)\} = \{V(t)\}^T[B_c]^T[B_c]\{V(t)\} \tag{40}$$

Noting that

$$\|V(t)\|^2 = \{V(t)\}^T\{V(t)\} \le \{f_c(t)\}^T\{f_c(t)\}/\lambda_{\min}([B_c]^T[B_c]) \tag{41}$$

and combining equation (39), the following criterion can be developed

$$J_a = \min_{S_{an}(n=1,2,\cdots N)} R_1\lambda_{\max}\{([B_c]^{-1})^T[B_r]^T[B_r][B_c]^{-1}\} - R_2\lambda_{\min}([B_c]^T[B_c]) \tag{42}$$

where R_1 and R_2 are the weighting coefficients. This criterion can be used to find the optimal locations and sizes of the piezoelectric actuator patches.

5.3 An Illustrative Example

As an illustrative example, consider a 1m×0.7m×1mm simply supported rectangular plate bonded with four PZT actuator patches and six sensor patches. The material properties of the host plate and PZT are the same with those in section 3.3. The sensor patches are made of PVDF whose piezoelectric stress constant is 0.06 N/Vm. The thickness of each actuator patch is 0.4mm and that of sensor patch is 0.1mm. When each sensor patches is constrained to be not larger than 6cm×4cm and not smaller than 2cm×2cm and only the first 10 modes are considered, the optimized locations and sizes of the sensor patches can be obtained using criterion (33). The optimal locations of these six sensor patches are (0.7527m, 0.2404m), (0.4926m, 0.2315m), (0.25m, 0.469m), (0.2498m, 0.2471m), (0.4882m, 0.4788m) and (0.7414m, 0.4671m), and the optimal sizes are 3.45cm×3.27cm, 5.8cm×2.73cm, 4.42cm×3.67cm, 5.54cm×2.51cm, 2.58cm×4.00 cm and 4.69cm×3.92cm respectively.

Unlike the sensor patches, the dimensions of the four actuator elements are fixed as 0.04m×0.015m, 0.03m×0.0125m, 0.02m×0.05m and 0.025m×0.025m respectively. When the weighting coefficients are chosen as R1=$10^5 / c_n^2$, R2= 0.1, Q= 105, the obtained optimal locations of the four actuators are (0.6315m, 0.3926m), (0.6295m, 0.2095m), (0.3398m, 0.3428m) and (0.7257m, 0.3582m) respectively.

With the optimum placement of the sensor and actuator elements, one can control the vibration of the plate caused by the sudden removal of a force of 3.5N acting on the point (0.5,0.4). To control the first four modes, design the modal control force in equation (36) by [11]

$$f_{ci}(t) = g_i \xi_i(t); \quad \ddot{\xi}_i(t) + 2\zeta_{ci}\omega_i \dot{\xi}_i(t) + \omega_i^2 \xi_i(t) = \omega_i^2 \eta_i^*(t), \; i = 1,4 \qquad (43)$$

in which the damping ratios ζ of the compensators are all taken to be 0.5. The modal control is performed by choosing the control gains in equation (43) as g1=600, g2=700, g3=1000, and g4= 1000. During the control process, the observed modal coordinates by the quasi-modal sensor and the control voltage on the actuator patches are also shown in Figure 5 and 6. As shown in Figure 5, the optimized quasi-modal sensor can give a good estimation of modal coordinates, and the remaining high frequency components in the estimated modal coordinates can be further removed by the compensators in equation (43). It is also shown that all the target modes are effectively controlled within 1 seconds. Figure 6 shows that the control voltage can be equally distributed among the actuator patches by optimizing their locations.

6 Conclusions

In this paper, three modal sensor/actuator designing methods for vibration control of smart plates are briefly introduced. In the first method, by modulating the thickness of the piezoelectric actuator layer, the modal actuators can be realized to actuate several modes simultaneously by using single voltage input. However, the thickness of actuator/sensor must be changed with the change of thickness of plate and the target modes. The second method, the DPE method, is suitable for change of plate thickness and controlled modes by changing software, but a multi-input multi-output control with great degree of freedom (DOF) is required. The third method, quasi-modal actuator/sensor developed by using the discretely distributed piezoelectric patches, can reduce the DOF of control system.

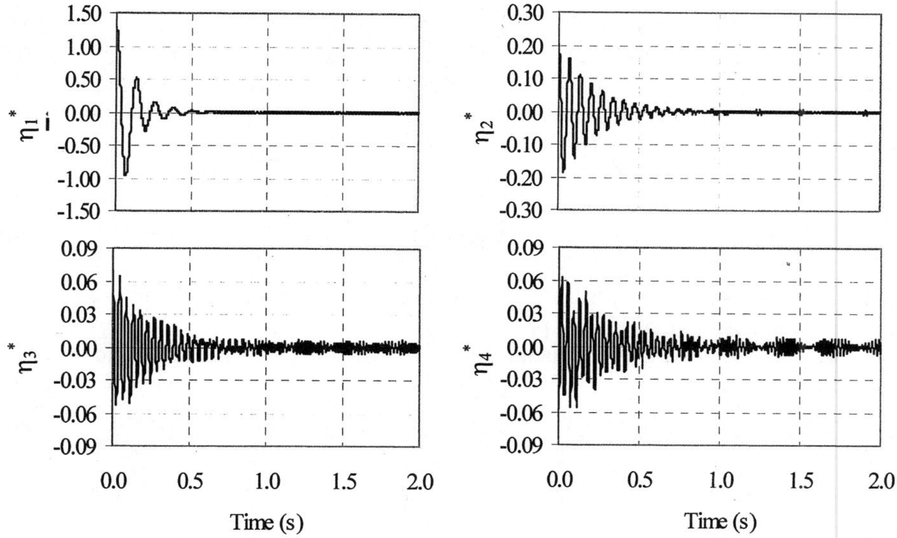

Figure 5 Observed modal coordinates by quasi-modal sensor [11]

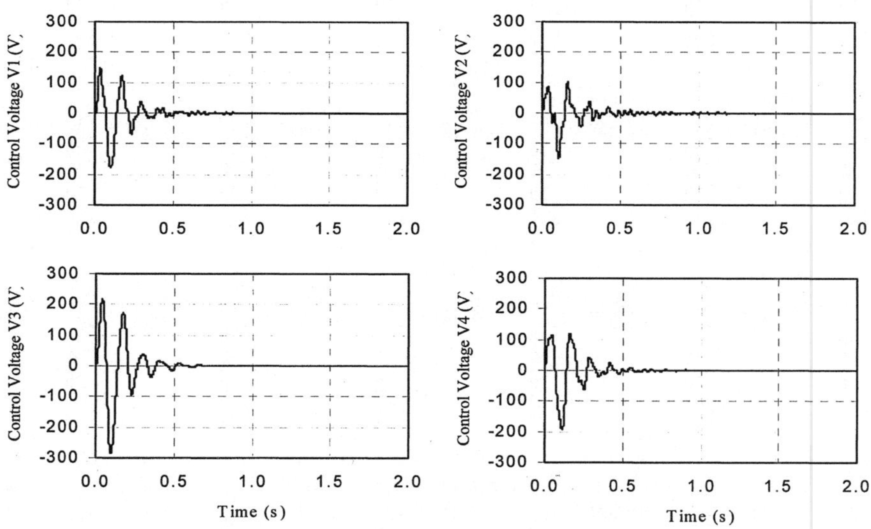

Figure 6 Control voltage distribution on actuator patches [11]

Acknowledgements

The authors would like to thank the support by the Australian Research Council under a Large Research Grant (Grant No. A89905990) and by the National Science Foundation of China (Grant No. 19802016).

References

[1] C. –K. Lee and F. C. Moon. "Modal Sensors/Actuators". Journal of Applied Mechanics, Vol. 57, 1990, pp434-441.

[2] C. –K. Lee, W. –W. Chiang and T. C. O'Sullivan. "Piezoelectric Modal Sensor/Actuator Pairs of Critical Active Damping Vibration Control". Journal of Acoustics Society of America, Vol. 90, 1991, pp374-384.

[3] Y. Gu, R. L. Clark and C. R. Fuller. "Experiments on Active Control Plate Vibration Using Piezoelectric Actuators and Polyvinylidene Fluoride (PVDF) Modal Sensors". Journal of Vibration and Acoustics, Vol. 116, 1994, pp303-308.

[4] H. S. Tzou, J. P. Zhong and J. J. Hollkamp. "Spatially Distributed Orthogonal Piezoelectric Shell Actuators Theory and Applications". Journal of Sound and Vibration, Vol. 177, 1994, pp363-378.

[5] S. E. Burke and J. M. Sullivan. "Distributed Transducer Shading via Spatial Gradient Electrodes". SPIE Proceedings, Vol. 2443, 1995, pp716-725

[6] S. A. Collins, D. W. Miller and A, H. Von Flotow. "Distributed Sensors as Spatial Filters in Active Structural Control". Journal of Sound and Vibration, Vol. 173, 1994, pp471-501.

[7] J. –K. Ryou, K. –Y. Park and S. –J. Kim. "Electrode Pattern Design of Piezoelectric Sensors and Actuators Using Genetic Algorithms". AIAA Journal, Vol. 36, 1998, pp227-233.

[8] H. S. Tzou, V. B. Venkayya and J. J. Hollkamp. "Orthogonal Sensing and Control of Continua with Distributed Transducers-Distributed Structronic System". In H. S. Tzou and L. A. Bergman (eds.), Dynamics and Control of Distributed Systems, Cambridge University Press, Cambridge, 1998, pp304-370.

[9] D. C. Sun, D. J. Wang and Z. L. Xu. "Distributed Piezoelectric Segment Method for Vibration Control of Smart Beams". AIAA Journal, Vol. 35, 1997, pp583-584.

[10] D. C. Sun, D. J. Wang, Z. L Xu, and H.X. Wu. "Distributed Piezoelectric Element Method for Vibration Control of Smart Plates". AIAA Journal, Vol. 37,1999, pp1459-1463.

[11] D. C. Sun, L. Tong and D. J. Wang. "Vibration Control of Plates Using Discretely Distributed Piezoelectric Quasi-modal Actuators/sensors". AIAA Journal, Vol. 39, No. 9, 2001, pp1766-1772.

Overview of Nonlinear Finite Element Modelling of Flexible Manoeuvring Aerospace Systems and Structures

PAVEL TRIVAILO and TRENTON GILBERT

Department of Aerospace Engineering, RMIT University
GPO Box 2476V, Melbourne, 3001, VIC, Australia,
trivailo@rmit.edu.au

Summary

This paper presents an overview of research into the use of a nonlinear co-rotational finite element technique for modeling the motion of flexible manoeuvring systems and structures. The applications include robotics and aircraft wings. In the case of robotics, modeling of manipulators, which are capable of being reconfigured to perform various tasks, is carried out and results from simple models are compared with experimental data. In the case of aircraft, results from simulations involving an aerodynamically loaded aircraft wing model are presented.

1 Introduction

Designers of aerospace systems and structures are forced by various constraints to reduce overall weight and improve performance. It is therefore inevitable that in some cases, the inherent flexibility of this equipment will increase. Two areas in which this phenomenon can manifest itself are robotics and aircraft structures.

When considering manoeuvring structures, the nature of the motion experienced by any flexible appendages (e.g. manipulator links, aircraft wings and fuselage) is typified by large rigid body rotations and translations, and significant elastic deformations, mainly due to inertial effects. This motion is highly nonlinear and thus modelling these structures requires the use of unconventional techniques. A co-rotational finite element method has been utilised for this task.

The co-rotational finite element method utilises multiple co-ordinate systems (a co-rotational co-ordinate system which rotates and translates with each element, and a global co-ordinate system) to cope with the geometric non-linearity present in either very flexible, or rotating systems. It was first proposed by Belytschko[1], and has since been the subject of many refinements, improvements and extensions [4, 8, 9, 5, 3, 10]. However, there is little evidence of the application of the co-rotational technique to problems involving reconfigurable robotic or aerodynamically loaded systems, as is presented here.

This paper therefore details the results of research performed into the use of the co-rotational finite element method for prediction of the motion of flexible manoeuvring structures. This work has included two main components; the modelling of flexible reconfigurable robotics[6], where it is possible to simulate manoeuvres involving connection to and disconnection from other manipulators or fixed points within the manipulators environment, as well as failure of the system due to high stresses; and aircraft structures[7], where the aerodynamic load on the structure (which can be modified to reflect control inputs), is evaluated using a lifting line analysis. In all cases, strains can be monitored throughout the simulation. Also discussed will be a comparison of simulation results with experimental data.

2 Methodology

2.1 Co-rotational Finite Element Method

A co-rotational finite element method [4, 8] is used to model the structures discussed here. The technique is suited to modeling geometrically nonlinear structures which experience large deformations (static or dynamic cases) and/or large rotations and translations (dynamic cases). In most cases presented here, deformations are relatively small, and the nonlinearity present is largely due to rotation of the models.

The characteristic of the technique, which allows modelling of this type of motion, is the fact that each element within the system is assigned a unique set of co-ordinates, which is allowed to rotate and translate with the element. All deformations and therefore strains within the element are described relative to the co-rotational co-ordinates, and providing element sizes are chosen appropriately, can be said to be linear relative to this co-ordinate system. Nodal forces due to deformation are therefore evaluated relative to these co-ordinates. System velocities and accelerations, and thus nodal forces proportional to these, are described relative to a global co-ordinate system. This leads to a non-standard form of the system's equation of motion,

$$M\ddot{\mathbf{d}} + C\dot{\mathbf{d}} + \mathbf{f} = \mathbf{F} \tag{1}$$

$$\mathbf{f}^i = Q^i\bar{\mathbf{f}}^i = Q^i\bar{K}^i\bar{\mathbf{d}}^i \tag{2}$$

where M, C and K are the mass, damping and stiffness matrices respectively, $\mathbf{f}$ is the nodal force vector due to deformation, Q is a transformation matrix, the superscript i represents elemental quantities, and an overbar ($\bar{\ }$) represents quantities expressed relative to the co-rotational co-ordinates. The quantity $\bar{\mathbf{f}}^i$ is evaluated relative to the co-rotational co-ordinates for each element, then transformed to the global co-ordinates for inclusion in the global deformation force vector. Rotation of the co-rotational co-ordinates, however, also requires that mass and stiffness matrices be transformed from co-rotational to global co-ordinates at each timestep. Proportional damping is used to evaluate C, where $C = \alpha M + \beta K$, and α and β are chosen to be 0.01 and 0.05 respectively.

Strains are evaluated using the standard expression for bending strain,

$$\varepsilon_{\bar{x}} = -\bar{y}\frac{\partial^2 v}{\partial \bar{x}^2} \tag{3}$$

where $\varepsilon_{\bar{x}}$ is the strain, $\bar{y}$ is the distance from the neutral axis of the beam element at which the strain is to be evaluated (the upper surface of the beam element in all cases presented here), v is the deformation of the element in the direction of the y axis of the co-rotational co-ordinates and x represents distance along the x axis of the co-rotational co-ordinates.

The equations of motion are solved at each timestep using the Newton-Raphson iterative procedure and Newmark integration[2]. All models consist of two-dimensional, two-noded, six degree-of-freedom, linear beam elements, which simplifies the modelling procedure and significantly reduces the computational cost of performing analyses.

2.2 Robotics - Manipulator Reconfiguration

A number of innovative techniques have been developed which allow a reconfigurable manipulator to be modelled. Simulations have been performed for manipulators of varying configurations, which are being attached to and disconnected from either locations within the manipulator's operating environment or other manipulator components. This facilitates either transfer of the manipulator from one location to another, or reconfiguration of the manipulator to form manipulators consisting of any number of links.

Reconfiguration of the manipulator was achieved by making changes to the degrees of freedom within the finite element model over time. For instance, in the case of a separation manoeuvre or failure of the system, the beam is effectively broken at the prescribed location by the addition of extra degrees of freedom at that location to allow the previously joined ends to move independently. An example of this alteration of the degrees of freedom can be seen for a simple two element system in Figure 1.

Connection of the manipulator to, or disconnection from, a particular location was achieved by altering the boundary conditions of the system at the required time. For instance, fixing the degrees of freedom at the end of a beam to a particular location enables connection to its current location,

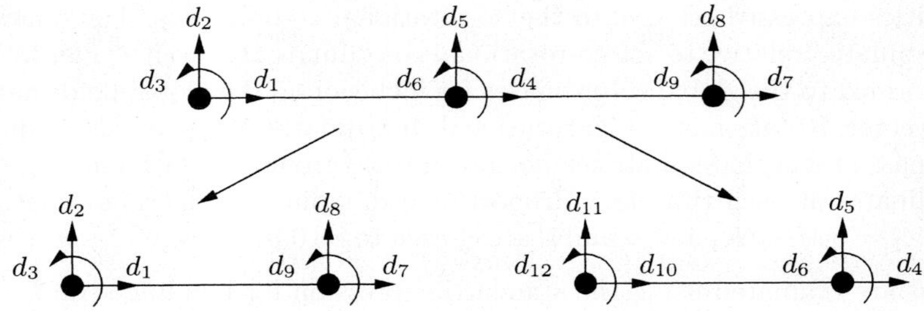

Degrees of freedom are deleted, added and re-ordered

Figure 1 Method of separating elements

whilst eliminating these boundary conditions provides the ability to disconnect that end of the manipulator. Figure 2 shows a pictorial representation of the changes in the mass and stiffness matrices as this occurs.

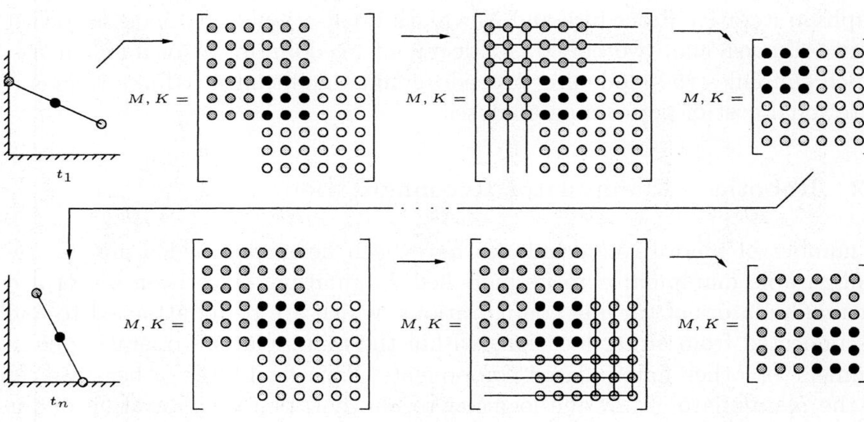

Figure 2 Variation of boundary conditions

Performing these alterations is simplified by the fact that the mass and stiffness matrices are recalculated at each time step, and the solution procedure utilised remains effective through these abrupt changes in system characteristics.

2.3 Aircraft - Aerodynamic Loads

Modeling a manoeuvring aircraft structure required the use of a technique for aerodynamic load estimation. A lifting line analysis[11] was used as this provided a satisfactory means of estimating the load, whilst being responsive to changes in angle of attack caused by vertical velocity at various locations along the wing and twisting of the wing under load. It was also possible to incorporate the change in aerodynamic load as a result of a control surface deflection using this technique.

3 Manipulator Examples

3.1 Transfer Manoeuvres

The following examples show a robotic manipulator connecting to and disconnecting from arbitrary locations such that the manipulator can be moved from one location to another.

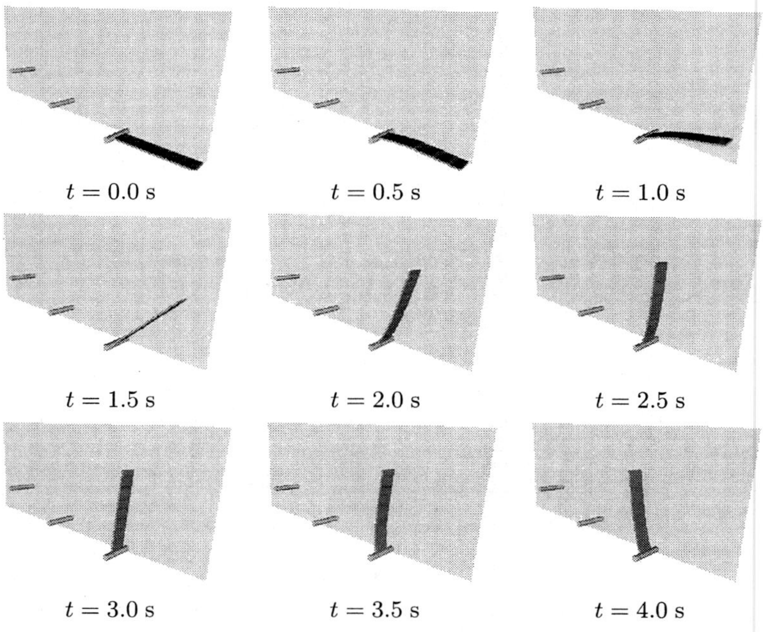

Figure 3 Displacement during transfer of single-body manipulator ($t = 0$–4 s)

Single-Body Manipulator: Figures 3 and 4 show a displacement driven single-body manipulator, of length 1 m, which is rotated about its root with an angular velocity of 0.5 rad/s. At $t = 6.0$ s and $t = 12.0$ s the free end of the manipulator is fixed at its current location, and the connected end is released. The newly connected end is then rotated with the same angular velocity (i.e.

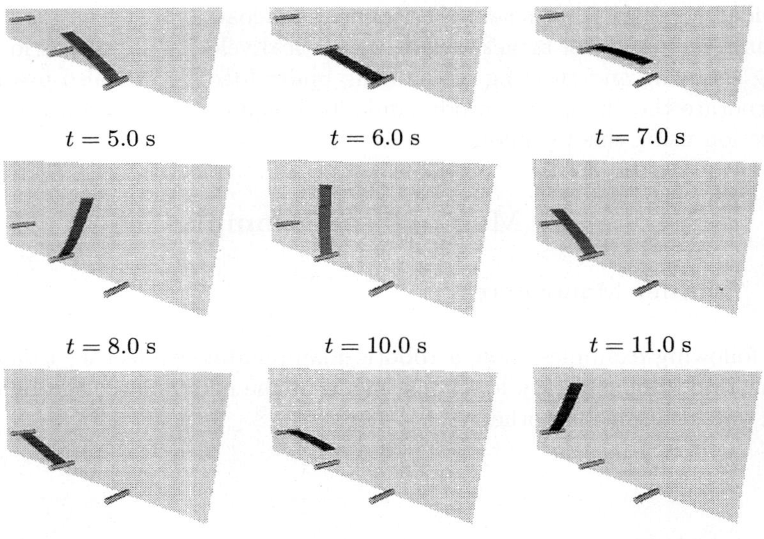

Figure 4 Displacement during transfer of single-body manipulator (t = 5–14 s)

0.5 rad/s). The strain time history for this case can be seen in Figure 5, which shows strains at the root, mid-point and tip of the manipulator, where the root of the beam is deemed to be the initially connected end.

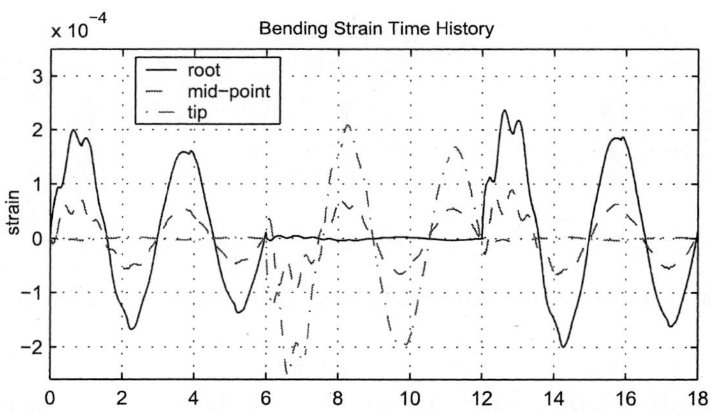

Figure 5 Bending strain during transfer of single-body manipulator

This example demonstrates the ability of the simulation to connect to and disconnect from different locations through variation of the model's boundary conditions. It is also evident that the flexibility of the system has a significant effect on the nature of its motion. The fact that the free end (or tip) of the manipulator oscillates relative to the root (or connected end), means that its path can be difficult to ascertain. Also, it can be noted that the highest strains are experienced at the connected end of the manipulator, whilst at the free end, strains are almost zero (as is expected). The strains (and the manipulator itself) oscillates with a dominant frequency equivalent to the first mode of the system, however the second mode is excited at key points in the simulation, noticeably at the start of the simulation, and when connection/disconnection occurs.

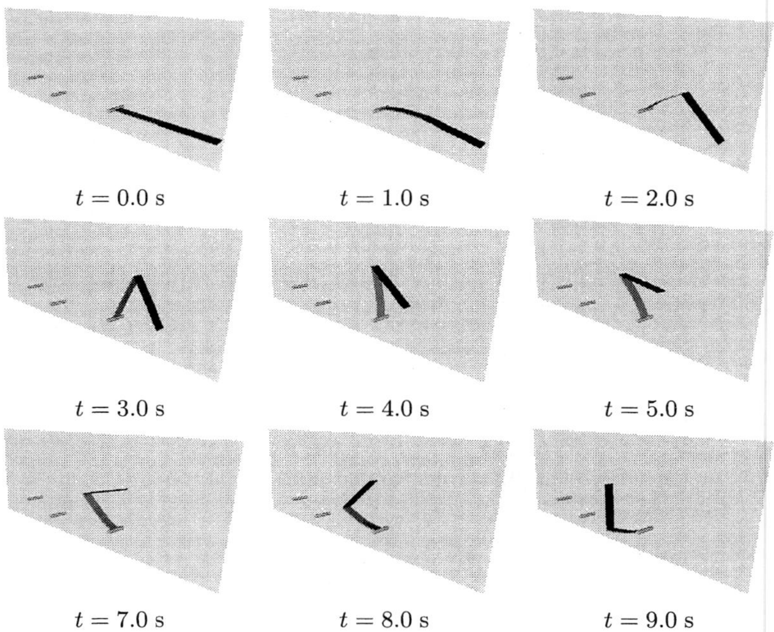

Figure 6 Displacement during transfer of multi-body manipulator ($t = 0$–9 s)

Multi-Body Manipulator: Figures 6 and 7 show a similar example to that described above, except that in this case the manipulator consists of multiple bodies. This two link manipulator experiences the same connection/disconnection sequence as the previous example, except in this case the times when the changes in boundary conditions occur are $t = 10$ s and $t = 16$ s. The length of each member is 1 m, and the angular velocity is 0.4 rad/ s. A strain time history can be seen in Figure 8, for the root, mid-

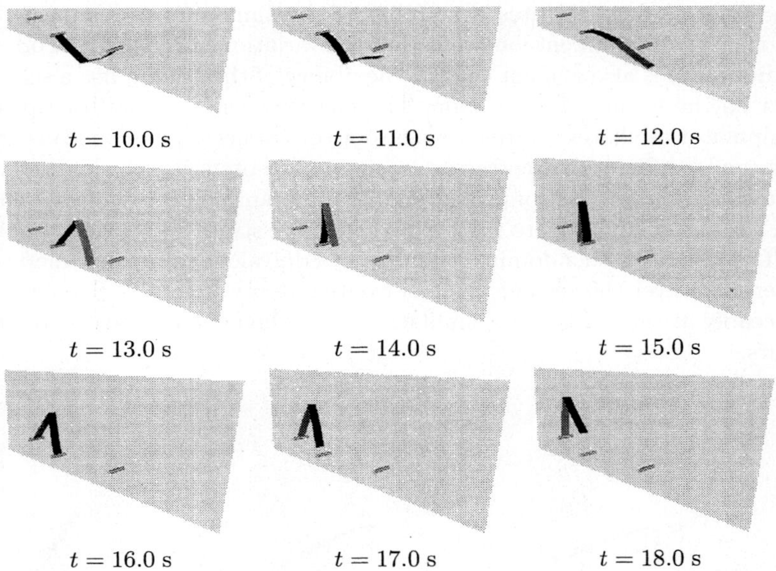

Figure 7 Displacement during transfer of multi-body manipulator (t = 10–18 s)

point and tip of each member of the manipulator, where the root of each member is the connected end.

In this example, the complex nature of the motion experienced by this type of system is again evident. The strain time histories show similar trends to the single-body case, with excitation of the first mode occurring predominantly, and excitation of the second mode occurring at points throughout the simulation where the disconnection or connection occurs.

3.2 Separation Manoeuvre

In this case, a multi-body manipulator, with each member of length 1 m, is rotated for 10 s and then connected via its free end to an arbitrary location. The manipulator is then separated at its centre at 14 s, effectively generating two new manipulators. The angular velocity of the manipulator(s) in all configurations is 0.4 rad/ s. For $t = 0 - 10$ s, the results from this simulation are the same as the previous example and can therefore be seen in Figure 6. Figure 9 shows the position of the manipulator(s) for $t = 10 - 20$ s.

This example emphasises the ability of the simulation to model the manipulator through changes in configuration involving the addition of degrees of freedom. This technique also allows modelling of systems experiencing failure.

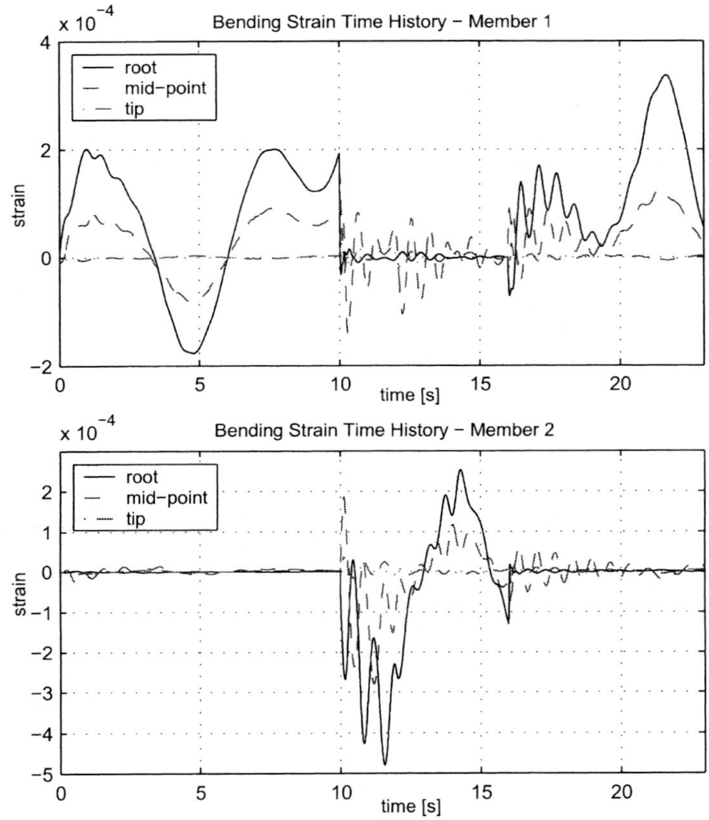

Figure 8 Bending strains during transfer of multi-body manipulator

4 Aircraft Example

This example shows a wing constructed from 4 beam elements with arbitrary material properties (chosen to emphasise effects of flexibility), undergoing a reversed rolling manoeuvre. The wing is pinned at its centre, and is therefore free to rotate about this point in response to the aerodynamic load applied to it. It is at first accelerated to a velocity of 100 kts, subjected to an aileron deflection of 30° from $t = 3 - 5.5$ s, then subjected to a deflection of $-30°$ from $t = 5.5 - 9$ s. A positive aileron deflection implies right aileron down, and the strains shown are for the wing's lower surface. Four elements are used here to represent the wing.

The deflection time history is shown in Figure 10, whilst lift coefficient and strain time histories can be seen in Figures 11 and 12 respectively. Evident in these figures, is the response of the wing to the deflection of the ailerons.

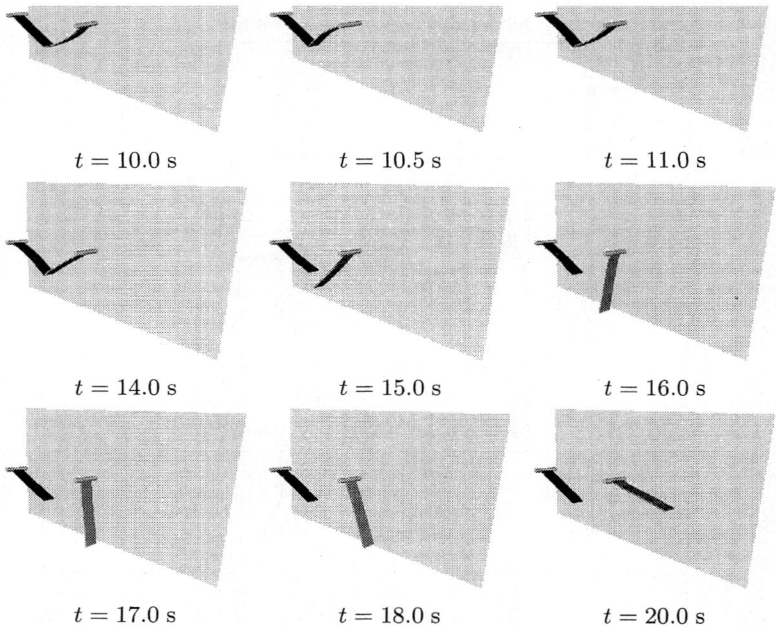

Figure 9 Displacement during separation of multi-body manipulator (t = 10–20 s)

At the instant the ailerons are deflected, there is a noticeable increase in aerodynamic load, and a subsequent increase in strain. The strain (and also the position of the wing) shows an oscillation as a result of the control input, then eventually reaches an equilibrium position. This emphasises that the strain within the aircraft wing is greater during both the application of control inputs and manoeuvring than when the wing reaches an equilibrium position.

5 Experimental Validation

An experimental setup was developed which enabled verification of results obtained using the finite element code used for the simulations presented previously. The model consisted of an extremely flexible aluminium beam, attached at one end to a DC motor which was capable of providing a constant torque for a specific time period. Strain was measured at 4 locations on the beam using strain gauges, and it was these values which were used for comparison with results from the FE model. The strains presented here are for the gauge located closest to the root of the beam.

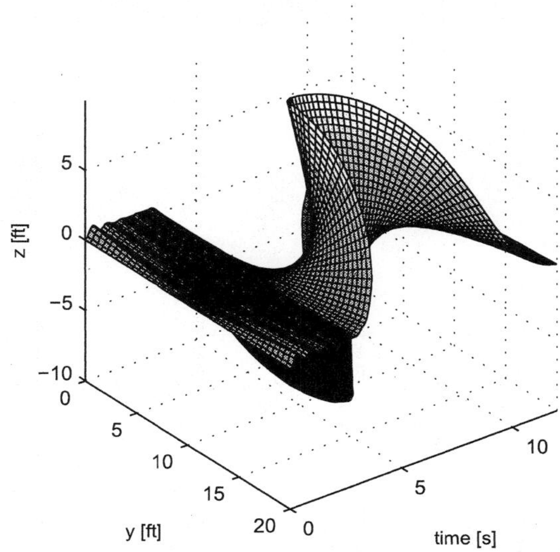

Figure 10 Deflection time history for wing during reversed rolling manoeuvre

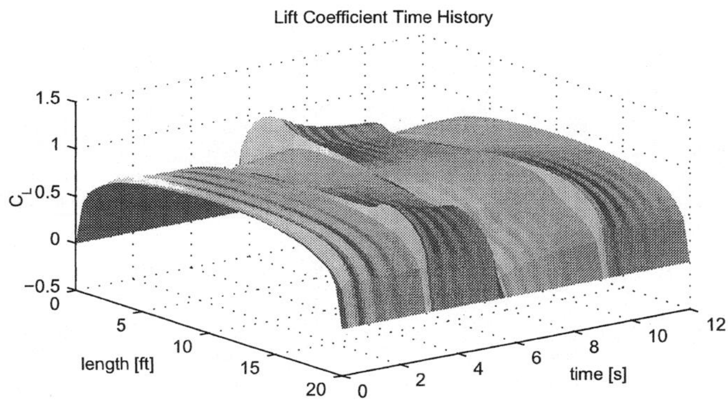

Figure 11 Lift coefficient time history for wing during reversed rolling manoeuvre

As the setup was created as simply as possible, using available equipment, tests were limited to very high rotation rates, and were not completed in a vacuum. Due to this, and the fact that the beam was extremely flexible, ($EI \approx 0.015$ Nm^2), it was found that aerodynamic drag on the beam had a significant effect on the results. A method of accounting for this was thus introduced into the model, and found to markedly improve comparisons.

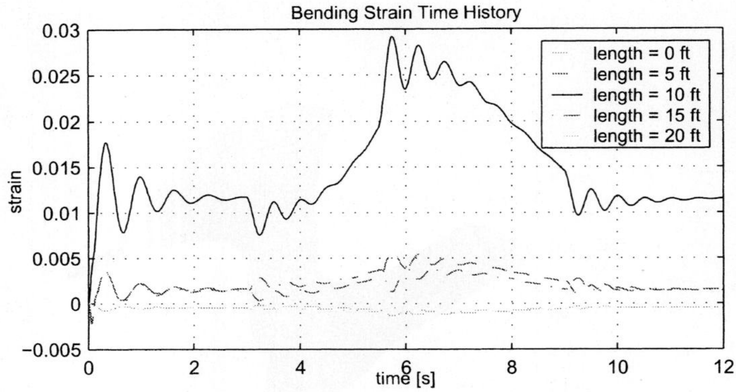

Figure 12 Bending strain time history for wing during reversed rolling manoeuvre

5.1 Applied Impulse

In these tests, a 200 ms pulse was applied to the motor, causing rapid rotation of the beam. The strains measured and calculated using the FE model can be seen in Figure 13.

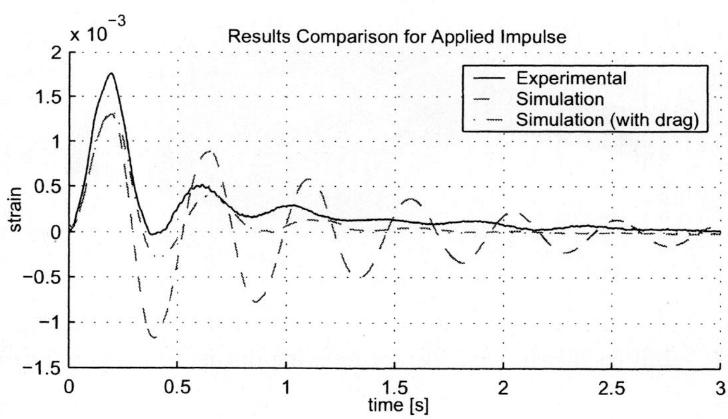

Figure 13 Comparison of results for applied impulse manoeuvre

It can be noted that the inclusion of drag significantly improved the simulation results, however there are still some discrepancies in the peak strain experienced at the beginning of the simulation and in decay rates. These errors could possible be attributed to the fact that the aerodynamic drag was estimated by treating the beam as a flat plate moving through fluid (neglecting any complex aerodynamic effects), and that proportional damping was used to predict the structural damping.

5.2 Halted Rotation

In this example, a pulse was applied to the motor, again causing rapid acceleration, however, after a short period an impulse is applied in the opposite direction to halt rotation. The strains can be seen in Figure 14.

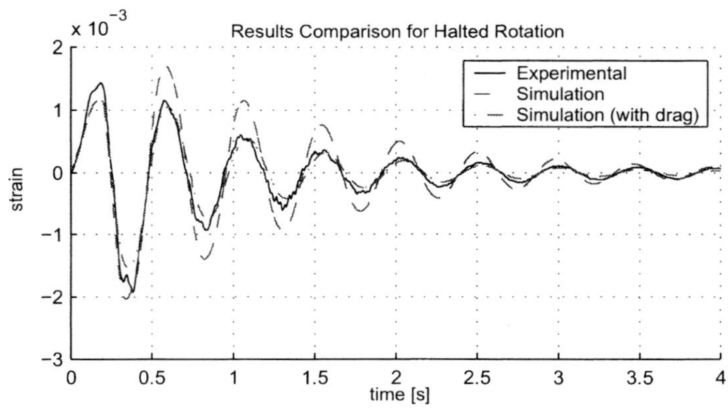

Figure 14 Comparison of results for halted rotation manoeuvre

The comparison of results for this case, show that the strains from both experiment and simulation correlate well. It is likely that this is due to the reduction in the significance of the aerodynamic drag, as the beam is stopped before it can obtain a high enough velocity for the drag to be significant.

6 Conclusions

A dynamic simulation capability has been developed, using a nonlinear corotational finite element method, which is capable of modeling the motion of flexible manoeuvring systems and structures undergoing large rigid body rotations and translations. The software developed is used to model aerospace systems and structures, including reconfigurable robotics and flexible aircraft wings. The technique implemented gives a useful indication of the nature of the motion experienced, as well allowing the calculation of strains as the simulation progresses. A comparison of the finite element results with experimental data has shown good agreement.

7 Acknowledgments

The authors would like to acknowledge the support of the DSTO/RMIT Centre of Expertise in Aerodynamic Loading (CoE-AL) and the Department of Aerospace Engineering, RMIT. Thanks also go to Askar Kadykov for his work on obtaining the beam experimental results used here.

References

[1] T. Belytschko, and B. Hasieh. "Non-linear transient finite element analysis with convected co-ordinates". In International Journal for Numerical Methods in Engineering, Volume 7, 1973, pp. 255–271.

[2] R. Cook. "Concepts and applications of finite element analysis", 3rd edition. Wiley, New York, 2001.

[3] Crisfield, M., Moita, G. "Unified co-rotational framework for solids, shells and beams". In International Journal of Solids and Structures, Volume 33, 1996, pp. 2969–2992.

[4] H. Elkaranshawy and M. Dokainish. "Corotational finite element analysis of planar flexible multibody systems". In Computers & Structures, Volume 54, 1995, pp. 881–890.

[5] U. Galvanetto and M. Crisfield. "Energy conserving co-rotational procedure for non-linear dynamics with finite elements". In International Journal for Numerical Methods in Engineering, Volume 39, 2265-2282.

[6] T. Gilbert and P. Trivailo. "Non-linear dynamic modelling of a reconfigurable flexible multi-link robotic manipulator arm, using a co-rotational finite element method". In 51st International Astronautical Congress, Rio de Janeiro, Brazil, AIAA Inc. (IAF-00-A.3.08), 2000, pp. 1–9.

[7] T. Gilbert and P. Trivailo. "Non-linear dynamic modelling of elastic wing models using a corotational finite element method". In Proceedings of the Fourth Biennial Engineering Mathematics and Applications Conference, Melbourne, Australia, Institution of Engineers, Australia, 2000, pp. 143–146.

[8] K. Hsiao, H. Horng and Y. Chen. "A corotational procedure that handles large rotations of spatial beam structures". In Computer Methods in Applied Mechanics and Engineering, Volume 27, 1987, pp. 769–781.

[9] K. Hsiao, R. Yang and C. Lee. "A consistent finite element formulation for non-linear dynamic analysis of planar beam". In International Journal for Numerical Methods in Engineering, Volume 37, 1994, pp. 75–89.

[10] K. Hsiao, J. Lin and W. Lin, W. "A consistent co-rotational finite element formulation for geometrically nonlinear dynamic analysis of 3-d beams". In Computer Methods in Applied Mechanics and Engineering, Volume 169, 1999, pp. 1–18.

[11] E.L. Houghton and P.W. Carpenter. "Aerodynamics for Engineering Students", 4th edition. Edward Arnold, London, 1993.

Inverse Problems in HF Radar

S J ANDERSON

Defence Science and Technology Organisation,
P.O. Box 1500, Salisbury, South Australia 5108, Australia.
Stuart.Anderson@dsto.defence.gov.au

Summary

The extraction of useful information from radar echoes cannot be accomplished solely within the domain of 'direct' problems, that is to say, the solution of Maxwell's equations with prescribed boundary conditions. In general, the interpretation of radar returns involves the solution of inverse problems, which are essentially ill-posed, even in the linear case, so special procedures are necessary in order to obtain meaningful solutions.

In this paper we illustrate the diversity of inverse problems which arise in HF radar by examining three examples : (i) the estimation of the ocean directional wave spectrum from the Doppler spectrum of scattered radar signals, (ii) the determination of an unknown S-matrix observed through a polarisation-transforming medium, and (iii) the determination of the temporal signatures of a class of wave-wave interactions in the ionosphere.

1 Introduction

Radar technology, in its diverse forms, constitutes a singularly effective means for probing our geophysical environment, and for detecting and classifying discrete man-made scatterers. For example, HF skywave radar echoes from (i) the distant ocean surface, and (ii) elevated plasma structures in the ionosphere, can be analysed to yield respectively (i) estimates of sea state, wave spectra and ocean currents, and (ii) profiles of electron density, characteristics of atmospheric gravity waves, the spatial and spectral properties of magnetic field line resonances, and meteor fluxes.

In order to extract this information, mathematical models of the associated physics must be used, in conjunction with appropriate mathematical techniques, to map the experimental data into the space of physical parameters being sought. 'Inverse' problems of this type are essentially ill-posed, even in the linear case, so special procedures are necessary in order to obtain meaningful solutions. In this context, it transpires that solution of the corresponding 'direct' problem is of interest not only for simulation and modelling, its natural domain, but as part of an iterative approach to the 'inverse' problem.

The primary purpose of this paper is to illustrate the variety of inverse problems which arise in HF radar physics by means of examples taken from DSTO's research. We emphasise cases where innovative techniques have been developed and applied to problems which have hitherto resisted attack but which are central to the objective of extending the scope and fidelity of radar-derived information.

2 Direct and Inverse Problems

Direct problems of mathematical physics are problems oriented along a cause-effect sequence, i.e., problems of finding out the consequences of given causes. For example, (i) the determination, using Maxwell's equations, of electromagnetic fields in time and space from given charge and current sources (ii) the calculation of the reaction of an electrical circuit to a specified input signal.

Inverse problems are those associated with the reversal of the chain of causally related effects, i.e. problems of finding the unknown causes of known consequences [1]. For example, (i) the determination of the characteristics of the sources of a field from the values of the field at certain points or in certain regions of space (ii) the reconstruction of the input signal from the reaction observed at the output of a device.

In the radar domain, several problems have acquired canonical status :
(i) the inverse obstacle problem - typically, estimate target shape and constitutive parameters from scattering measurements over some range of aspect and frequency
(ii)the synthetic obstacle problem - typically, minimise target radar cross section over some sector
(iii) the inverse profiling problem - typically, determine the height variation of some property of a stratified medium by multifrequency sounding
(iv) the source synthesis problem - typically, design an antenna to achieve a specified radiation pattern
In addition, a number of other radar-related problems can be viewed from an inverse problem perspective, including adaptive beamforming, channel equalisation (and calibration), together with system identification and blind deconvolution.

Most previous work on electromagnetic inverse scattering in the context of these canonical problems has used linearising assumptions to simplify the models of the physics. For example, we can cite formulating the diffraction tomography problem in terms of the Born approximation for weakly scattering dielectric objects, or the Kirchoff approximation (physical optics) for perfectly conducting targets [2]

In general, inverse problems are ill-posed in the Hadamard sense : the problem of solving an operator equation $Kf = g$ is called well-posed on a pair of metric spaces (F, U) if
(i) for any element $g \in U$ there exists a solution f from the space F (the solvability condition)
(ii) the solution is unique in F (the uniqueness condition)
(iii) the solution is stable on the spaces (F, U) (the stability condition)
If K is bijective and the inverse operator $K^{-1} : U \to F$ is continuous, then the problem is well-posed. If at least one of these conditions is violated, the problem is said to be ill-posed. Singularity of K or the existence of nontrivial solutions to the homogeneous equation $Kf = 0$ are sufficient conditions for ill-posedness.

3 Linear Inverse Problems

The classical continuous linear inverse problem $Kf = g$ can be formulated as a linear Fredholm equation of the first kind,

$$\int K(x,y)f(y)dy = g(x) \tag{1}$$

In practice the measurement set $g(x_i)$ is necessarily discrete, which immediately introduces non-uniqueness for all kernels (Green's functions) which are physically realizable (integrable), even in the absence of data noise. (This follows from the fact that the null space of K is nonempty, as implied by the Riemann-Lebesgue lemma.) But this is only the first difficulty. To proceed, we must discretise the equation (1) by one or other standard technique - matrix quadrature, expansion in (orthogonal) special functions, product integration or expansion in terms of the singular functions of the kernel :

$$\sum_{j=1}^{n} K_{ij} f_j = g_i \qquad i = 1, m \tag{2}$$

where f_j and g_i are the corresponding linear functionals of the functions $f(y)$ and $g(x)$.

We find that, in each case, the linear system becomes increasingly poorly conditioned as we increase the number of samples in the representation of the source function. In other words, the price of trying to map f in more detail is that the solution f_j develops an extremely strong dependence on variations of the inhomogeneous term, as well as to numerical round-off errors and inexactitude in the specification of the elements of K. Thus the system is neither unique nor stable.

4 Methods for Linear Inverse Problems

(i) Classical Methods

Traditional methods for 'solving' (1) include (i) rank reduction by parameterisation, (ii) expansion of the inverse operator in a Neumann series when f can be assumed highly smooth, and (iii) direct solution of the linear system (2) for a sufficiently low order n at which instability has not yet emerged.
Often the system of equations is over-determined, and a solution minimizing some quadratic functional of the residuals is selected. If the system is under-determined, other options exist. But in all cases, these methods exhibit considerable arbitrariness, and offer little reassurance in the form of meaningful error bounds. In fact, successive approximations do not in general satisfy $||f_{n+1} - f_n|| \to 0$ as $n \to \infty$. Further, finding an f which makes $||Kf - g||$ negligibly small is NOT sufficiently restrictive to ensure a source function which bears any resemblance to reality.

(ii) Non-classical Methods

The modern approach recognises that 'optimum inversion depends on properties of the unknown solution which cannot be determined form the data alone' [3]. Information which is needed to stabilize the solution is applied in the form of structural constraints on the source function. Provided that this /itala priori information is soundly based, for example from a deep understanding of the physics, the ill-posed problem can be transformed into a well-posed problem.

The best known method embodying these ideas is the method of regularization. Here the extra information is introduced as a 'smoothness condition' on the source function. A solution is then obtained by bounding an appropriate linear functional of the source function, Hf say, subject to the classical constraint of minimizing $||Kf - \hat{g}\ ||$

$$|| \ Kf - \hat{g} \ ||^2 + \lambda \ || \ Hf \ ||^2 = \min \tag{3}$$

where λ, the regularization parameter, appears as a Lagrange multiplier. A wide range of options for H exists (see eg. [4]). The practical problem then becomes that of determining the regularization parameter which optimally balances residual error against smoothness of the solution. We note that regularisation methods can be formulated for either deterministic or statistical data. When implemented in a Bayesian framework, the technique achieves its greatest efficacy, but the computational load of calculating prior distributions can be excessive.

5 Methods for Nonlinear Inverse Problems

(i) Directed Search Methods

When the operator K is nonlinear, the problems associated with ill-posedness are still present. Suppose we reduce our operator equation to a system of nonlinear equations of the general form

$$\vec{\rho}^{\,*} = \vec{\rho}(\mathbf{C}) \tag{4}$$

where $\{\vec{\rho^*}\}_j = \rho_j^* = \rho^*(\omega_j)(j = 1, 2, ...m)$ is a set of measured quantities and $\{\mathbf{C}\}_i = C_i(i = 1, 2, ...n < m)$ is the set of parameters to be retrieved. The dependence $\vec{\rho}(\mathbf{C})$ is given by the model of the corresponding physical process, now nonlinear. Following [5], we note that most approaches to the solution of (4) are based on the minimisation of a quadratic functional of the form

$$M(\mathbf{C}) = \sum_{j=1}^{m} \{\rho_j^* - \rho_j(\mathbf{C})\}^2 \tag{5}$$

using an iteration procedure such as Gauss-Newton

$$\mathbf{C}^{p+1} = \mathbf{C}^p - (K_p^T K_p)^{-1} K_p^T \Delta\vec{\rho}^{\,p} \tag{6}$$

where $\Delta\vec{\rho}^{\,p} = \vec{\rho}(\mathbf{C}^p) - \vec{\rho}^{\,*}$, and K_p is an $m \times n$ matrix with elements

$$\{K_p\}_{ji} = \frac{\partial \rho_j(\mathbf{C})}{\partial(C_i)} |_{\mathbf{C}^p} \tag{7}$$

To enhance search efficiency, and to counter ill-posedness, we can employ a modified search algorithm, such as those of Levenberg and Marquardt :

$$\mathbf{C}^{p+1} = \mathbf{C}^p - \eta_p(K_p^T K_p + \mu_p D_p)^{-1} K_p^T \Delta\vec{\rho}^{\,p} \tag{8}$$

where

$$D_p = \{ \begin{matrix} I_n & \text{in Levenberg's method} \\ \text{diag}(K_p^T K_p) & \text{in Marquardt's method} \end{matrix} \tag{9}$$

The parameters $\eta_p > 0, \mu_p > 0$ are introduced to optimise the convergence of the procedure. η_p imposes a limitation on the step size $\Delta \mathbf{C}^p$ and is introduced to ensure monotonic variation of the functional,

$$M(\mathbf{C}^{p+1}) < M(\mathbf{C}^p) \tag{10}$$

while the diagonal loading weighted by μ_p,

$$K_p^T K_p \rightarrow K_p^T K_p + \mu_p D_p \tag{11}$$

is a simple form of regularisation used here to prevent loss of convergence of the iteration procedure should $K_p^T K_p$ approach degeneracy. The formulation (8) can be improved using the full statistical regularisation ,

$$\mathbf{C}^{p+1} = \mathbf{C}^p - (K_p^T C_\rho^{-1} K_p + C_C^{-1})^{-1}(K_p^T C_\rho^{-1} \Delta \vec{\rho}^{\,p} + \mathbf{Y}^p) \tag{12}$$

Here C_ρ is the covariance matrix of errors $\vec{\rho}^{\,*}$ and C_C is an a priori covariance matrix of the sought parameter set **C**. The term $\mathbf{Y}^p$ is introduced for the nonlinear case in order to satisfy the convergence conditions [5]. It can be written as $\mathbf{Y}^p = K_p^T C_\rho^{-1} K_p(\mathbf{C}^p - \mathbf{C}^{p-1})$ or as $\mathbf{Y}^p = K_p^T C_\rho^{-1} K_p(\mathbf{C}^p - \mathbf{C}^0)$.

(ii) Stochastic Search Methods

In recent years, stochastic methods based on evolutionary algorithms have found widespread application in optimisation problems. We have developed a general-purpose genetic algorithm (GA) code which is well-suited to inverse problems and used it in lieu of the deterministic search approaches mentioned above. In addition to the well-known ability of GA's to avoid being trapped on local extrema, the attribute which makes them especially effective for inverse problems of the classes encountered in HF radar applications is the ability to combine a diverse set of physical criteria in the metric used for ranking solutions.

Our investigations of GA convergence, supported by extensive animated graphics developed to monitor the solution space, have convinced us that the most efficient approach is to employ hybrid methods with adaptive handover to multiple directed search branches at the point where the entropy of the population crosses a threshold. Otherwise the final stages of optimisation need to be driven by mutation, with a difficult trade-off between preservation of desirable traits and convergence time.

6 The Radar Process Model Formalism

At the most fundamental level, the radar process consists of driving a current i through a (transmitting) antenna and observing the subsequent voltage V generated across the output feed of a (receiving) antenna,

$$V = Zi \tag{13}$$

where the 'impedance' operator Z contains all the physics of radiation, propagation, scattering and so on. To serve any practical purpose, we must decompose Z into factors which embody the various physical mechanisms involved, so that we can identify the known and the unknown parameters and hence derive the equations to be solved. This operator decomposition is in many cases straightforward, even trivial, but in some cases of interest the interaction of antennas and scatterers with their respective environments prevents any simple partitioning. Here we shall proceed on the basis of a conventional decomposition [6], though we extend the representation to cover the waveform selection and signal processing stages of the radar process :

$$s = \tilde{P}\tilde{R}\tilde{M}_S^R\tilde{S}\tilde{M}_T^S\tilde{T}w + \tilde{P}\tilde{R}\tilde{M}_N^R n + m + \text{higher order terms} \tag{14}$$

where
w represents the selected waveform
$\tilde{T}$ represents the transmitting complex, including transmitters and antennas
$\tilde{M}_T^S$ represents propagation from transmitter to target scattering region
$\tilde{S}$ represents all scattering processes in the target region
$\tilde{M}_S^R$ represents propagation from target scattering region to receiver
n represents an external noise source
$\tilde{M}_S^R$ represents propagation from noise source to receiver
m represents internal noise
$\tilde{R}$ represents the receiving complex, including antennas and receivers
$\tilde{P}$ represents the signal processing
s represents the signal decomposition after processing

We note that the dimension of the signal space may vary along the process; all desired physical attributes may be assigned appropriate dimensions, and integration over intermediate domains is implicit. Alternatively, it may be useful, or necessary on occasion, to expand these primary level operators into concatenations of secondary operators, and so on, to isolate specific stages within a hardware subsystem or algorithmic procedure, or to expand the dimensionality of models accounting for the physical processes which are being taken into account. One advantage of this formalism is that it provides a framework for constructing abstract representations of arbitrary signal paths. We can then work backwards, proposing any consistent operator expression and exploring its realisability, its identifiability and its potential for exploitation. In the case studies discussed in this paper, we shall show how specific problems are embedded in this process model.

7 Example I : Directional Wave Spectrum Estimation from HF Sea Clutter

As first demonstrated by Barrick [7], the measured HF radar Doppler spectrum of the sea surface $D(\omega; k_0)$ is related to the directional wave spectrum of ocean gravity waves $S(\boldsymbol{\kappa})$ through an expansion of the form :

$$D(\omega; \boldsymbol{k_i}, \boldsymbol{k_s}) = \int F^A(\boldsymbol{k_i}, \boldsymbol{k_s}, \boldsymbol{\kappa_1}; \omega) S(\boldsymbol{\kappa_1}) d\boldsymbol{\kappa_1} + \int\int G^A(\boldsymbol{k_i}, \boldsymbol{k_s}, \boldsymbol{\kappa_1}, \boldsymbol{\kappa_2}; \omega) S(\boldsymbol{\kappa_1}) S(\boldsymbol{\kappa_2}) d\boldsymbol{\kappa_1} d\boldsymbol{\kappa_2} + ... \quad (15)$$

where k_i is the wavevector of the incident field, k_{is} is the wavevector of the scattered field and F^A,G^A, etc are kernels embodying the scattering mechanisms. This expression serves as the $\tilde{S}$ operator in the process model.

Now, for HF surface wave radars, only vertically polarised fields are significant, so the problem is a scalar one. For HF skywave radars it is necessary to take account of polarisation, but owing to the facts that (i) the incident skywave polarisation correlation length is less than the resolution cell size, (ii) the antennas employed are single polarisation antennas (almost always linear), and (iii) the VV-polarised scattering coefficient greatly exceeds the other elements of the scattering matrix, we can replace the model's polarisation space structure throughout by a simple 3 dB scalar loss term. Further, from the saturation properties of the high wavenumber directional spectrum, we can usually calibrate the echo approximately, thereby removing the need to know the magnitudes of the $\tilde{M}_S^R$, $\tilde{M}_T^S$, $\tilde{T}$ and $\tilde{R}$ operators. When these simplifications are valid, the fundamental inverse problem of radio oceanography consists of solving (15) for $S(\boldsymbol{\kappa})$.

The nonlinear and nonlocal structure of (15) have lead researchers to examine a number of approaches :
(i) linearisation by making assumptions about the high wavenumber regime of $S(\boldsymbol{\kappa})$, followed by application of the linear inversion methods described earlier
(ii) reduction to a low-rank parameter estimation problem by adopting a simple parametric form for $S(\boldsymbol{\kappa})$
(iii) application of the nonlinear inversion methods described earlier, with various forms of regularisation incorporating the physics of wave-wave interactions
When the radar system is multistatic, or has the capability to employ multiple frequencies, we can extend the inversion problem to exploit additional information. For example :
(i) Given a measured spectrum , $D(\omega; \boldsymbol{k_i}, \boldsymbol{k_s})$, $\boldsymbol{k_s} \neq -\boldsymbol{k_i}$, estimate $S(\boldsymbol{\kappa})$ - the bistatic case
(ii) Given measured spectra $D(\omega; \boldsymbol{k_i}, \boldsymbol{k_s})$, $\boldsymbol{k_s} = -\boldsymbol{k_i}$, $\boldsymbol{k_i} = 2\pi f_j \hat{k_0}/c, j = 1, n$, estimate $S(\boldsymbol{\kappa})$ - the multi-frequency case

(iii) Given measured spectra $D(\omega; \boldsymbol{k_i}, \boldsymbol{k_s})$, $\boldsymbol{k_s} = -\boldsymbol{k_i}, \boldsymbol{k_i} = 2\pi f_j \hat{k_0}/c, j = 1, n$, estimate $S(\boldsymbol{\kappa})$ - the bistatic multi-frequency case

The independent samples afforded by multi-frequency and multi-geometry observations reduce or even eliminate the classical left-right ambiguity problem, and in addition, dramatically accelerate the convergence of the inversion procedure.

8 Example II : S-Matrix Estimation Through Polarisation-Transforming Media

In a magnetised plasma such as the ionosphere, the polarisation state of the signal varies along the propagation path. Accordingly, the standard lexicographic technique for measuring the scattering matrix (S-matrix) of a scatterer is frustrated. The fundamental question is, can the S-matrix be determined in spite of the unknown transformations which occur en route to and from the target ?

Modelling the ionosphere as a plane-stratified, magnetised, cold, collisionless plasma, Maxwell's equations can be written as a set of four coupled linear differential equations,

$$\boldsymbol{e}' = ik\tilde{C}\boldsymbol{e} \tag{16}$$

where the vector $\boldsymbol{e}$ is given by

$$\boldsymbol{e}' = [E_x, -E_y, H_x, H_y]^T \tag{17}$$

and $\tilde{C}$ is the usual coupling matrix from magneto-ionic theory. Let $\tilde{V}$ be the matrix whose columns are eigenvectors of $\tilde{C}$. Then

$$\tilde{C}\tilde{V} = \tilde{V}\tilde{Q} \tag{18}$$

where the eigenvalues q_i, the entries along the diagonal of $\tilde{Q}$, are the roots of the Booker quartic, that is, the generalised form of the dispersion relation for a cold uniform magnetoplasma. Now, for the slowly-varying plasma, we can decompose the total field in terms of the eigenvectors corresponding to a fictitious homogeneous medium which matches the local medium properties (the WKB approximation) :

$$\boldsymbol{e} = \tilde{V}\boldsymbol{f} \tag{19}$$

where $\boldsymbol{f}$ is the column vector of amplitudes of the characteristic wave modes. Substituting in (16),

$$f' + i\tilde{Q}f = \tilde{\Gamma}f \tag{20}$$

When the coupling can be neglected, which at VHF (30 - 300 MHz) is for all normal ionospheric conditions except when intense sporadic E layers are present, this equation integrates directly to yield solutions

$$f_j = K_j \exp(-ik \int q_j dz), \qquad j = 1,..,4 \tag{21}$$

which each contribute to the total field an amount

$$\boldsymbol{e}_j = K_j \boldsymbol{s}_j \exp(-ik \int q_j dz) \tag{22}$$

The eigenvalues q_j appear in the ionospheric case as two complex conjugate pairs, for which the sign of the imaginary part determines whether the wave is upgoing or downgoing. Thus, provided that the coupling term Γ is small, we can partition the field into upgoing and downgoing waves,which enables us to exploit the partitioning invoked in the radar process model. In other words, the only coupling occurs at the scatterer, where it is embodied in the target S-matrix. We can therefore model the (unitary) propagation operators $\tilde{M}$ using only a 2×2 representation for the coherent case, that is, the classical Jones calculus of optics.

With reference to our model equation (14), measurements are conducted by radiating in turn at two independent (generally orthogonal) transmission polarisation states (represented by $\tilde{T}$), and collected each time for two orthogonal receive polarisation states (represented by $\tilde{R}$).

We depart from the standard data requirements by demanding access to measurements at a set of points z_i for which differential propagation effects are non-negligible,

$$\tilde{J}_i = \tilde{M}_S^R(z_i)\, \tilde{S}\, \tilde{M}_T^S(z_0) \qquad i = 1, \ldots, n \tag{23}$$

where z_0 corresponds to the transmitter→target separation. The $\tilde{M}(z)$ matrices model the progressive transformation of the transverse electric polarisation state $\boldsymbol{a} = [E_x, E_y]$:

$$\frac{d\boldsymbol{a}}{dz} = \tilde{A}\boldsymbol{a} \tag{24}$$

where

$$\tilde{A} = \begin{pmatrix} j2\beta & \alpha \\ -\alpha & 0 \end{pmatrix}$$

combines the effects of conventional Faraday rotation and ellipticity conversion, resulting from birefringence and other mechanisms. Integrating,

$$\boldsymbol{a} = \tilde{M}(z)\, \boldsymbol{a}_0 \tag{25}$$

so

$$\tilde{M}(z) = \exp z\tilde{A} \tag{26}$$

As demonstrated in [8], this constraint on the form of $\tilde{M}(z)$, together with the constraint that the radar observations are monostatic, leads us to an equation linking the observational data $\tilde{J}_i$ to the eigenstructure of $\tilde{M}(z)$; this in turn sets up matrix equations for $\tilde{S}$ which must be solved using one of the inverse methods described earlier. It transpires that for all propagation conditions with non-zero Faraday rotation, the scattering matrix $\tilde{S}$ is identifiable given an unambiguous receiving antenna array. Of course, the disposition and scaling of the array must be consistent with the differential propagation effects. It is a long way from this precursor result to the conclusion that HF skywave radar polarimetry is a realistic prospect, but experiments are underway to assess this ambitious goal.

9 Example III : Estimation of Radiowave Modulation by Magnetohydrodynamic Processes in the Ionosphere

One common type of wave modulation which impacts significantly on HF radar performance is the result of a complex chain of interactions between the solar wind and the magnetosphere-ionosphere system. Hydromagnetic turbulence generated by the Kelvin-Helmholtz instability at the plasmasphere boundary propagates as Alfven and fast mode magnetohydrodynamic waves to the ionosphere, perturbing it and exciting field line resonances. HF radiowaves traversing the ionosphere interact with the MHD waves; this interaction manifests itself as a phase modulation which results from the dependence of the refractive index on the magnetic field, the advection of the ionospheric plasma, and the compression of the plasma along the radiowave propagation path [9].

The modulation effects can be expressed in terms of the equivalent Doppler velocity $\hat{V}$, defined for the vertical incidence case as :

$$V^* = \frac{d}{dt}\int_0^{Z_R} \mu[B_L(z,t), B_T(z,t), N(z,t)]dz \tag{27}$$

The physical mechanisms responsible for nonzero $V*$ can be associated with the terms resulting from carrying out the differentiation,

$$V = V_1 + V_2 + V_3 \tag{28}$$

where

$$V_1 \simeq -2i\omega k \int_0^{Z_R} \left[\frac{\partial \mu}{\partial B_L} b_z + \frac{\partial \mu}{\partial B_T} b_x \right] ds \tag{29}$$

(plasma remagnetisation)

$$V_2 = -2k \int_0^{Z_R} \left[\frac{\partial \mu}{\partial N} \frac{\partial N}{\partial z} v_Z \right] dz \tag{30}$$

(plasma advection)

$$V_3 = -2k \int_0^{Z_R} \left[\frac{\partial \mu}{\partial N} N(\nabla \cdot \boldsymbol{v}) \right] dz \tag{31}$$

(plasma compression)
In practice it has been shown in [10] that the advection term V_2 usually dominates in the oblique F-layer reflection case. As for E-region measurements, and Sporadic-E, the calculations reported in [10] indicate that the Doppler velocity magnitudes are an order or two smaller than those at F-region heights, with Pc-3 and Pc-4 oscillations frequently not detectable with conventional techniques. The essence of the method described here (and first published in [11])lies in enhancing sensitivity by exploiting the conserved rank property of the echoes scattered from the distant terrestrial surface.

With reference to the radar process model (14), we are concerned with modulations which, to a good approximation, affect all polarisation components equally, except at ranges very close to the skip distance. Further, the interaction is linear in the transmitted signal so the modulation is independent of radiated power, and is usually of low azimuthal wavenumber, so it is more or less independent of antenna patterns. Accordingly the process model reduces to the scalar form :

$$s(t) = e^{i\phi_S^R(t)} \tilde{S}(t) e^{i\phi_T^S(t)} \tag{32}$$

where

$$\tilde{M}_S^R(t) = e^{i\phi_S^R(t)} \tilde{M}_S^R(0) \tag{33}$$

and

$$\tilde{M}_T^S(t) = e^{i\phi_T^S(t)} \tilde{M}_T^S(0) \tag{34}$$

The associated inverse problem is to estimate $e^{i[\phi_S^R(t)+\phi_T^R(t)]}$ from measurements of $s(t)$.

Modulation Detection

In view of the diversity of wave processes in the ionospheric plasma, the first step in estimating the MHD wave modulation signature is to show that such modulation is indeed present. The importance of this step is a consequence of the existence of alternative wave processes in the ionosphere with different physical mechanisms but which cannot be distinguished from MHD modulation in the power spectrum domain. This can be accomplished by exploiting the fact that the sample covariance matrix of phase-modulated clutter preserves the eigenstructure of the unaffected clutter, with the eigenvectors of the respective matrices linked by unitary transformation. The noise subspace dimension of the sample covariance matrix is thus independent of the modulation. Accordingly, nonstationarity of the signal arising from wave modulation may be tested by examining whether the eigendecomposition preserves the symmetries associated with stationarity. Specifically, if the given clutter time sequence belongs to the family of stationary processes, then the noise subspace eigenvectors should be orthogonal (in the stochastic sense) to the signal subspace of the time-reversed (averaged) matrix. If this condition does not hold, then modulation is present [11].

Modulation estimation

Having shown that modulation is present, the inverse problem of estimating the modulation function could, in principle, be achieved by direct estimation of the unitary diagonal transform which provides time reversibility, using an appropriate regularising factor.

The computational burden of performing this inversion in real-time is excessive, so we have developed a simpler approach which can be adopted when the instantaneous frequency of modulation may be assumed constant over a few repetition periods, as is the case for virtually all the ionospheric wave phenomena of interest. Under this assumption we may assume that any two subintervals, consisting of these few repetition periods, experience the frequency shift caused by the phase modulation. This frequency shift, *not the signal instantaneous frequency as in [10]*, may be estimated by a MUSIC-type approach, regardless of the spectral content of the short-interval covariance matrix of the electromagnetic field.

More specifically, the method may be summarised as follows :
(i) for each subinterval form an estimate of the clutter covariance matrix using a sliding window with forward and backward averaging as well as averaging over the set of available clutter sequences with "identical" a priori power spectra (e.g. adjacent range cells, subarray outputs, etc.)
(ii) for each subinterval i form the eigendecomposition of the associated covariance

matrix

$$\hat{R}_i = \hat{U}_i \hat{\Lambda}_i \hat{U}_i^H \tag{35}$$

and define the signal and noise subspaces $\hat{S}_i$, $\hat{G}_0$
(iii) form the function $\mu_i(\Delta)$:

$$\mu_i(\Delta) = \sum_{n=1}^{M-r} \sum_{m=1}^{r} \hat{S}_{i+1,m}^H D(\Delta) \hat{G}_{in} \hat{G}_{in}^H D(\Delta) \hat{S}_{i+1,m} \tag{36}$$

where

$$D = \mathrm{diag}(e^{j\omega T_R}, e^{j2\omega T_R}, \ldots, e^{j(N-1)\omega T_R}) \tag{37}$$

$\omega = 2\pi\Delta$ and T_R is the repetition period
(iv) estimate the instantaneous frequency shift, $\hat{\Delta}_i$: by minimising $\mu_i(\delta)$:

$$\hat{\Delta}_i = \mathrm{argmin}\ \mu_i(\Delta) \tag{38}$$

(v) estimate the phase sequence by integration of the instantaneous differential frequency $\hat{\Delta}_i$.

The resulting function is an accurate representation of the temporal structure of the MHD wave integrated, with appropriate weighting, along the propagation path [12].

10 Conclusion

The case studies described in this paper illustrate the diversity of inverse problems which have arisen in our studies of HF radar :
(i) The estimation of the directional wave spectrum of the ocean surface from the Doppler spectrum of scattered radar signals is based on a nonlinear, nonlocal integral formulation of the scattering process. We have shown through modelling that augmenting the data space by means of multi-frequency and multi-static observations leads to accelerated and more reliable convergence.
(ii) The prospect of estimating a target's scattering matrix through a polarisation-transforming medium has been shown to be possible under physically plausible conditions, and an inversion method constructed.
(iii) A method for determining the characteristics of MHD waves in the ionosphere from their modulation phase signatures on skywave radio signals has been has been established by an efficient step-wise reduced rank inversion technique.

Much of the early research addressing these problems has been carried out without due regard to the radar process context, or the implications of implicitly-adopted process models. This situation is now changing as we begin to re-examine the interdependence of these elemental problems within a global process model. Questions such as how to maintain consistent fidelity of the physics whilst preserving mathematical integrity of the solutions can then assume their due importance. Ultimately we expect that questions such as how to optimise utilisation of radar resources, and the converse issue of optimising radar design, will be able to be addressed in this general framework, using the methods of modern inverse theory to avoid the pitfalls of more simplistic approaches.

References

[1] V.F. Turchin, V.P. Kozlov and M.S. Malkevich, "The use of mathematical-statistics methods in the solution of incorrectly posed problems", Soviet Physics Uspekhi, Vol.13, May-June 1971, pp.681-703

[2] K.J. Langenberg, T. Gurke, and M. Brandfass, "Multidimensional electromagnetic vector inverse scattering", in W-M Boerner et al (eds.)'Direct and Inverse Methods in Radar Polarimetry', Part 2, Kluwer, 1992, pp.1065 - 1074

[3] I.J.D. Craig and J.C. Brown, "Inverse problems in Astronomy", Adam Hilger, 1986

[4] A.N. Tikhonov, A.S. Leonov and A.G. Yagola, "Nonlinear Ill-posed Problems", Vols. 1 and 2, Chapman and Hall, 1998

[5] O.V. Dubovnik, S.L. Oshchepkov and T.V. Lapyonok, "Iteration-regularisation method of solution of nonlinear inverse problems and its application to interpretation of the spectra of brightness coefficient of a layer of water", Physics of the Atmosphere and Ocean (English trans.), Vol.30, No.1, August 1994, pp.103 - 110

[6] S.J. Anderson, "The challenge of signal processing for HF over-the-horizon radar", Proc. Workshop on Signal Processing and Applications, QUT Signal Processing Research Centre, Brisbane, December 1993, pp.1-14

[7] D.E. Barrick, "Remote sensing of sea state by radar", in *Remote Sensing of the Troposphere*, V.E.Derr (Ed.), Ch.12, US Govt. Printing Office, 1972

[8] S.J. Anderson, Yu.I. Abramovich and W-M. Boerner, "On the solvability of some inverse problems in radar polarimetry", Proc. SPIE Conference on Wideband Interferometric Sensing and Imaging Polarimetry, San Diego, July 1997

[9] P.R. Sutcliffe and A.W.V. Poole, "The relationship between ULF geomagnetic pulsations and ionospheric Doppler oscillations : model predictions", Planet. Space. Sci. 38(12), 1990, pp.1581-1589

[10] J. Parent and A. Bourdillon, "A method to correct HF skywave backscattered signals for ionospheric frequency modulation", IEEE Trans. Ant. Propag., AP-36, January 1988, pp.127-135

[11] Yu.I. Abramovich, S.J.Anderson, G.J.Frazer, and I.S.D.Solomon, "Measurement and interpretation of phase fluctuations in HF radar echoes propagated via sporadic E layers", Proc. Conf. on Solar Physics, Adelaide, 1995, pp.152-155

[12] S.J. Anderson and Yu.I. Abramovich, "A unified approach to detection, classification and correction of ionospheric distortion in HF skywave radar systems", Radio Science, Vol.33, No.4, July-August 1998, pp.1055-1067

Antenna Integration on Aircraft: Needs and Methodology

CHRISTOPHE CALNIBALOSKY
DASSAULT AVIATION Direction Générale Technique Défense
78, quai Marcel Dassault – Cedex 300 92552 St Cloud Cedex France

1 Introduction

Modern combat aircraft increasingly incorporate more and more operational functions. These functions use a wide range of frequencies over the entire electromagnetic spectrum. They are essentially related to three functional areas, which are :

- Radar function
- Electromagnetic Countermeasures (ECM)
- Communication - Navigation - Identification (CNI).

Each of these functions calls upon a whole suite of equipment units, to ensure the necessary processing, and of antennas, to ensure the associated radio-electric coverage. Each function has its own antenna, or set of antennas, which, more particularly in the case of the ECM and CNI functions, makes it obvious that :

- such antennas form protuberances, becoming less and less compatible with the aircraft's global performance characteristics (drag, RCS,...)
- the great number of them makes installation, more and more costly.

Also, in the case of the Radar function, the antenna, whether of parabolic, mechanical scan or of active, electronic scan type, is generally positioned on the aircraft forebody, with the consequence that this restricts the obtained coverage and furthermore, wastes a "precious" location which could be profitably used for other functions.

2 What New Generation Antennas Should Be

When seeking to remedy these impediments on the performance capabilities of the aircraft and its systems (optronics, air data system, ...) by use of new solutions, it becomes obvious that our efforts should be directed towards :

- *Fewer antennas*, by use of compatible radioelectric grouping methods, so as to satisfy the needs of the growing number of operational functions, whilst maintaining the number of antennas to be fitted within reasonable limits; this corresponds to concept of multi-function antennas.

To be noted in this context is - taking the example of CNI functions - that if previous generation aircraft (roughly 20 years ago) often followed the "1 function = 1 antenna" principle, nowadays some degree of function grouping, such as in VHF/UHF/MIDS/TACAN antennas has already been achieved, although the involved dimensions of such blade antennas sometimes prove penalizing for their integration with the aircraft (Figure 1).

- *Antennas that better conform to the structure*, so as to overcome the drawbacks of the protuberances caused by present-day antennas ; this corresponds to concept of <u>conformal antennas</u>, antennas that fit the aerodynamic shape.

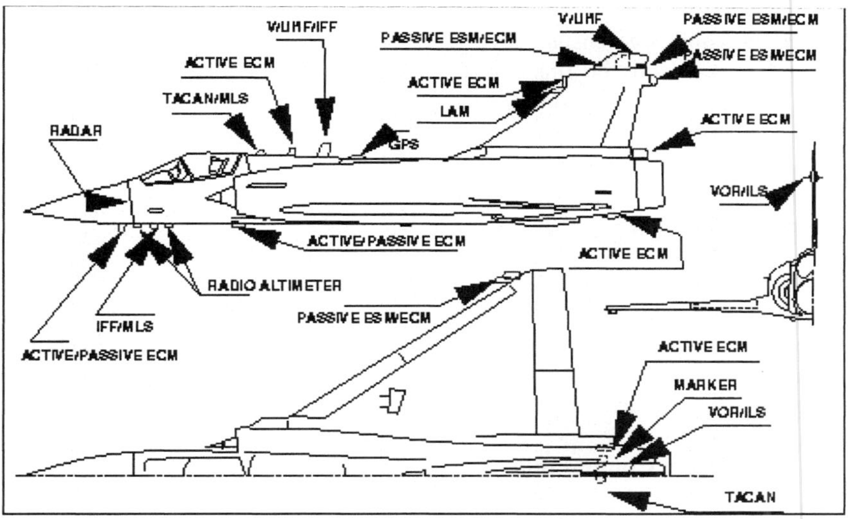

Figure 1 Military Antennas Proliferation

3 Operational Interests

Use of such conformal multi-function antennas should contribute towards improving:

- the performance characteristics of the operational functions making use of such antennas
- the global performance capabilities of the aircraft and their systems
- the maintainability/ reliability aspects.

3.1 Operational Function-Related Aspect

Radio-Electric Coverage

Radio-electric coverage improvements may arise from a better distribution of such antennas. These antennas are aircraft structure members and, then, must be able to be fitted to the spine as well as to the leading edge of the wing, to the lower surface of the forward fuselage or other possibilities.

Three aspects amongst these needs for improvement are currently clearly identified :

- coverage of the radar surveillance and IFF interrogation function, to provide for surveillance of an area of a surface larger than what is currently possible
- possibility to ensure, for firing control purposes, radar target tracking and aircraft/ missile links over larger sectors and not, as currently feasible, exclusively limited to the aircraft front sector
- possibility to have larger antenna bases so as to increase the ESM/ECM functions accuracy.

Emission Control

For most operational functions, it is furthermore desirable to endow such antenna types with a <u>coverage</u>, and hence, scanning <u>adaptativity</u> function which will provide for improvement of discretion during transmission (via direction, beam provision), or even, for protection against jamming during reception (implying, in this case, via antenna milling provision).

Inter-Function Compatibility

The problems relative to the compatibility between the different operational functions have to be resolved by use of communal, elementary antennas in conjunction with the necessary components, whether radio-electric (multiplexers, filtres,...) or processing sequencing (temporal comparability) units. Here again, two examples may be given ; the compatibilities between (1) radar/ ECM-jammer and (2) IFF interrogator - responder/ MIDS Data Link.

3.2 Aircraft and Systems Global Performance Characteristic Aspect

Aerodynamic Characteristics

Elimination of the current antenna-generated protuberances (essentially in the ECM and CNI function case) will result in less aircraft drag.

Electromagnetic Signature

Use of conformal, multi-function elements is expected to result in reduced global aircraft RCS, since this would mean fewer antennas. This will however, require particular care with such new antenna installations. The apparent RCS reduction that would result from the absence of protuberances must not be cancelled out by new problems which are bound to arise, such as gaps or impedance disruption at the aircraft/ antenna interface.

Global Aircraft System Aspect Integration Constraints

Use of conformal antennas spread over the whole aircraft surface is expected to contribute to global improvement of the airborne system. As an example and as already mentioned, the benefits of replacing the current antennas ensuring the radar function on the aircraft forebody by conformal antennas spread over the aircraft surface would be :

- liberation of installation areas, thus making them available for accommodation of other sensors (optronics, for example)
- new possibilities to improve the aircraft's air data system performance capabilities (today we remove the pitot tube so as not to interfere with the radar performance capabilities).

Maintainability/ Reliability Aspects

Potential benefits for these aspects would be :

- the need of fewer antennae to be integrated with the aircraft and hence, ultimately, of reduced associated maintenance action. To be noted also is that the conformal antenna aspect will also improve their erosion resistance, which again will contribute to reduced maintenance
- to offer the possibility of using the operational functions in degraded modes without, in the case of active antennas, complete loss of the involved function. This would arise from the adaptativity notion which would imply the use of a great number of active modules. Failure of the modules may tolerate progressive, although not necessarily total, function degradation.

4 Role of Modelling Tools

Significant progress has been yielded over recent years by both the processors and the software used for electromagnetism modelizations. Numerical simulation plays an increasingly important role which is complementary to that of experimentation. Measurements are a pre-requisite for identification of the elementary constituents' characteristics, for analyses and for validation of electromagnetic systems. Computations offer essential additional functionalities for designing antennas that fit the aircraft.

4.1 Hybrid Methods

The computation code SPECTRE, developed by DASSAULT AVIATION involves numerical and asymptotical electromagnetic computation methods. The numerical methods use finite-element field techniques. These offer great accuracy and robustness for computations on complex structures composed of different dielectric materials which may be prone to attenuation losses. These methods are, however, limited by the size of the matrix problems that can be handled by the processors. On current computers, it is challenging to conduct numerical method computations on the whole aircraft when the frequency exceeds the GHz range.

Asymptotical methods are the numerical methods' natural complement to conduct higher frequency computations. These methods describe the electromagnetic field propagation along the ray traces, by use of analytical reference solutions. Figure 2 shows the ray tracing taken into account for the computation of a fraction of the antenna pattern. As asymptotical formulations show serial $1/K^{\alpha+N}$ developments, the method applies best to the higher frequencies, but is not suited for geometrical parts which are of small size compared to the wavelength.

Such methods require continuous validation efforts for generic cases.

In order to associate the different methods' respective contributions, the SPECTRE code uses hybrid methods. By combining numerical methods, by means of which are computed the fields in small-size complex areas, with asymptotical methods which take into account larger areas, a great variety of problems can be solved. An example of the use of these methods is given in Figure 3. The computation of a blade antenna placed on a circular plate breaks down into 2 parts :

1. Numerical method modelling of a blade antenna placed on a plane gives the field in every point of its surface and allows definition of electric and magnetic currents equivalent to those of an antenna with respect to electromagnetic emission
2. Asymptotical method modelling of the equivalent current distribution radiation gives the reflections on the plate and the scattering by the circular rim, to determine the far field pattern.

These methods although very attractive by their potential capabilities require full mastery in their employment. The choice of how a problem is to be broken down into different sub-domains according to the different associated computation methods calls upon a physical sense of appreciation of the encountered phenomena.

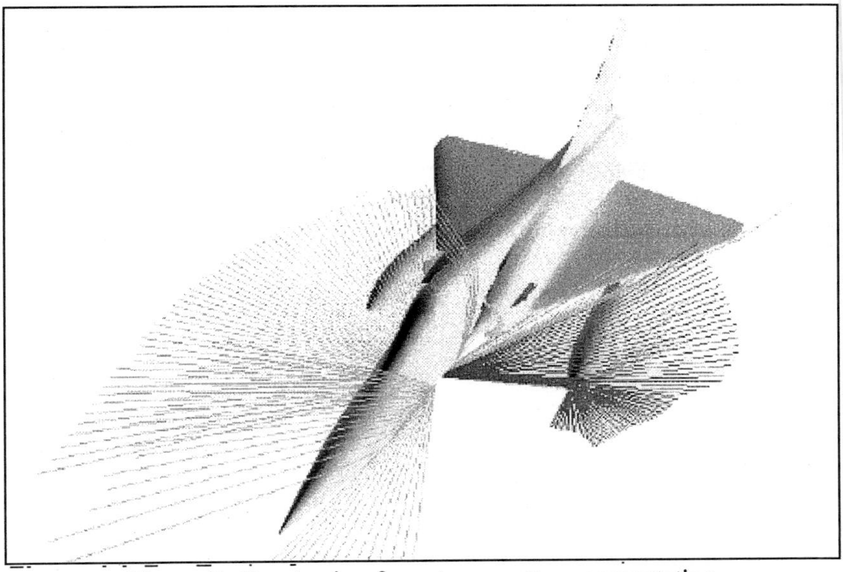

Figure 2 Ray tracing for antenna pattern computation

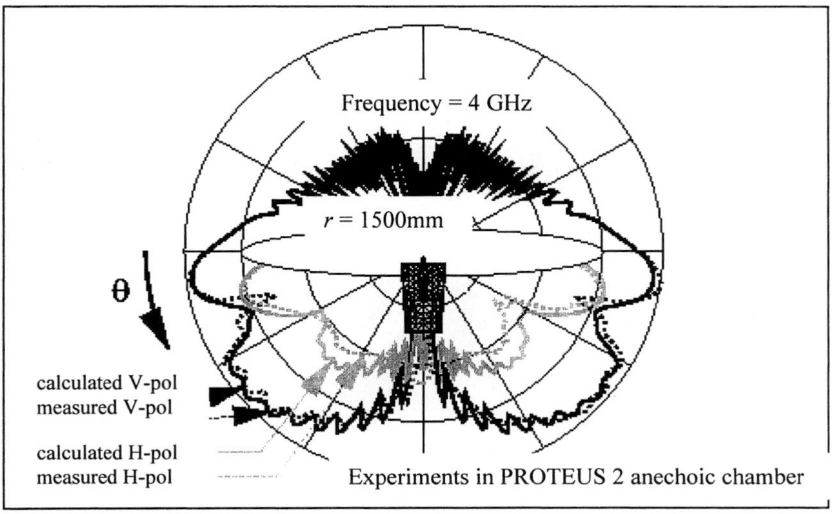

Figure 3 Blade antenna on a circular plate, comparison of numerical and experimental patterns

4.2 Optimization of Antenna Locations

Determining preferred antenna locations is a task becoming increasingly complex on account of the number of antennas and the number of constraints that come into play. From the radioelectrics point of view, such optimization must take into account the sought coverage pattern, as well as the compatibility between the different equipment. Such optimization must also account for the thermal and chemical aeromechanical environment constraints, notably with respect to structure deflections liable to reach amplitudes that risk jeopardizing the systems' performance capabilities.

4.3 Antenna Field Isolation

With the near field computations can be obtained the impedance inputs and the mutual impedance between antennas. The computations require detailed modelling of the geometry around the excitation accesses, together with a precise description of the dielectric materials.

The elementary result leads to the coupling power between 2 antennas. For more global analysis, the decoupling level can be represented in each point of the surface vis-à-vis one antenna of a given type. A presentation example is given in Figure 4 showing the decoupling vis-à-vis an antenna fitted to the fuselage upper surface.

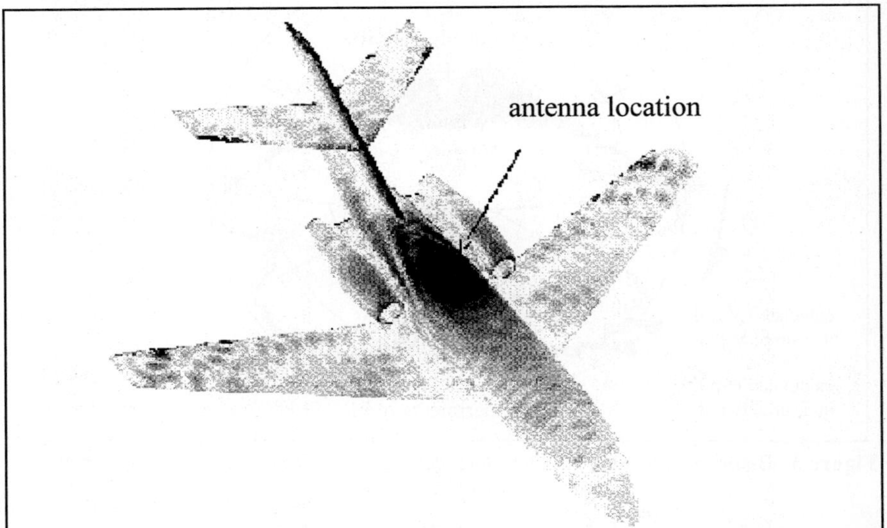

Figure 4 Mutual coupling with an antenna in a dorsal location

4.4 Designing Electromagnetic Windows

Design of radomes has to take into account a great number of constraints :

- aeromechanical environment (vibrations, deflections)
- thermal and chemical environment
- low observability (radar and improved)
- radioelectric transparency in the used frequency bandwidths.

Electromagnetic computations are used to determine the geometry and materials that offer on acceptable compromise. Figure 5 shows an S-band transparent leading edge structure that allows integration of radiating elements.

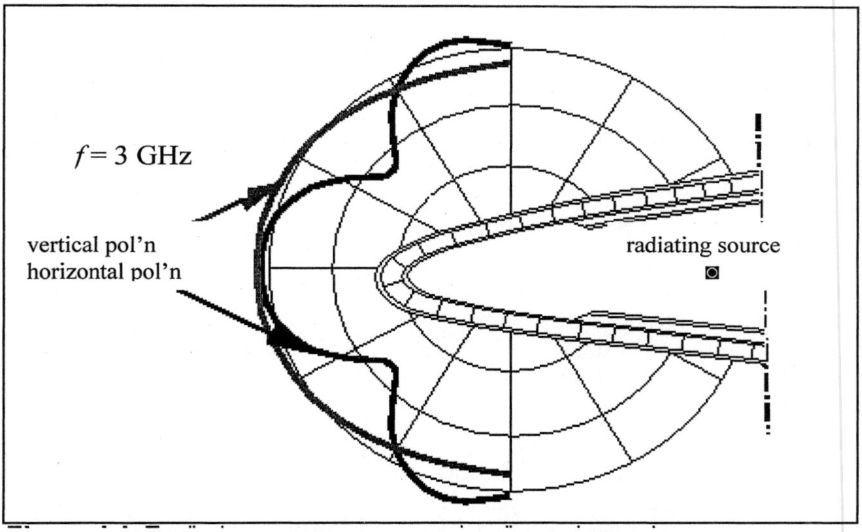

Figure 5 Radiation pattern across a leading edge radome

On account of the stealth constraints, the radome transmission bandwiths have to be limited to those of the used frequencies. V/UHF antennas, for example, have RCS of considerable levels. One solution consists in using a radome whose transmission bandwidth is limited to frequencies of less than 1 GHz, with low transmission above such levels.

This characteristic property can be achieved by certain frequency-selective surfaces (FSS). Figure 6 shows the transmission factor, via FSS composed of 3, very dense, cross dipole or tripole type layers. This allows us to achieve low transmission factors between 5 and 20 GHz.

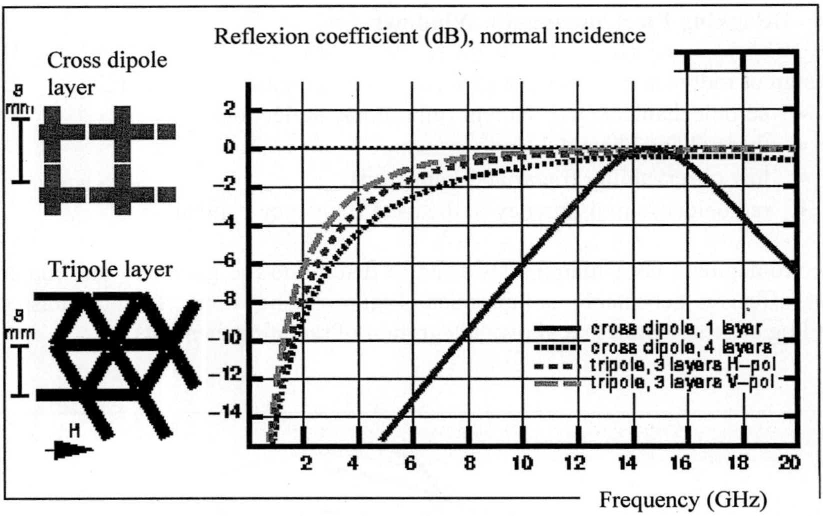

Figure 6 Reflectivity of frequency selective surface, normal incidence. Comparison of three different configurations

5 Integration Methodology

The modelling tools mentioned above provide for the use of new methodologies to determine the locations of antennas aboard the aircraft. The object is to bring into play an equipment model and an aircraft model so as to derive, as early as possible, the behaviour of antennas that are to be integrated with the aircraft.

The **equipment model** accurately represents the antenna (power supply, shapes, material...) and gives the radiated fields on a fictitious surface area around the antenna, fitted to a locally representative surface of the aircraft shape (typically, over a few tens of centimeters). The **aircraft model** gives a fine-grid representation of what is a fictitious interface (surface-external) and indicates the electric currents on this electrically bonded surface for each incidence direction.

By the reciprocity principle, the data of the two models (fields and currents) can be simply connected on the interface surface and thus give the radiated far field, where :

- $E^{\alpha}(x)$ is the equipment model's electric fields on the interface surface, antenna emitting
- $J^{s}(x)$ is the aircraft model's electric current on the bonded interface surface, the aircraft being exposed to a unitary plane wave in $\hat{\imath}$-direction
- $G(\hat{\imath})$ is the antenna gain in $\hat{\imath}$-direction.

$$\sqrt{G}(\hat{i}) = \frac{\mathrm{j}\omega\mu}{4\pi} \int_{\mathrm{Sint}} \mathrm{E}^{\alpha}(\mathrm{x}) . \mathrm{J}^{\mathrm{s}}(x) dS(x)$$

This principle is classically used for the hybrid methods presented above. There are several variants to this principle, which allow adaptations to the cases under study in accordance with the computation methods that are used.

A validation of a simple case consisting in a monopole antenna mounted on a parallelepiped box is shown in Figure 7, 8 and 9. The validity of the approach is demonstrated by the compared results of the models coupled with direct computations and measurement.

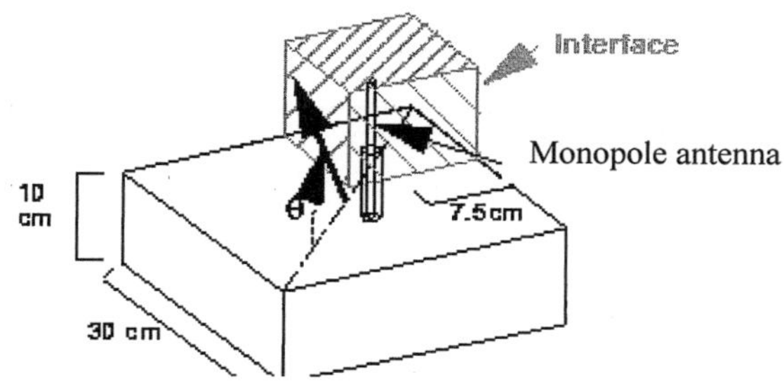

Test case, geometry

Figure 7 Monopole mounted on a box

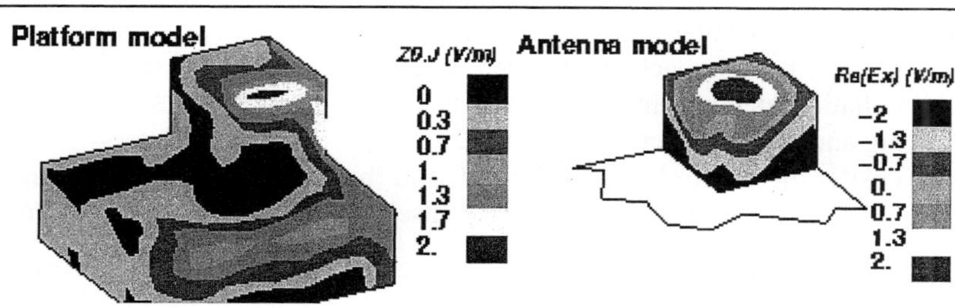

Data : electric current on conducting interface Data : Electric field on fictitious interface

Figure 8 Illustration of the platform and antenna models

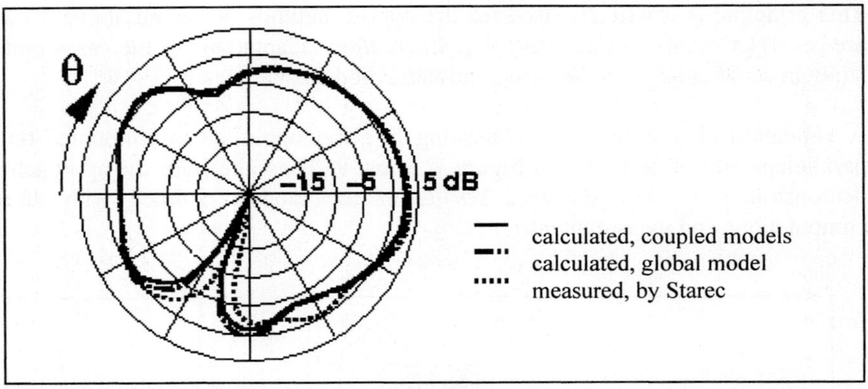

Figure 9 Radiation pattern (data from M. Jelioui, *Radiation of wire antenna mounted on complex perfectly conducting structure, J INA 88)*.

6 Recommended Solutions

- It is at present obviously not conceivable to have, for the envisaged applications, wide band antennas that can globally cover the entire radioelectric spectrum used on combat aircraft. This means that choices have to be made.
- Also, given that for certain functions radiation pattern adaptativity is wanted (the latter, here again, depending on the covered spectrum), the possibilities for achieving such adaptativity within one single unique radiation element must not result in elementary antenna sizes that are too big and hence difficult to integrate.
- Lastly, the functions to be integrated in these new multi-function antennas must be compatible amongst themselves, as regards the performance characteristics to be ensured.

A first line of thought thus leads to:

Recommendation 1: Elimination of the low frequency HF and VHF bands and grouping of the CNI functions, from the IFF to the GPS band within one and the same elementary antenna

A first analysis, essentially focused on printed circuit or wave guide-type antennas, has, indeed, shown that the required sizes of a radiating surface covering the considered bandwidth (200 to 1500 MHz) and providing a potential antenna pattern modulation minimum are in the order of the 700 mm diameter and 100 mm depth.

Given such sizes and the difficulty to integrate them with the aircraft, the design of conformal antennas of lower bandwidth (for a civil non-crypted voice transmission VHF function, for example) and of the same performance level as current solutions (faired printed circuit-type antenna in fin tip) cannot be envisaged. Also, the vast majority of CNI functions require omnidirectional coverage types (voice, IFF response, data transmission...), which must be compatible amongst themselves.

Recommendation 2: Grouping the Radar/ESM/ECM/IFF Interrogation functions, in conjunction with using a lower radar band

These three functions must, indeed, have common coverage areas and common scanning domains, especially if the latter are to be more extensive than in current practice. Furthermore, the use of common radioelectric scanning axes (in surveillance areas) will operationally lead to a better association of the tracks provided by these different sensors. Also, in the case of the radar function, distributing several (conformal) antennas over a greater aircraft surface than current practice will provide for lowering the working frequency (> = L band) and thus improve the detection of stealthy targets and enhance anti-jammer protection, both currently non-existent in such frequency domains.

As a conclusion, there are two equipment families that appear to emerge for future aircraft which should be endowed with conformal multi-function antennas :

- one "low band" family (essentially grouping the CNI functions)
- one "high band" family (countermeasures and radar, amongst others).

An industrial example of a coupled simulation illustrates these recommended solutions. The previously described methodology was successfully applied between DASSAULT AVIATION and THALES Airborne Systems during a feasibility study to install CNI conformal and multi-function antennas on Rafale.

The GPS, IFF, MIDS and TACAN functions were grouped within the same antennas block, located on the Rafale spine behind the canopy. This development was performed to achieve the following specifications :

- conformal antennas
- adaptative patterns for MIDS and GPS to ensure resistance to jamming.

Figure 10 shows the magnetic currents, induced by the radiation of a MIDS element, on the fictitious interface, computed by THALES AS, whilst Figure 11 shows the Rafale fine grid mesh used by DASSAULT AVIATION to compute the radiated pattern of this element, installed on the aircraft.

Improved antenna integration with the aircraft may furthermore arise from exchanges with partner contractors involved in such new design methodologies.

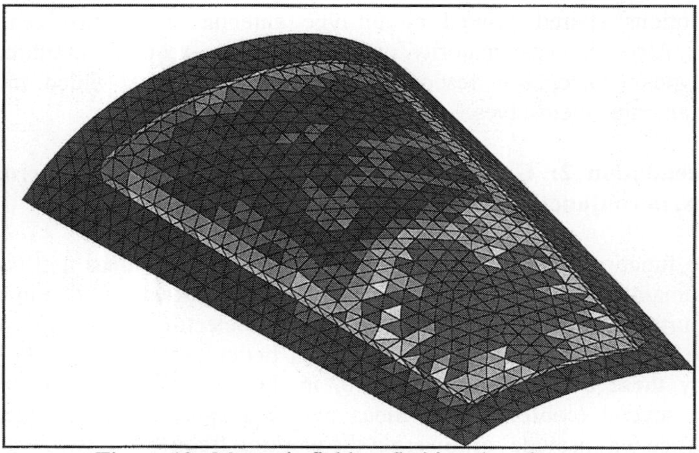

Figure 10 Magnetic field on fictitious interface

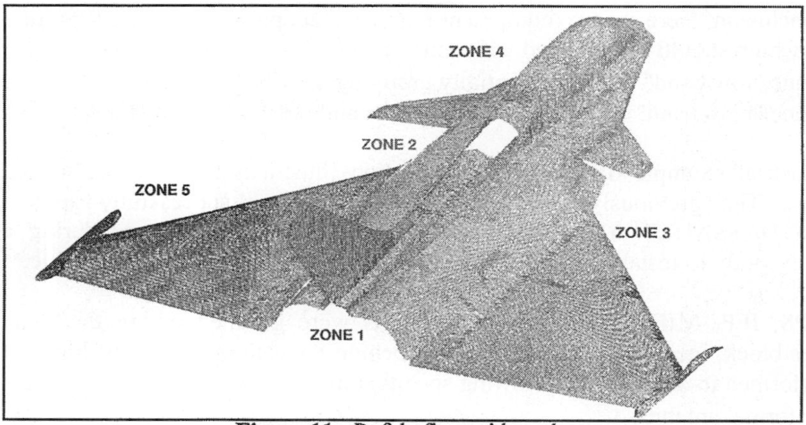

Figure 11 Rafale fine-grid mesh

An Iterative Method for Time-harmonic Integral Maxwell's Equations

Francis Collino[1] and Bruno Després[2]

[1] INRIA, Domaine de Rocquencourt 78153 Le Chesnay, France
[2] Laboratoire JLL, Université Paris VI, 175 rue du Chevaleret, 75013, Paris, France

Summary

We present an iterative algorithm for time harmonic Maxwell's equation, using a particular integral formulation proposed in [13], [8]. We discuss the rate of convergence of the algorithm. In the case of of sphere scatterer the condition number of the method is adressed by means of Fourier transform.

1 Introduction

The design of efficient iterative algorithms for the numerical solution of linear systems for time harmonic 3D scattering problem with small wavelength is still challenging, mainly because of the very large number of unknowns and the bad conditioning of the matrix. In the general case it is known that the lack of coercivity of classical integral equations can result in very poor convergence rates for iterative methods, at least for general geometries and with small wavelengths : we refer to [23], [25], [20] for a presentation of iterative methods for classical integral equations. This paper is devoted to the presentation of the iterative solution of a new system of integral equations for solving obstacle scattering by time-harmonic electromagnetic waves, with a particular emphasis on impedance boundary conditions : see to [2], [4], [1] for some recent works about scattering with impedance boundary conditions. We think that the method presented in this work could have some advantages from the computational point of view with respect to more classical integral equations like Electric Field Integral Equation (EFIE) or Magnetic or Combined Field Integral Equation (MFIE, CFIE), [10], [22]. The reason is the particular algebraic structure of the system : it allows us to obtain convergence theorems for iterative algorithms (we give a proof for one of them in this paper), what is known to be difficult to get with classical integral equations.
Let $D^- \subset R^3$ be a bounded domain with regular boundary Γ (at least of class C^2), $n(x)$ be the unit normal vector to Γ directed into $D^+ \subset R^3$, the exterior domain of D^-. The problem we address is the numerical computation of the outgoing electromagnetic field scattered by Γ by means of itrative algorithms. Let $k > 0$ be the wavenumber and Z_0 the vacuum impedance. The outgoing electromagnetic field is

a solution to (1) where

$$\begin{cases} \nabla \wedge E^+ - ikZ_0H^+ = 0, & \text{in } D^+ \\ \nabla \wedge H^+ + ikZ_0^{-1}E^+ = 0, \text{in } D^+ \end{cases} \tag{1}$$

is the Maxwell system and $\lim_{|x|\to\infty} |x| \left(Z_0H^+ \wedge \frac{x}{|x|} - E^+ \right) = 0$ is the Silver-Müller radiation condition at infinity. We assume an impedance boundary condition of the type $n(x) \wedge (E^+_{/\Gamma}(x) \wedge n(x)) + Z_0\, \mathcal{Z}_r(H^+_{/\Gamma}(x) \wedge n(x)) = F$, where $\mathcal{Z}_r$ is some impedance operator that we assume symmetric with a positive real part, i.e. $(\Re e \mathcal{Z}_r J, J) \geq 0, \quad \forall J \in D(Z_r)$. We associate to $\mathcal{Z}_r$ its reflection coefficient operator $\mathcal{R} = (Id - \mathcal{Z}_r)(Id + \mathcal{Z}_r)^{-1}$, which satisfies $||\mathcal{R}||_{\mathcal{L}(TL^2(\Gamma))} \leq 1$.

In [8] it is proved that the original Maxwell's equation can be recast as a system of integral equations with strong coercivity properties. The system is

$$\begin{cases} (1+\beta)\mathbf{u} + (\mathbf{A}^\infty)^*\ \mathbf{A}^\infty \mathbf{u} - \mathbf{T}^*\mathbf{v} + \mathbf{N}_\mathcal{R}\ \mathbf{u} + i\beta\mathbf{v} = \mathbf{f} \\ \mathbf{T}\mathbf{u} - i\beta\mathbf{u} + \beta\mathbf{v} + (\mathbf{A}^\infty)^*\ \mathbf{A}^\infty\mathbf{v} = 0. \end{cases} \tag{2}$$

Here $\beta > 0$ is a coefficient ($\beta = 1$ in practice). The unknown is

$$\mathbf{u} = \begin{bmatrix} \sqrt{iZ_0}\ n \wedge H^+_{/\Gamma} \\ -\sqrt{iZ_0}^{-1}\ n \wedge E^+_{/\Gamma} \end{bmatrix}, \tag{3}$$

$\mathbf{v} = i\mathbf{u}$ is a Lagrange multiplier. Operator $\mathbf{A}^\infty$ is smooth. Operator $\mathbf{T}$ is a matrix integral operator obtained as a combination of classical integral operators with singular kernels. Operator $\mathbf{N}_\mathcal{R}$ depends on the boundary condition. A consequence of $||\mathcal{R}||_{\mathcal{L}(TL^2(\Gamma))} \leq 1$ is $||\mathbf{N}_\mathcal{R}|| \leq 1$. The main question adressed in this paper is the rate of convergence and condition number of an iterative method, based on (2), for the numerical solution of Maxwell's equation.

2 The Iterative Algorithm and its Convergence

To solve (2), the proposed iterative algorithm is a relaxed Jacobi method. Of course, this Jacobi algorithm is not necessarily the best one to solve *in practice* our integral equation system. But what we intend to do is to show that the structure of the system is exploited to get convergence results for iterative algorithms, which is difficult to obtain with the other classical formulations.

Let $1 > r > 0$ be some relaxation parameter. The algorithm reads

- initialization: $\mathbf{u}^0 = \mathbf{v}^0 = 0$
- loop over p

- solve

$$\begin{cases} (1+\beta)\tilde{\mathbf{u}}^p + (\mathbf{A}^\infty)^* \, \mathbf{A}^\infty \tilde{\mathbf{u}}^p - \mathbf{T}^* \tilde{\mathbf{v}}^p = \mathbf{f} - \mathbf{N}_{\mathcal{R}} \mathbf{u}^{p-1} - i\beta \mathbf{v}^{p-1} \\ -\mathbf{T}\tilde{\mathbf{u}}^p - \beta \tilde{\mathbf{v}}^p - (\mathbf{A}^\infty)^* \, \mathbf{A}^\infty \tilde{\mathbf{v}}^p = -i\beta \mathbf{u}^{p-1}. \end{cases} \tag{4}$$

- relax

$$\begin{cases} \mathbf{u}^p = (1-r)\mathbf{u}^{p-1} + r\tilde{\mathbf{u}}^p \\ \mathbf{v}^p = (1-r)\mathbf{v}^{p-1} + r\tilde{\mathbf{v}}^p. \end{cases} \tag{5}$$

Here we assume that solving the discrete counterpart of (4) is easy due to the strong coercivity of the operator on the left hand side. For example discrete conjugate gradient might be used for this calculation. In section 4, the condition number of the system is studied in the special case of a sphere. Concerning the convergence of the algorithm (4) one has the following result :

Theorem 1 *Assume* $||\mathcal{R}|| < 1$ *then* $(\mathbf{u}^p, \mathbf{v}^p)$ *converges strongly to* $(\mathbf{u}, \mathbf{v})$ *in apropriate functionnal spaces.*

In the case $||\mathcal{R}|| = 1$ (which covers perfect conductor boundary conditions) we get weaker results.

For simplicity, we give only a formal proof more adapted to the understanding of algorithmic issues, without mentioning the functional spaces. Let us consider the algorithm for the error. It means that the source term is set to 0, $f = 0$: the algorithm is initialized by $(\mathbf{u}^0, \mathbf{v}^0) = -(\mathbf{u}, \mathbf{v})$. We begin by multiplying the first (resp. second) equation of (4) by $\mathbf{u}^p$ (resp. by $\mathbf{v}^p$) then we subtract the real part of the two results. Both terms involving $-\mathbf{T}^*\mathbf{u}^p$ and $\mathbf{T}\mathbf{u}^p$ cancel each other and we get

$$\begin{aligned} &(1+\beta)||\mathbf{u}^p||^2 + \beta||\mathbf{v}^p||^2 + ||\mathbf{A}^\infty \mathbf{u}^p||^2 + ||\mathbf{A}^\infty \mathbf{v}^p||^2 \\ &\quad = -\Re e(i\beta \mathbf{v}^{p-1}, \mathbf{u}^p) + \Re e(i\beta \mathbf{u}^{p-1}, \mathbf{v}^p) - \Re e(\mathbf{N}_{\mathcal{R}} \mathbf{u}^{p-1}, \mathbf{u}^p). \end{aligned} \tag{6}$$

This is the basic *a priori* estimate for the algorithm. Recall that $||\mathbf{N}_{\mathcal{R}}|| \leq 1$, so the right-hand side of the inequality controls the left-hand side at iteration $p-1$. The remaining part of the proof of convergence is just technical, details are in [8]. Let us use the notation $\mathbf{u}^p = \begin{bmatrix} J_1^p \\ M_i^p \end{bmatrix}$, and define

$$\mathcal{V}^p = |\mathbf{A}^\infty \mathbf{u}^p|^2 + |\mathbf{A}^\infty \mathbf{v}^p|^2 + \frac{\beta}{2}||i\mathbf{u}^{p-1} - \mathbf{v}^p||^2 + \frac{\beta}{2}||i\mathbf{v}^{p-1} + \mathbf{u}^p||^2 \tag{7}$$

$$+\frac{1}{2}||i\tilde{J}_1^p + n\wedge \tilde{M}_1^p + \mathcal{R}(iJ_1^{p-1} - n\wedge M_1^{p-1})||^2 + \frac{1}{2}(1-||\mathcal{R}||^2)||iJ_1^{p-1} - n\wedge M_1^{p-1}||^2.$$

Iterating from (6) we find out that

$$||v^q||^2 + \sum_{p=0}^{q} r(1-r)||v^p - v^{p-1}||^2 + \sum_{p=0}^{q} (1-r)\mathcal{V}^q \leq ||\mathbf{u}||^2 + ||\mathbf{v}||^2. \tag{8}$$

This estimate implies that $||v^p||^2$ is bounded in an appropriate functionnal space and that the two series are convergent. So the generic term approaches 0 as p tends to infinity. Thus $\mathcal{V}^p$ tends to 0. By hypothesis $||\mathcal{R}|| < 1$, then $iJ_1^{p-1} - n \wedge M_1^{p-1} \to 0$, so $i\tilde{J}_1^p + n \wedge \tilde{M}_1^p \to 0$. Thus the error $\mathbf{u}^p$ tends to 0 and the approximate solution tends to the exact solution.

Note that if we have readily obtained the strong convergence when the reflection coefficient is strictly less than 1, there is no reason for the error to be a geometric decreasing function of the iterations (even if numerical evidence shows it is probably the case as in Figure 3). Indeed, all that we have obtained is that a series is convergent and consequently its generic term must goes to 0. However if $\beta = 0$ then it implies a geometric convergence to 0. At the theoretical level it is worthwhile to note that the system has much more stronger coercivity properties with $\beta > 0$. This is why it is probably preferable to use $\beta > 0$.

Finally estimate (8) clearly shows the interest of the under relaxed algorithm, $0 < r < 1$, versus the non relaxed algorithm $r = 1$. This is confirmed by numerical experiments.

3 Fourier Decomposition for the Sphere

We specialize the study to the special case of a spherical scatterer where all the calculations can be done analytically. We study the condition number of the left-hand side of the iterative system (4). Just recall that we need to inverse the left-hand side at each iteration of the global algorithm. Thus the condition number of this left-hand side system is an important issue for practical algorithms. This section is dedicated to show that the left hand side of (4) is, mode by mode in the case of a sphere, equivalent to (24) where all entries of the matrix are explicitly known in terms of Bessel functions. We define

$$\begin{cases} \tilde{Y}_n^m(\hat{r}) = P_n^{|m|}(\cos\theta)\, e^{im\varphi}, \quad n \geq 0, |m| \leq n \\ d_{n,m} = \frac{1}{4\pi} \frac{(n-|m|)!\,(2n+1)}{(n+|m|)!\,n(n+1)}, \end{cases} \tag{9}$$

$$\begin{cases} u_{nm}^{(+)}(\hat{r}) = d_{n,m}^{1/2}\, \nabla^t \tilde{Y}_n^m(\hat{r}) \\ u_{nm}^{(-)}(\hat{r}) = d_{n,m}^{1/2}\, \hat{r} \wedge \nabla^t \tilde{Y}_n^m(\hat{r}), \end{cases} \tag{10}$$

where (θ, φ) are the usual spherical coordinates and ∇^t is $\frac{1}{a}\hat{\theta}\frac{\partial}{\partial\theta} + \frac{1}{a\sin\theta}\hat{\varphi}\frac{\partial}{\partial\varphi}$. Any tangential field on a sphere of radius a may be decomposed into

$$J(\hat{r}) = \sum_{n=1}^{\infty} \sum_{m=-n}^{+n} \sum_{\varepsilon=\pm} J_{mn}^{(\varepsilon)}\, u_{nm}^{(\varepsilon)}(\hat{r}). \tag{11}$$

The $u_{nm}^{(\varepsilon)}$'s form a complete set of orthonormal functions in the space of the square integrable complex fields on the sphere. We have

$$\|J(\hat{r})\|^2_{TL^2(S^2_a)} = \sum_{n=1}^{\infty} \sum_{m=-n}^{+n} \sum_{\varepsilon=\pm} |J_{mn}^{(\varepsilon)}|^2. \tag{12}$$

In [19], Hsiao and Kleinman used this decomposition to study two potentials denoted here by T_{HS} and K_{HS} and which are related to ours by the relationships

$$\begin{cases} T = T_r + T_i = i\hat{r} \wedge T_{HS} \\ K = K_r + K_i = -\hat{r} \wedge K_{HS}. \end{cases} \tag{13}$$

It is shown in this paper that if j_n and y_n are the spherical Bessel functions and

$$\begin{cases} J_n = kaj_n(ka) \ J'_n = (kaj_n(ka))' \\ Y_n = kay_n(ka) \ Y'_n = (kay_n(ka))', \end{cases} \tag{14}$$

then, (see (78), (79), (85), (86) of [19])

$$\begin{cases} (T_{HK} u_{nm}^{\varepsilon})(\hat{r}) = T_n^{\varepsilon} \, u_{nm}^{-\varepsilon}(\hat{r}) \\ (K_{HK} u_{nm}^{\varepsilon})(\hat{r}) = K_n^{\varepsilon} \, u_{nm}^{\varepsilon}, \end{cases} \tag{15}$$

with

$$\begin{cases} T_n^+ = -J'_n(J'_n + iY'_n), \qquad T_n^- = J_n(J_n + iY_n) \\ K_n^+ = -\dfrac{1}{2} - iJ_n(J'_n + iY'_n), \ K_n^- = -\dfrac{1}{2} + iJ'_n(J_n + iY_n). \end{cases} \tag{16}$$

(Note: let us remark that formulae (16) differs from those given in Hsiao *et al.*'s paper by a change of sign. The reason is that there is an error of sign in formulas (68) and (69) of this paper: $-ik$ and $-ik^2$ must be changed into ik and ik^2. But, except this point all the remaining calculations are valid hence the only change of sign in the result). From this and after some algebraic manipulations we easily deduce for T

$$\begin{cases} (T_r u_{nm}^+(\hat{r}) = -J'_n Y'_n \, u_{nm}^+(\hat{r}), \ (T_i u_{nm}^+)(\hat{r}) = (J'_n)^2 \, u_{nm}^+(\hat{r}) \\ (T_r u_{nm}^-)(\hat{r}) = -J_n Y_n \, u_{nm}^-(\hat{r}) \ (T_i u_{nm}^-)(\hat{r}) = J_n^2 \, u_{nm}^-(\hat{r}), \end{cases} \tag{17}$$

while for K

$$\begin{cases} (K_r + \varepsilon \dfrac{\hat{r}}{2} \wedge) u_{nm}^{\varepsilon}(\hat{r}) = -J'_n Y_n \, u_{nm}^{-\varepsilon}(\hat{r}), \ (K_i u_{nm}^{\varepsilon})(\hat{r}) = J'_n J_n \, u_{nm}^{-\varepsilon}(\hat{r}) \\ (K_r - \varepsilon \dfrac{\hat{r}}{2} \wedge) u_{nm}^{\varepsilon}(\hat{r}) = -J_n Y'_n \, u_{nm}^{-\varepsilon}(\hat{r}). \end{cases} \tag{18}$$

Our aim is to apply this harmonic analysis for the Jacobi algorithm described in section 2. At each step of the induction, the solution of the linear system (4) is required. It reads

$$\left\{\begin{array}{l}\begin{bmatrix}(1+\beta)+T_i & K_i\\ K_i & (1+\beta)+T_i\end{bmatrix}\begin{bmatrix}J_1\\ M_1\end{bmatrix}\\ \qquad -\begin{bmatrix}T_r & K_r+\frac{\hat{r}}{2}\\ K_r+\frac{\hat{r}}{2} & T_r\end{bmatrix}\begin{bmatrix}J_2\\ M_2\end{bmatrix}=\begin{bmatrix}G_1\\ H_1\end{bmatrix},\end{array}\right. \tag{19}$$

and

$$\left\{\begin{array}{l}-\begin{bmatrix}T_r & K_r-\frac{\hat{r}}{2}\\ K_r-\frac{\hat{r}}{2} & T_r\end{bmatrix}\begin{bmatrix}J_1\\ M_1\end{bmatrix}\\ \qquad -\begin{bmatrix}\beta+T_i & K_i\\ K_i & \beta+T_i\end{bmatrix}\begin{bmatrix}J_2\\ M_2\end{bmatrix}=\begin{bmatrix}G_2\\ H_2\end{bmatrix}.\end{array}\right. \tag{20}$$

This system can be split into two blocks

$$\mathbf{M}^{\pm}\begin{bmatrix}J_1\pm M_1\\ J_2\pm M_2\end{bmatrix}=\begin{bmatrix}G_1\pm H_1\\ G_2\pm H_2\end{bmatrix}, \tag{21}$$

with

$$\mathbf{M}^{\pm}=\begin{bmatrix}(1+\beta)+T_i\pm K_i & -T_r\mp(K_r+\frac{\hat{r}}{2})\\ -T_r\mp(K_r-\frac{\hat{r}}{2}) & -\beta-(T_i\pm K_i)\end{bmatrix}. \tag{22}$$

Using the decomposition with the basis functions, we have a block structure for $\mathbf{M}^{\pm}$, each block being 4 $\times$4 (caution, the sign $\pm$ has not the same meaning as the one in (10))

$$\mathbf{M}^{\pm}=\bigoplus_{n,m,\,|m|\leq n}(\mathbf{M})^{\pm}_{mn} \tag{23}$$

with

$$\mathbf{M}^{\pm}_{mn}=\begin{bmatrix}1+\beta+(J'_n)^2, & \pm J_nJ'_n, & J'_nY'_n, & \pm J_nY'_n\\ \pm J_nJ'_n, & 1+\beta+J_n^2, & \pm J'_nY_n & J_nY_n\\ J'_nY'_n, & \pm J'_nY_n & -\beta-(J'_n)^2 & \mp J_nJ'_n\\ \pm J_nY'_n, & J_nY_n & \mp J_nJ'_n & -\beta-J_n^2\end{bmatrix}. \tag{24}$$

Remark that $\mathbf{M}^{\pm}_{mn}$ is independent of m as we can expect since the problem is invariant under any rotation.

4 Condition Number for the Sphere

It is now an easy matter to study the condition number of all four by four matrices given in (24). In Figures 1 and 2 are depicted the 4 real eigenvalues of $(\mathbf{M})^{+}_{nm}$ (or those of $(\mathbf{M})^{-}_{nm}$ as they coincide) as function of n. Coefficient β was chosen equal to 1. Figure 1 corresponds to a sphere of moderate size with respect to the wavelength ($ka = 20\pi$). In Figure 2, ka equals 200π and the sphere is large with a radius

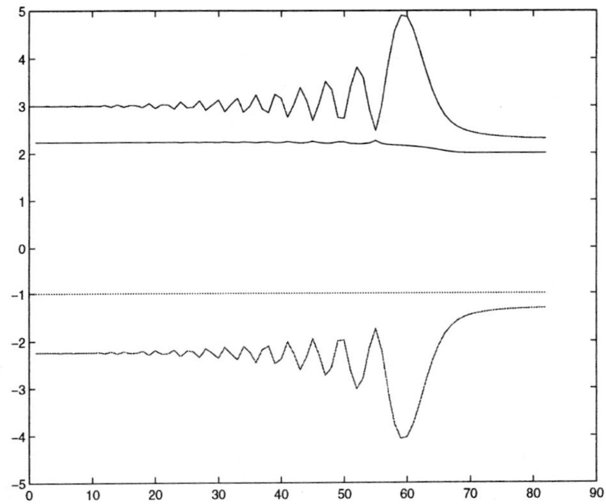

Figure 1 The 4 eigenvalues of the matrix $\mathbf{M}^{\pm}_{mn}$ as function of n for $ka = 20\pi$

of 100 wavelengths. Note that the spectrum is real, as expected, and the range of the eigenvalues does not change so much between the two experiments. Extensive calculations with Matlab have shown that the maximum modulus of the eigenvalues (it is the pike on both Figures 3 and 4) in the area $n \leq ka + 10\log(ka + \pi)$ is a low increasing function of the frequency. This area corresponds to the excited modes for incident plane waves. For larger n (not displayed on 3 and 4), the asymptotic behavior of Bessel functions enables us to obtain $-n$, $-\beta$, $1 + \beta$ and $+n$ as asymptotic eigenvalues. Figure 3 shows the spectrum of the matrix associated to the error of the Jacobi Algorithm for the large sphere. This spectrum is composed of the set of all the eigenvalues of the iteration matrices (Id is the 2×2 identity matrix)

$$\mathbf{E}^{\pm}_{mn} = (1 - r)\begin{bmatrix} Id, & 0 \\ 0, & Id \end{bmatrix} - ir\beta\left(\mathbf{M}^{\pm}_{mn}\right)^{-1}\begin{bmatrix} 0, & Id \\ Id, & 0 \end{bmatrix}. \tag{25}$$

In that example, the reflexion coefficient is equal to zero (model problem) and the relaxation parameter is 0.7. All the eigenvalues are found located inside a circle

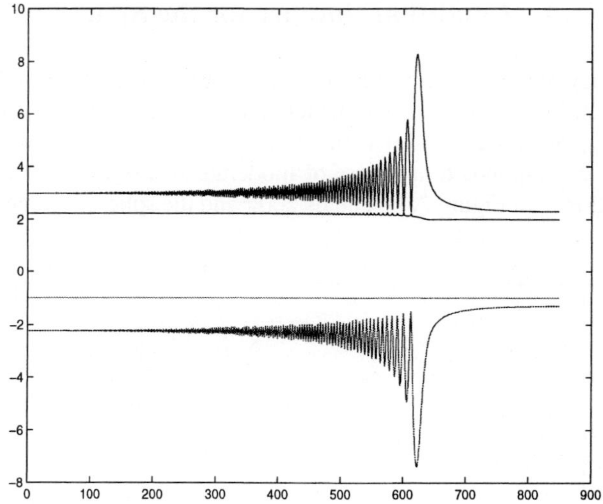

Figure 2 The 4 eigenvalues of the matrix $\mathbf{M}^{\pm}_{mn}$ as function of n for $ka = 200\pi$

of radius 0.782 and geometrical convergence occurs. Other algorithms than Jacobi can be used to solve the problem. They can be constructed from our linear system in its coercive version. Figure 4 shows the spectrum for the non iterative system where we solve directly (2). Parameters β, R and ka are unchanged with respect to the previous examples: $\beta = 1$, $R = 0$, $ka = 200\pi$. The coercivity property can be checked on the Figure 4 : all eigenvalues are located to the right of a line $\Re z = c$ with $c \approx 0.41$.

5 Conclusion

Due to the strong coercivity properties of our integral systems, they are well suited for the coupling with interior problems treated for example with domain decomposition algorithms [17], [16], [9].

A very promising method for the numerical calculation of time-harmonic obstacle scattering solutions is the multipole method, see [18], [27], [11] among others. Multipole methods is a way to speed up the matrix vector product for linear systems coming from integral equations. For those who are familiar with these methods, it is well known that the usual integral equations require at least 2 scalar (*i.e.* with scalar far and near fields) multipole computations, and even more when impedance problems are considered. For our system, 4 multipole computations are enough to compute $\mathbf{Tu}, \mathbf{T}^*\mathbf{v}$ and 4 other multipole computations (with a very simple translation function) provide $(\mathbf{A}^\infty)^*(\mathbf{A}^\infty)\mathbf{u}$ and $(\mathbf{A}^\infty)^*(\mathbf{A}^\infty)\mathbf{v}$. Furthermore the near interaction matrix required by the method for $\mathbf{T}$ is now real and so is halved in stor-

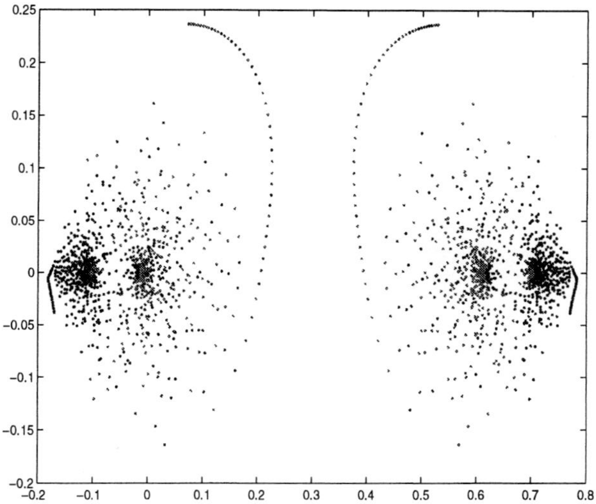

Figure 3 Spectrum of the iteration matrix for the Jacobi algorithm applied to a sphere ($ka = 200\pi$). Coefficient β is 1, relaxation parameter is $r = 0.7$ and reflexion coefficient is $R = 0$ (model problem)

age with respect to the classical ones. It is reasonable to think that coupling multipoles methods and our integral system should give a good compromise in terms of an accurate, fast, robust and reliable algorithm.

References

[1] T. Abboud and T. Sayah, *Potentiels retardés pour les Equations de Maxwell avec condition d'impédance généralisée*, Tech. Report R.I. 387, Ecole Polytechnique, France, 2000.

[2] X. Antoine, H. Barucq and A. Bendali, *Bayliss-Turkel-like Radiation Conditions on Surfaces of Arbitrary Shapes*, Journal of Math. Anal. and Appl., 229(2000), pp. 184–211.

[3] N. Bartoli and F. Collino, *Integral Equations via Saddle Point Problems for the 2-D electromagnetic Problems*, M2AN, 34(2000),no 5, pp. 1023–1050.

[4] A. Bendali, *Boundary Element Solution of Scattering Problems Relative to a Generalized Impedance Boundary Condition*, proccedings of the Boca Raton Cerfacs conference, London-New-York, 1999.

[5] F. Brezzi and M. Fortin, *Mixed and Hybrid Finite Element Methods*, no. 15 in Springer series in Computational Mathematics, Springer-Verlag, 1991.

[6] M. Cessenat, *Mathematical Methods in Electromagnetism*, no. 41 in Series on advances in Mathematics for Applied Sciences, World-Scientific, 1996.

[7] G. Chen and J. Zhou, *Boundary Element Methods*, Academic Press, 1992.

[8] F. Collino and B. Despres, *Integral Equations via Saddle Point Problems for Time Harmonic Maxwell's Equations .*, to appear in Journal of Ap. and Comp. Math.

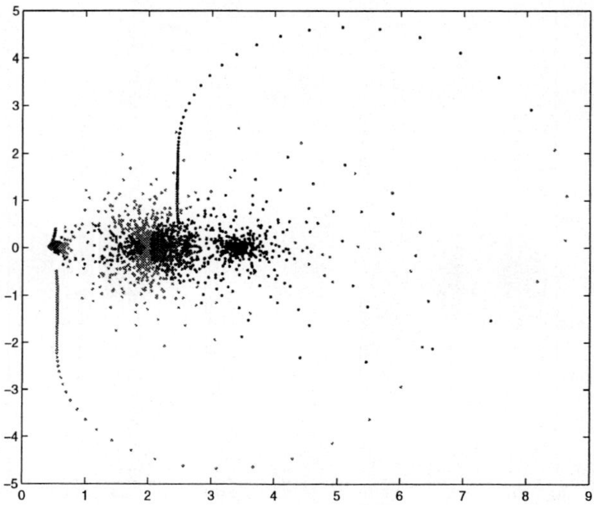

Figure 4 Spectrum of the coercive matrix for the model problem. Case of the sphere with $ka = 200\pi$. Coefficient β is 1

[9] F. Collino and S. Ghanemi and P. Joly, *Domain Decomposition Method for the Helmholtz Equation: A General Presentation.*, Comput. Methods Appl. Mech. Engrg, 184 (2000), no. 2-4, pp. 171–211.

[10] D. Colton and R. Kress, *Inverse Acoustic and electromagnetic scattering theory*, Applied Mathematical Sciences 93, Springer-Verlag, 1992.

[11] E. Darve, *The fast multipole method: numerical implementation.* , J. Comput. Phys. 160 (2000), no. 1, pp. 195–240.

[12] R. Dautray and J. L. Lions, *Analyse mathématique et calcul numérique pour les sciences et les techniques*, no. 2, Masson, 1985.

[13] B. Després, *Fonctionnelle quadratique et équations intégrales pour les équations de Maxwell en domaine extérieur*, Comptes Rendus de l'Académie des Sciences, Paris, Série I, 323 (1996), pp. 547–552.

[14] B. Després, *Fonctionnelle quadratique et Équations Integrales pour les problèmes d'onde harmonique en domaine extérieur*, M2AN, 31 (1997), pp. 679–732.

[15] B. Després , *Quadratic functional and integral equations for harmonic wave equations*, in Mathematical and numerical aspects of wave propagation (Golden, CO, 1998),SIAM, Philadelphia, 1998, pp. 56–64.

[16] B. Després and B. Stupfel, *A domain decomposition for the solution of large electromagnetic scattering problem*, J. of Electromagnetic Waves and Applications, vol. 13-number 11 (1999), pp. 1553-1568.

[17] B. Després and P. Joly and J. E. Roberts, *A domain decomposition method for the harmonic Maxwell's equations.*, in Proceedings of the IMACS international symposium on iterative methods in linear algebra, North Holland, 1990.

[18] M. Epton and B. Dembart, *Multipole translation theory for the three-dimensional Laplace and Helmholtz equations.*, SIAM J. Sci. Comput., 16 (1995), pp. 865–897.

[19] G.C. Hsiao and R.F. Kleinman, *Mathematical foundations for error estimations in numerical solutions of integral equations in electromagnetism* , IEEE Trans. Antennas and Propag., no. 3 45(1997), pp. 316–328.

[20] R. E. Kleinman and P. M. Van den Berg, *Iterative methods for solving integral equations.*, Elsevier, 1988.

[21] P. Lascaux and R. Theodor, *Analyse numérique matricielle appliquée à l'art de l'ingénieur*, Masson, 1987.

[22] J. C. Nédelec, *Cours de l'école d'été d'analyse numérique*, tech. report, CEA-EDF-IRIA, 1977.

[23] M. D. Pocock and S. P. Walker, *The complex bi-conjugate gradient solver applied to large electromagnetic scattering problems, computational costs, and costs scaling*, IEEE Trans. Antennas and Propag., no. 1 45(1997), pp.140–146.

[24] M. Reed and M. Simon, *Scattering Theory*, Academic-Press, New York, 1979.

[25] T. K. Sarkar, *Application of conjugate gradient method to electromagnetics and signal analysis*, in From reaction concept to conjugate gradient: have we made any progress?, Elsevier, 1988.

[26] T. B. A. Senior and J. L. Volakis, *Approximate Boundary conditions in Electromagnetism*, no. 41 in IEEE Electromagnetism Waves Series, The institution of Electric Engineers, 1995.

[27] J. Song and C. C. Lu and W. C. Chew, *Multilevel Fast Multipole Algorithm for Electromagnetic Scattering by Large Complex Objects*, IEEE Trans. Antennas Propagat., 45 (1997), pp. 1488–1493.

[28] B. Stupfel and B. Despres, *A domain decomposition method for the solution of large electromagnetic scattering problems*, Journal of Electromagnetic waves and applications, 13 (1999), pp. 1553–1568.

High Performance Computation and Visualization of EMs Using an Integrated Computation System

JUNWEI LU
School of Microelectronic Engineering, Faculty of Engineering and Information Technology, Griffith University, Nathan, Qld 4111 Australia

Summary

This paper demonstrates a basic concept of high performance computational and visual electromagnetics (EMs) using an integrated computation system, and discusses several different computational methods with emphasis on different problems requiring different computational methods. Several application problems of computational electromagnetics (CEMs) in the areas of electrical and electronic engineering and communications systems are presented. The paper also presents computational and visual electromagnetics (CVEM) using an integrated programming, integrated computation system and high performance parallel computer. The future trend of high performance computing and collaborative visualization for coupling problems is also described.

1 Introduction

Many practical electromagnetic field problems in electrical and electronic engineering and communications systems are extremely difficult or impossible to solve by purely analytical methods. Analytic methods involve finding and solving mathematical equations that define the required variables. These different problems require different methods of approach, but have in common a high complexity making them insolvable by analytic means. The most complicated three-dimensional problems e.g, electromagnetic field, eddy-current, coupled field, wave propagation, radiation and scattering require numerical techniques for solution determination. With the associated boundary value conditions required for low and high frequency structures, required techniques are often the only means by which solutions can be obtained with the required degree of accuracy.

In the past, it has been common practice to simplify such complex problems to a point where an analytical solution could be obtained which, it was hoped, would bear some resemblance to the solution of the real problem. On the other hand, experimental procedures were often used to measure the characteristics of devices. The very expensive experimental "cut and try" technique is still predominant in many industrial design projects. With rapidly developing computer technology, the emphasis in engineering analysis has moved towards versatile numerical methods or computational methods. Indeed in the past decade, computing has become an essential tool for advancing various sciences and engineering disciplines [1]. With the advent of parallel computers, it has become clear that parallel computing has become an integral part of many aspects of these disciplines [2]. Modern high-performance computing including low cost and high performance cluster based parallel computing and Internet computing has become a very important computation technique for computational electromagnetics. Conventional computers are not fast enough to do what we need to do in computational electromagnetics. The price also puts them out of range in any case. The importance of using low cost and high performance computational techniques to solve large-scale electromagnetic field problems has been recognized by scientists and engineers. Unfortunately, most commercial software packages for electromagnetic design and simulation are single processor based (for PC or workstations) packages. They are limited to small-scale computation problems, and often take a long time to compute a real scientific or engineering problem. Sometimes it is very difficult to solve a large-scale 3D nonlinear and dynamic complex problem. Software for large-scale 3-D nonlinear electromagnetic, coupled field, wave propagation, radiation and scattering problems is still in the development stages and will become very popular with the rapid development of the electronic and communication industries.

2 Integrated Computational EM Systems

2.1 Effective Computational EM System

Various numerical computation methods are used in commercially available electromagnetic software packages. Among them, the most popular numerical methods are FD-TD, FEM, MOM and BEM. These numerical methods theoretically can be used for either high frequency or low frequency structure simulation. In practice, only limited applications are selected for a particular frequency range as the generalized packages often exhibits some weakness in solving a problem with complex structural, coupling or nonlinear problems resulting in lack of accuracy for engineering design. Also, these software packages do not have optimal design functions. The following examples of computational electromagnetics present how to utilise the effective computational methods for different engineering design and problem solving.

1) Optimal design of smart mobile terminal antenna using FD-TD and the adaptive simulated annealing optimization algorithm [3]

a. Dielectric embedded antenna structure: In mobile terminals the size of the antenna is important, as portability is a key requirement. The rapid growth of the wireless market has resulted in new technologies being investigated to improve performance and usage of the available spectrum in the most efficient way. Smart antennas with controllable directionality are one promising candidate as they allow higher re-use of channels and increased system performance. The electronically switched multiple beam (ESMB) structure lies on a finite ground plane and the cross section of the FD-TD simulation domain is shown in Figure 1.

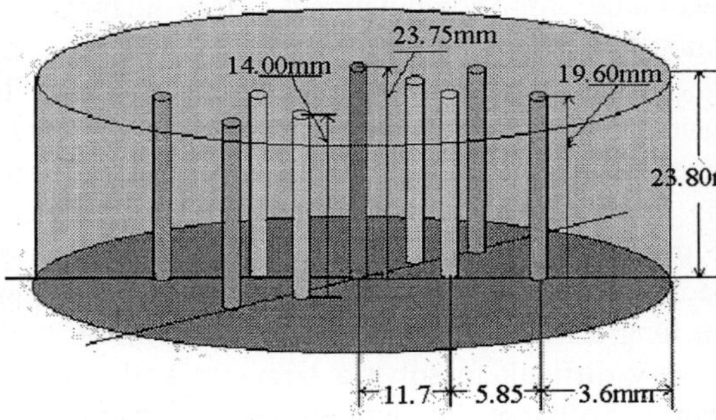

(a)

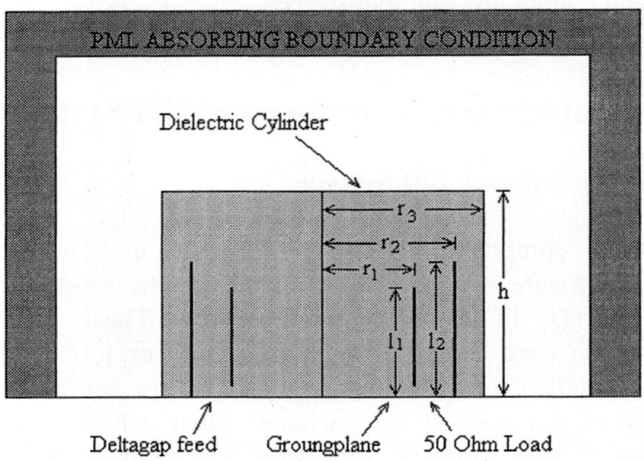

(b)

Figure 1 Geometry of the dielectric embedded switching antenna array, (a) configuration of DE-ESMB antenna array, (b) cross section of the FD-TD simulation domain.

A nine-element ESMB antenna can produce two beams simultaneously, separated by 180 degrees, with four possible beam directions at 90 degree increments through the azimuthal plane. The centre element is slightly longer than a resonant quarter wavelength element and is shorted to the ground plane giving the array Yagi-Uda characteristics. The centre element is surrounded by elements placed along two radially concentric circles. These elements can be switched to a number of possible different states. The inner circle of elements may be either active elements, passive elements shorted to ground or passive elements isolated from the ground plane. The outer circle of elements are always parasitic but can be switched from open circuit to short circuit. Each circle has the same number of elements. The switching can be achieved electronically using active semiconductor devices (p.i.n. diode or FET switches) or by RF MEMS switches.

It is well known that by embedding a wire in a dielectric medium, the electrical length of a wire antenna can be reduced [3]. If the dielectric is of infinite extent, then the wavelength of radiation in the medium is given by $\lambda_r = \lambda_0 / \sqrt{\varepsilon_r \mu_r}$, where ε_r is the relative permittivity and μ_r is the relative permeability of the material. If the antenna elements are embedded in a high permittivity medium (non-magnetic material) of infinite extent, the wavelength in the medium can be rewritten as $\lambda_r = \lambda_0 / \sqrt{\varepsilon_r}$ and the size reduction factor F_r for the length of antenna elements will be $F_r = \sqrt{\varepsilon_r}$. To reduce the total volume and weight of the DE-ESMB array, the radius of the dielectric from outer element to air interface should be smaller than the height of the dielectric. In a practical antenna, the reduction factor will not be as great as F_r as the embedding dielectric will not be of infinite extent. Clearly, the reduction factor will be a function of the shape, size, electrical characteristics of the dielectric material and the location of each wire element in the material. Thus the actual reduction factor, F_{ra}, for the size of antenna will be

$$F_{ra} = f(\sqrt{\varepsilon_r}, h, r_1, r_2, r_3, l_1, l_2) \qquad (1)$$

where h is the height of dielectric, r_1 is the radius of inner element circle, r_2 is the radius of outer element circle, r_3 is the radius of embedding dielectric, l_1 the length of the elements on the inner circle and l_2 the length of the elements on the outer circle respectively, as shown in Figure 1b.

b. FD-TD based computation method: Unlike free space ESMB antenna arrays, the switched parasitic and active elements of the DE-ESMB array are embedded in a cylinder of high permittivity dielectric material. The physical parameters must be optimized due to complex wave interaction of radiation between elements and the dielectric/air interface of the cylinder. Since there are no design rules for such DE-ESMB antenna arrays, the only way to achieve an optimal antenna is to use a full wave numerical solution and control it with a robust optimization algorithm. In this design, we used the FD-TD method, as it is well suited to modeling complex objects, being able to include finite dielectric media, thin wires and lumped loads. Parameters of interest that can be derived from FD-TD simulations include broadband input impedance, radiation patterns and time harmonic field strengths. The FD-TD method is based on Maxwell's curl equations as shown in equation (2). A system of six coupled partial differential equations in the Cartesian coordinate system forms the basis of the FD-TD numerical algorithm for electromagnetic wave interactions with general three-dimensional objects [4]. The size of the simulation domain was 120x120x60 cells with a cubic space discretization of 1mm per cell. The time step was set to the Courant stability limit.

$$\frac{\partial \vec{H}}{\partial t} = -\frac{1}{\mu}\nabla \times \vec{E} - \frac{\rho'}{\mu}\vec{H}$$
$$\frac{\partial \vec{E}}{\partial t} = \frac{1}{\varepsilon}\nabla \times \vec{H} - \frac{\sigma}{\varepsilon}\vec{E} \qquad (2)$$

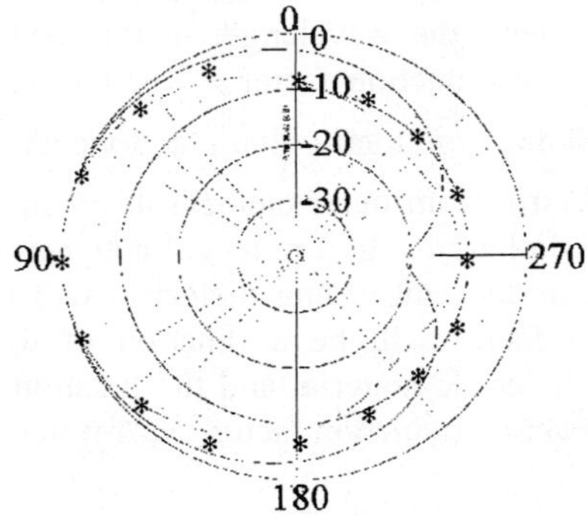

(a)

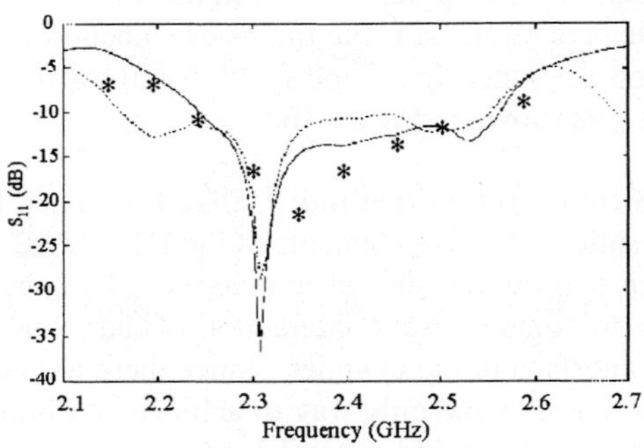

(b)

Figure 2 Radiation pattern and return loss for 9-element DE-ESMB antenna,

(a) the *H* plane radiation pattern at 2.4 GHz, where ----- line indicates the free space antenna and — line indicates the dielectric embedded antenna. **** line indicates the simulation result of DE-ESMB antenna,

(b) the return losses, where ----- line indicates the experiment result of free space ESMB antenna, — line indicates experiment result of DE-ESMB antenna, **** line indicates the simulation result of DE-ESMB antenna.

Berenger's original split-field PML Absorbing Boundary Conditions (ABC) were implemented with separate memory allocation domains between the FDTD [5] and PML regions to reduce memory requirements since standard FDTD only requires 6 field components per cell whereas PML require 12 fields per cell. The PML ABC was 4 cells deep with a parabolic conductivity profile with $R(0)=10^{-5}$. A sub-cellular thin wire model was used to include the effects of thin wires in the simulation, the wire diameter was specified to be 0.9mm. To obtain broadband impedance results, a 40ps Gaussian pulse $p(\tau)$ was applied as a delta-gap excitation on the base of a resistively loaded monopole element.

c. Optimal design of DSMB antenna: An optimal design for antennas usually includes many factors that need to be considered simultaneously. In this case we wish to reduce the volume of the antenna, however the optimal electrical characteristics are probably not achievable if the only optimisation objective is size. The objective functions (our target result) and constraints (physical limits) should be carefully defined before applying the optimizing design process to DE-ESMB antenna array design.

Since the size reduction factor F_{ra} for the antenna controls the physical size of the antenna, and it also directly defines the antenna electrical characteristics. The dielectric scaling factor can be used to get an initial solution for these parameters. Manual optimization was used to gain a deeper understanding into the characteristics of the nine-element DE-ESMB antenna array. The main parameters of interest in this manual optimization run were front to back ratio R_{fb}, and return loss S_{11}. The material chosen for the simulations and experiment was a low loss L6G nylon rod of permittivity ε_r=4.4, lossless material was assumed for the FD-TD simulation. The hand optimized dimensions for the nine-element antenna array are as follows: h = 24mm, r_1 = 12mm, r_2 = 18mm, r_3 = 22mm, l_1=19mm and l_2=20mm. Figure 2 shows the calculated return loss compared with measured data. It was found that the use of a double sided PCB substrate in the experimental model resulted in a slightly increased electrical length for the active element, resulting in the shift in resonant frequency to 2.32GHz. We have achieved significant reduction in volume for the DE-ESMB antenna array as shown in Fig. 1a. Compared to the free space ESMB antenna array, about 50% footprint reduction and 80% volume reduction was achieved by embedding the array in a cylinder of high permittivity dielectric material with minimal effects on antenna performance.

From above investigations we found that the height of dielectric is basically dominated by the central grounding element that follows the reduction factor, thus we can simplify the optimization by fixing the height of dielectric and the length of the centre element. For a complete optimization, we can consider the return loss S_{11}, the front to back ratio for the radiation pattern, R_{fb}, antenna gain G_0, beamwidth f_{bw} and volume Vol, as objectives. From this, the following multiple objectives, can be defined as follows:

$$F_{obj} = f(S_{11}, R_{fb}, G_0, f_{bw}, Vol) \tag{3}$$

Based on the above investigation, we can define a constrained optimization procedure. The adaptive simulated annealing (ASA) optimization algorithm [6] was used with physically constrained input dimensions.

One problem that needs to be addressed with multiple objectives is that careful weighting of the importance of each objective is required. Our solution to this is to use a clamping function to dynamically reduce the weighting once a particular objective is acceptable. For example, clamping S_{11} from ***clamp*** = -15dB over a ***ydel*** = 10dB range, resulting in a minimum clamped S_{11} of –25dB, can be defined as follows:

$$S_{11}\cos t = clamp + (ydel * \tanh(ydel^{-1} * delta)) \tag{4}$$

where ***delta*** = S_{11} – ***clamp***. R_{fb} can be clamped in a similar manner giving $R_{fb}cost$.

2) Wave propagation in dispersive transmission line using FE-TD method [7]

Transmission lines are often used to describe the plasma problems in the solar system or space. The wave propagation of an arbitrary waveform on a dispersive transmission line is a time-dependent problem. The dispersive wave equation for lossless dispersive transmission line as shown in Figure 3 can be derived as

$$\frac{\partial^2 u}{\partial t^2} - C_S^2\frac{\partial^2 u}{\partial x^2} - \lambda_D^2\frac{\partial^4 u}{\partial x^2 \partial t^2} = 0 \tag{5}$$

where $C_s^2 = 1/LC_0$ and $\lambda_D^2 = C_k/C_0$.

In the case of $C_k \to 0$, λ_D in the third term of (5) tends to zero. The lossless wave equation is retrieved, there is no dispersion, and the initial waveform does not change during propagation. It is the well-known wave equation. However, when λ_D (which is called the Debye shielding length in plasma physics) is larger than zero, the wave is significantly dispersive and the waveform changes as the wave progresses, as shown in Figure 3. Using Galerkin method, equation (5) for one dimensional lossless dispersive transmission line can be written as

$$G_i = \int_0^\ell \left(\frac{\partial^2 u}{\partial t^2} - C_s^2\frac{\partial^2 u}{\partial x^2} - \lambda_D^2\frac{\partial^4 u}{\partial t^2 \partial x^2} \right) N_i dx = 0 \tag{6}$$

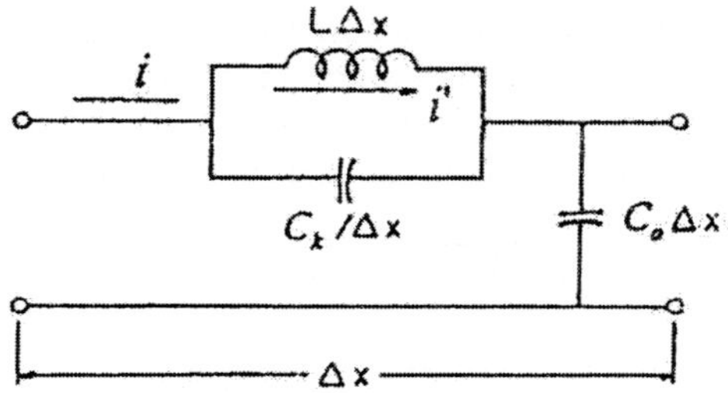

Figure 3 Lossless dispersive transmission line model

where N_i is a shape function, and the matrix equation for a single element is

$$[M]\{\ddot{n}_i\}+C_S^2[S]\{n_i\}+\lambda_D^2[S]\{\ddot{n}_i\}=\{0\} \qquad (7)$$

where (``) indicates the derivative with respect to time, [M] and [S] are

$$[M]^e=\frac{\ell^e}{6}\begin{bmatrix}2 & 1\\ 1 & 2\end{bmatrix}, \qquad [S]^e=\frac{1}{\ell^e}\begin{bmatrix}1 & -1\\ -1 & 1\end{bmatrix} \qquad (8)$$

where l^e is the length of element.

The system matrix equations can be solved by Newmark-B method in the time domain. The results have been shown in [7]. This method is also called the finite element time domain (FE-TD) method.

3 Integration of High Performance Computation and Visualisation System

Applied computational EM software packages are now widely used in engineering practice. With the rapid growth in the communications and high-speed electronics industries, electrical and electronic engineers demand more and more electromagnetic simulation software packages. It is very important to introduce the concept of computational electromagnetics in our engineering society. Without a detailed knowledge of computational electromagnetics, graduate engineers could not use these commercial available software packages to design the electromagnetic products efficiently. Unfortunately, most commercial software packages are "black-box" based packages and developed for EM product design and analysis. Vendors do not consider the need for teaching and learning of computation methods and algorithms in the classroom. Inexperienced engineers may need a few months or more to learn how to use these software packages properly. To overcome such problems, a new educational program in computational electromagnetics (CEMs) should be introduced into engineering courses, especially in undergraduate courses [8].

3.1 Modularized Computation Systems

Advanced Computational and Visual Electromagnetics (ACVEM) is an "open-box" based computer aided teaching and learning tool that is developed for the purpose of teaching and learning computational electromagnetics in the engineering courses. ACVEM is implemented by using MATLAB based integrated programming language [9-10]. To help the teaching and learning of numerical computation techniques and computation algorithm in electromagnetics, the following features have been considered: data input environment using Graphic User Interface (GUI), select different numerical methods, pre-process with mesh generation features, modularized numerical methods as PDE solvers, post-process and visualization functions. The GUI features in ACVEM allow the user to choose the particular numerical method from Numerical Techniques in GUI working platform. After the user selects a particular numerical method, the model attributes can be defined on the common GUI working platform. The following GUI features have been designed into ACVEM pre-process section: select the numerical methods, and define the model attributes such as geometry, material properties, constraints and boundary conditions, and excitations, mesh generation and model check.

ACVEM uses MATLAB M files for the common GUI features and a modularized numerical computation system, in which various numerical methods can be added or built into ACVEM system by the user. The user can open the "open-box" based MATLAB M files to modify the program codes and add more features into ACVEM. Under each numerical computation module, many other modularized computation features for different application problems such as static or quasi-static field problem, and time domain or frequency domain field problems can be added in. The flexibility of the modularized computation systems in ACVEM can be realized. Figure 4 shows a block diagram of modularized computation systems in ACVEM.

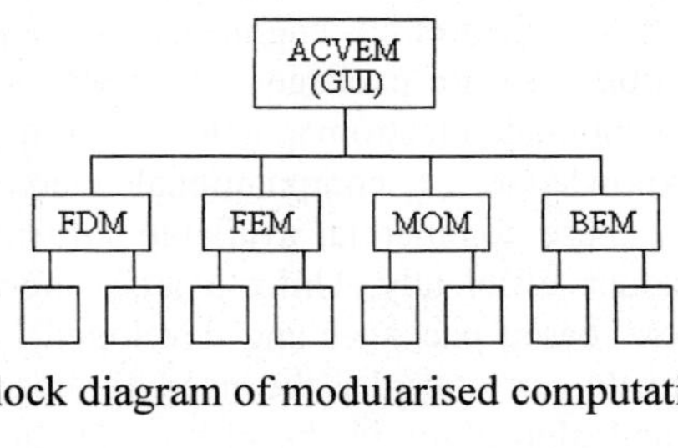

Figure 4 Block diagram of modularised computation system

ACVEM has the interactive GUI features for pre-processing. The user can add self-developed numerical methods or PDE solvers as a module into ACVEM for different applications under the GUI working platform. The GUI features allow the user to choose the particular numerical method from **Numerical Techniques** in GUI working platform. The user can also change the GUI features for different applications. After the user selects a particular numerical method, the model attributes can be defined on the GUI working platform. The detailed information can be found in [8]. The following GUI features have been designed into ACVEM pre-processing section:

- Select numerical methods
  FDM, FEM, BEM and MOM,
  Other numerical methods under each module
- Define the model attributes
  Geometry, material properties, boundary conditions, excitations
- Mesh generation
  mesh size refine, model check by rendering the element

Advances in electromagnetic computation are allowing mathematical model and simulations to become increasingly complex and detailed. This results in a closer approximation to reality thus enhancing the possibility of acquiring new knowledge and understanding. After the appropriate representations of the data have been constructed, information can be superimposed onto the surface, to provide further insight into the complex character of electromagnetic phenomena.

ACVEM visualisation system will simplify the task of interpreting results. MATLAB built-in plot functions are implemented into the ACVEM program. The following GUI features have been designed into ACVEM post-process section:

- Flux contour and vector field plot for field distribution.
- 3-D surface plots and rotation for field distribution.
- Animation for the time and frequency domain field distribution.

3.2 High Performance Parallel Computing

High performance parallel supercomputers have been widely used for scientific analysis and engineering design since the 1980's. Unfortunately, not many companies and universities could afford to use them freely because they cost millions of dollars and required considerable programming effort. This situation limited the scientists' research activities and engineers' productivity in the area of computational science and engineering for quite a time. For research, the important measure is high performance at low cost, while for practical engineering design, the focus is on fast and cheap product development, so application program packages and their efficient usage are the main factors [11]. The recent "home-made", cluster-based parallel computer could make the situation different because of its high performance and low cost. Various software packages have become available for such high performance, low cost cluster based parallel computing systems in recent years. There are even some commercially available packages, such as CFD, which can be run on cluster based high performance parallel computing (HPPC) systems, (e.g. SGI Beowulf Cluster) [12]. The new trend of HPPC is to develop high performance parallel CAD/CAE for high performance cluster computing systems.

1) Case Study of Parallel FD-TD Computation

The FD-TD method is a fully synchronous application and will perform best on a SIMD multicomputer. It can, however, also achieve good performance on cluster-based, distributed, parallel computers. In teaching practice, students are required to implement a parallel FD-TD program with MPI. The programming language used in the classroom is C++. MPI is used, as it is an industry standard focussing purely on message passing. Several free implementations of MPI are available, and many computer hardware vendors have produced their own versions [13]. Since MPI delivers great data-transfer speeds, it is generally preferred above PVM in this education program. The FD-TD method in 2-D electromagnetic wave propagation problems excited by a Gaussian pulse at the center of the structure, is implemented using the C programming language and MPI standard. The number of nodes in X direction is 2000, and in Y direction is 100, and the spatial increment is 1 mm in both directions. Absorbing boundary conditions are also used for this problem. A Gaussian pulse excitation is shown in (9).

$$E = E * \exp(-\alpha((\tau - \beta\delta t)^2)) \tag{9}$$

where τ = absolute time, $\alpha = (4/(\beta\delta t))^2$, β = number of time steps from peak to truncation. Figure 5 shows the speed-up versus number of nodes used in both the SP2 and cluster computing systems.

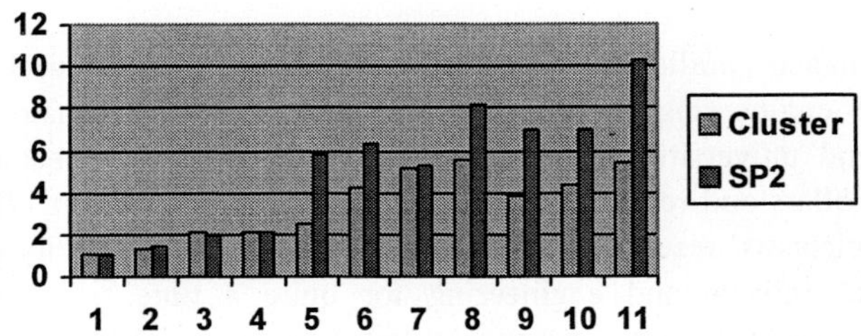

Figure 5 Comparison of speed-up of computation versus the number of nodes

The parallel speed-up is calculated by the following equation:

$$S = T_{1p} / T_{np} \tag{10}$$

where T_{1p} is single node computation time and T_{np} is ***n*** nodes computation time.

From the results we can see that the communication time of the cluster will degrade the total computation time if we use too many processors with the slow communications. The IBM SP2, with its high performance communication switch, shows better scalability. In real-world problem solving this issue might be of major significance, and thus justify the higher costs associated with acquiring and maintaining such a system. For the more modest demands of the classroom, we have found clusters to have adequate communication performance.

Table 1 Comparison of HPCC and SP2 parallel computing systems

Item for comparison	Clusters	IBM SP2
Performance	High	High
Communication speed	Slow	Fast
Computing speed	Fast	Slow (old SP2)
Maintenance cost	Low	High
Cost	Very low	Very high

3.3. Future Trend of Collaborative Visualization System

As computational techniques extend from numerical analysis to visualization analysis, the demand for innovative visualization techniques becomes higher and higher [14]. Interactive visualization is closely related to high performance computation using visualization techniques to deal with the high complexity of computation problems. It is a relatively young field in computational electromagnetics. Collaborative technologies will play a major role in interactive visualization, as most real world engineering designs are carried out as collaborative teamwork. A typical example of collaborative teamwork is the microelectromechenical system (MEMS) analysis and design, where the MEMS device requires multidisciplinary optimization from the FEM-based computation solutions of thermal, mechanical stress, electric field and so forth. Collaboration between groups of scientists and engineers has shown to be one of the most effective ways to achieve some of the best results. Therefore, technologies and systems for supporting collaborative teamwork have become a major area of research in computer science and software engineering [15]. However, teamwork has not yet been supported in any interactive visualization systems for product design or scientific analysis. Single user oriented CAD/CAE systems still dominant the scientific analysis and engineering design. The research on collaborative visualization aims to integrate the collaborative computing technologies and interactive visualization techniques to simulate, visualize, modify and redesign the same computation model by a group of scientists and engineers connected by the global Internet [16].

One of most promising systems is hybrid computation and visualization system. Hybrid architecture takes the elements from both centralised and replicated architecture [16]. The system uses a high performance supercomputer or cluster computer and uses the power of high performance computer for geometry calculation or computational steering, then copies this calculation representation to graphic workstations and produces the final image according to the user viewing perspectives. In addition, videoconference will be also integrated in this hybrid computation and visualization system to provide a completed high performance computation and visualization system with multimedia capabilities. This system is typically useful for coupling problems or multi disciplinary problems. Figure 6 shows the conceptual hybrid computation and visualization system.

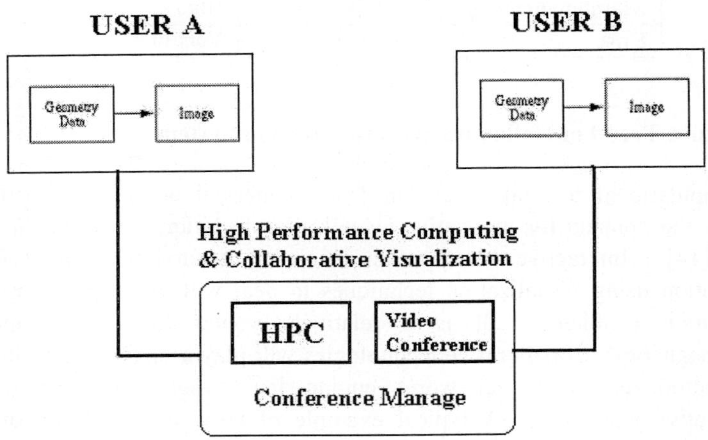

Figure 6 Hybrid computation and visualization system

4 Conclusion

The paper discussed high performance computational and visual electromagnetics using an integrated computation system for engineering design and analysis. The paper presented several major numerical methods for computational electromagnetics and its applications in relating to computational electromagnetics. The optimal design for smart mobile terminal antennas is also discussed in this paper. Finally, the paper addressed the challenging work in the area of high performance parallel computing and collaborative visualisation for coupling problems.

Acknowledgment

The author wishes to thank David Thiel, Seppo Sarrio and Denis Sweatman for their contributions to this paper.

References

[1] Abmed Sameb and Fobn Riganai: "Computational Science and Engineering", IEEE Computer,1994

[2] R. M. Hord, *Understudying Parallel SuperComputing,* IEEE Press, 1998

[3] J. Lu, D. Thiel and S. Saario: FD-TD Analysis of Dielectric Embedded Electronically Switched Multiple-Beam Antenna Array, IEEE Trans. On Magnetics, Oct., 2001 (in press)

[4] A. Taflove, "Computational Electromagnetics, The Finite-Difference Time-Domain method", Artech House, 1995

[5] J. P. Berenger, "Perfectly Matched Layer for the FDTD Solution of Wave-Structure Interaction Problems", IEEE Trans. Antennas Propagation, vol. 44, no. 1, pp. 110-117, 1996

[6] www.ingber.com, "Adaptive Simulated Annealing method"

[7] J. Lu, Y. Kagawa and David V. Thiel: Numerical Solutions of Wave Propagation in Dispersive and Lossy Transmission Lines Using a Finite Element Model, International Journal of Applied Computational Electromagnetics, Vol. 9 pp57-66, July, 1994

[8] J.Lu and D.Thiel: Computational and Visual Electromagnetics for Undergraduate Engineering Courses, IEEE Trans. on Magnetics, Vol. 36, No. 4, 7/2000

[9] J.Lu: *ACVEM-Applied Computational and Visual Electromagnetics for Computer Aided Engineering in Classroom,* Procedings of IEEE CEFC2000, June, 2000

[10] L. K. Kale, "Programming Languages for CSE: The State of the Art", IEEE Computational Science & Engineering, pp18-26, April-June, 1998

[11] U. Harms, "Clusters triumph over custom systems," Scientific Computing, pp14-14, Oct./Nov., 1999

[12] R.Buyya, *High Performance Cluster Computing*, PH PTR, pp350-374, 1999

[13] W. Gropp, E. Lusk and A. Skjellum, *Using MPI: Portble Parallel Programming with the Message-Passing Interface*, The MIT Press, 1992

[14] J.X. Chen and D. Rine: "Advancing Interactive Visualization and Computational Steering", IEEE Computational Science & Engineering, pp13-17, Winter 1996

[15] D. Chen, and C. Sun: "A distributed algorithm for graphic object replication in real-time group editors," Proceedings of ACM Conference on Supporting Group Work, pp. 121-130, Phoenix, Arizona, USA, Nov. 1999

[16] J.Wood, H.Wright and K.Brodlie. "Collaborative Visualization" in proceedings of Visualization '97, pp 253–259, (R.Yagel and H.Hagen, eds) IEEE Computer Society Press, 1997

Genetic Algorithm Optimisation of Mathematical Models — An Aircraft Structural Dynamics Case Study

S.A. DUNN

Airframes & Engines Division, Defence Science & Technology Organisation, 506 Lorimer St. Fishermans Bend, Victoria, Australia
Shane.Dunn@dsto.defence.gov.au

Summary

In this paper, a technique for tackling inverse problems in structural dynamics will be reviewed. The processes applied here involve using the artificial intelligence optimisation tool known as genetic algorithms (GAs) where the optimisation problem to be tackled involves creating a structural dynamic aircraft model that best matches the available experimental frequency response function (FRF) data. It will be demonstrated how the inverse structural problem is difficult for more traditional optimisation approaches and how the stochastic processes inherent in a GA allow the problem to be tractable. Unsteady aerodynamic modelling is then used in conjunction with the structural models for aeroelastic analyses to determine the airspeed at which instabilities (commonly known as flutter) will arise. The implications of such model optimisation on aeroelastic analyses, and the estimates of uncertainty that may then be made, will also be briefly discussed.

1 Introduction

With the increasing reliance on mathematical modelling in the design of aircraft, it is more important than ever that the mathematical models underpinning these processes are shown to be accurate. The best way of demonstrating this accuracy is to make use of well–targeted tests to assess how well the end product correlates with the behaviour predicted by the mathematical model. The question that then arises is: what is the best way of correcting the model when the test and model predictions do not agree? A circumstance where such checks are compulsory concerns the mathematical modelling that is performed to assess the aeroelastic stability of aircraft structures. If an aeroelastic instability, or flutter, were to arise on an aircraft in service, it can arise with little, or no, warning and can result in the loss of the aircraft. Mathematical models are typically used to estimate the airspeed at which such an instability will arise. Given the dire consequences that can result if the mathematical model is flawed, it is a requirement of the airworthiness standards that, where such modeling is used, the structural aspects of the model must be validated against experimental data from a ground vibration test (GVT). The standards, however, are silent on how such a validation should be carried out, or how the model should be improved if it is found to be deficient. A process which will be described here, uses a genetic algorithm (GA) to create a mathematical model by determining an optimal set of mass, stiffness and damping properties such that the best agreement between the experimental data and the model predictions is achieved.

When creating mathematical models of complex systems, many approximations are typically made. The hope is that these approximations will lead to a model that is both tractable, and accurate within acceptable levels. With the exponential growth in computational power observed over the past three decades, models can now be very large and complex and still be tractable — but are they more accurate? The complexity of even modern mathematical models typically does not approach the complexity of the real systems being investigated. The main gain in going to more complex models is that the degree of approximation required in creating the model is reduced, leading to increased confidence in the model's predictions. Such increased confidence, however, is not necessarily well founded.

In the case of aircraft structural dynamic models, the models of about 20 years ago required of order 10^3 unique pieces of information to describe them fully; a modern model may require 10^5 to 10^6 pieces of information. Both types of model, however, must be validated against experimental data and the question for both is: how do you fix such complex models when they are found to be inaccurate? The information content of the experimental data will not be sufficient to fully identify the 20 year–old model, let alone a modern one. One of the crucial features of the process that will be described here is that it isolates what are the truly important properties required of the mathematical model, and identifies these in such a way as to give a model of very high fidelity and relative simplicity.

An important outcome of such a model optimisation process is that an estimate can also be made of the uncertainty present in the final structural dynamic model. The impact of such an estimate for the model uncertainties on aeroelastic analyses will be briefly discussed.

2 The Inverse Structural Dynamic Problem

From a GVT, frequency response functions (FRFs) are often collected – an experimental FRF might be as shown in Figure 1. Such FRFs will be collected at many locations on the structure while the aircraft is being shaken. A typical model updating exercise, however, will use data only at the resonant frequencies, or eigenvalues. In this paper, more of the FRF data from the

GVT will be used to create an optimal physically–based model that gives an accurate representation of the data.

The object is to find the vector of unknown model properties, $\{\eta\}$, such that the error, ε, is minimised by comparing the model predictions, $\chi(\eta)$, and measurements, χ', for the n freedoms measured at the N selected frequencies, as shown in equation(1):

$$\min(\varepsilon(\eta)) = \sum_{j=1}^{N}\sum_{i=1}^{n} \left||\chi_{i,j}(\eta)| - |\chi'_{i,j}|\right| = \sum_{j=1}^{N}\sum_{i=1}^{n} \varepsilon_{i,j} \tag{1}$$

This cost function is demonstrated graphically in Fig. 1.

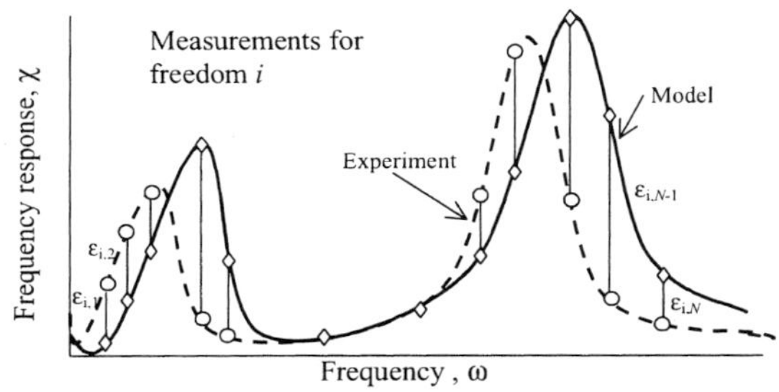

Figure 1 Graphical depiction of the cost function for the ith freedom of equation(1)

The model predictions are given by:

$$\chi_{i,j}(\mu,\kappa) = \left(\phi\left([K] + \mathrm{i}\omega[\Delta] - \omega^2[M]\right)^{-1}\right)_{i,j} \tag{2}$$

where ϕ is a vector of applied sinusoidal load amplitudes, K is the stiffness matrix, Δ is the damping matrix, M is the mass matrix, and ω is the frequency of loading. This forms the physical framework within which the optimisation is performed. The structure of the above matrices is dictated by the physics of mass/spring/damping systems, and within this optimisation procedure, this structure is maintained. If it were not maintained, better agreement between the model and the data could be achieved, but in losing the physics behind the model it would have no predictive capabilities (it is not unknown that following other model updating procedures which have less regard for the laws of physics, the final optimised model has negative masses).

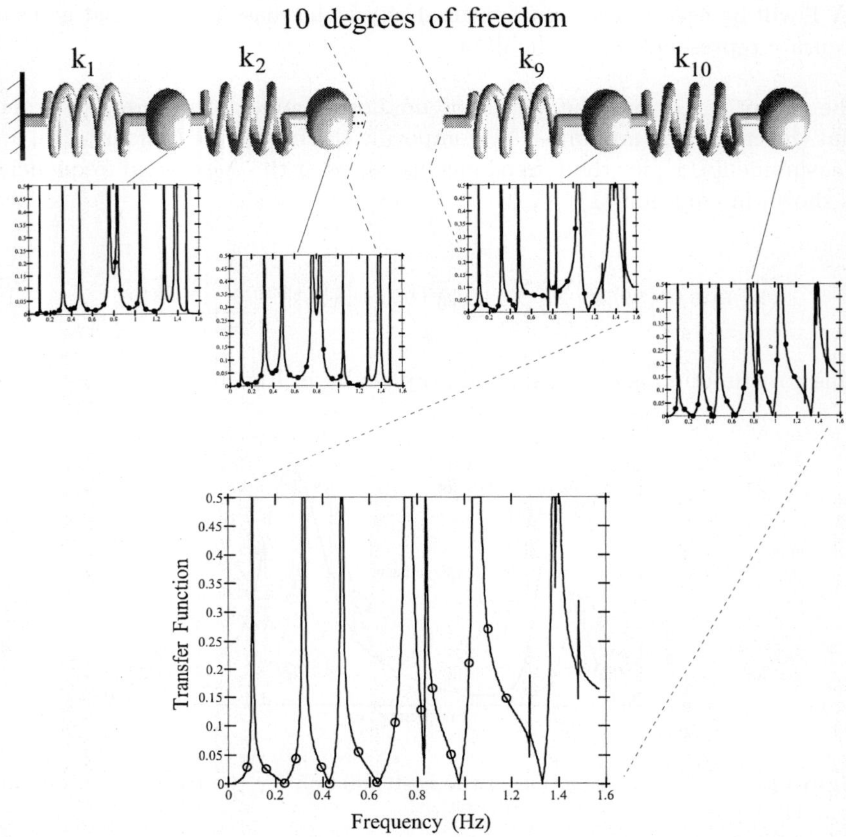

Figure 2 Simulated 10 degree–of–freedom mass/spring system and measurements.

2.1 10 Degree–of–Freedom Mass/Spring System

To demonstrate the difficulties involved with the optimisation described at equation(1), consider the simulated 10 degree–of–freedom undamped mass/-spring system shown in Figure 2.

Applying the cost function, as described in equation(1) for first one spring stiffness as an unknown, then with two spring stiffnesses unknown yields the search–spaces as shown in Figure 3. These results are based on solving the problem by enumeration (solving the cost function in equation(1) for all possible solutions); this is impractical for a real problem of typical size, but it does allow the illustration of the search landscape that must be traversed by an optimisation routine in order to find the best result. Standard optimisation techniques are based on starting somewhere within the search–space,

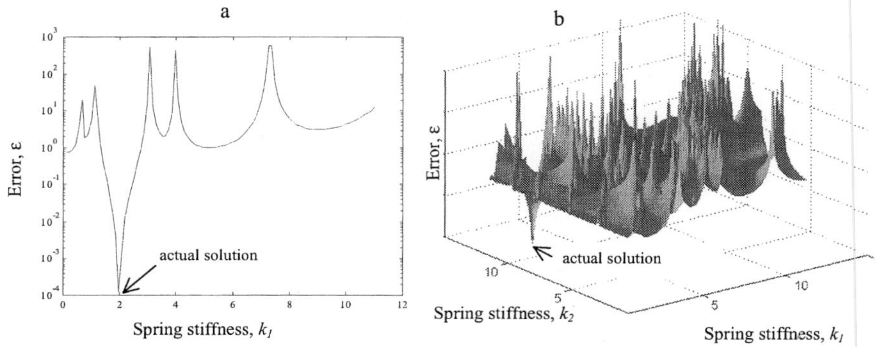

Figure 3 a & b, search landscapes for the mass/spring system shown in Figure 2 with one and two unknown spring stiffnesses respectively.

making an approximation to the local derivative, and heading downhill until no further improvement can be made. As can be seen in Figure 3, the search–space is very deceptive in that many starting points will lead to sub-optimal solutions, and the number of these deceptive solutions grows exponentially as the dimension of the search–space grows. The effects of this can be illustrated by applying a simplex algorithm[1] to the 10 degree–of–freedom problem first with one unknown, then two and so on. For the simplex algorithm, a random initial starting point within a defined search–space is chosen and the algorithm run until it has converged to a solution. Upon convergence, the process is started again and so on a number of times. The number of cost functions required to find the correct solution can then be estimated by dividing the total number of cost function evaluations performed by the number of times the correct solution was found. The results for this are shown in Figure 4. As can be seen, as the dimensionality of the search–space grows, the number of cost function evaluations grows exponentially, and, as a consequence, rapidly becomes effectively impossible.

3 Genetic Algorithms

The inspiration for GAs arose from the realisation that the result of the principles of Darwinian evolution in nature is a process that attempts to solve a vast optimisation problem. Those individuals that are best suited to their environment are more likely to breed and therefore pass their genetic material onto subsequent generations. Genetic algorithms are a computer–generated analogue of this process where the better individuals are chosen based on how they test against a specified cost function. For the case of using a GA to optimise mathematical models, the cost function is based on a measure of how well the model predicts actual measured data. The GA

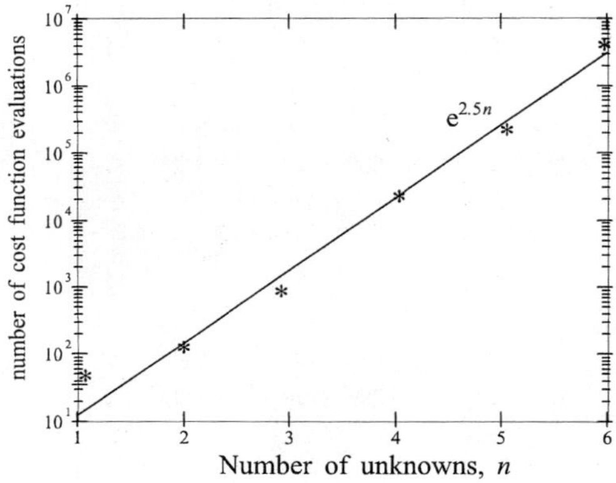

Figure 4 Exponential growth in the number of cost function evaluations required by a simplex algorithm as the number of unknown spring stiffnesses increases.

is started with a given population size of randomly generated individuals - where each individual is a potential complete solution to the problem. Based on equation(1), a value of *fitness*, $(1/\varepsilon)$, is then assigned to each individual. Breeding is done by selecting pairs from the original population in a weighted random process (where the weights are determined by the fitness such that the better solutions have a higher probability of breeding), and then swapping the properties of these pairs in a random manner to give rise to a new individual; this is done until a new population has been created and the processes of cost function evaluation and breeding *etc.* are repeated, over and over, until some stopping criterion is reached. There are many different ways of applying GAs; details of how they are applied for the case studied here can be found in [2], though a brief summary will be given as follows:

for $j = 1 : n$
    for $i = 1 : N_1$
        `select random population` — Easily parallelised through distributed computing environment – **Beehive** (see [3])
        `run GA for` g `generations`
        `append results to` R_j
    end for, i
    for $i = 1 : N_2$
        `select random population` — Serial process, best handled by processors dedicated to the task
        `include previous results in` R_j
        `run GA for` g `generations`
        `append results to` R_j
    end for, i
    `append final result to` R_f
end for, j
`Examine and compare all results in` $R_{1\,\mathtt{to}\,n}$ `and` R_f

This process of seeding the initial population with earlier solutions has been found to be more efficient than a standard 'linear' GA for these problems. Nevertheless, the computational requirements are still significant. The above process, in that it requires many independent solutions, the results of which are used as seeds for subsequent solutions, means that it is very amenable to parallel computation. A PC–based distributed computation system known as *Beehive* has been developed for these processes and is described in detail in [3].

Applying these GA processes to the 10 DOF problem considered in the previous section yields the results as shown in Figure 5. As can be seen in this figure, the required number of cost function evaluations still grows significantly as the dimensionality of the search–space grows, but the problem remains tractable, which is not the case for a more traditional approach.

There are many ways of applying GAs to the many and various optimisation problems that exist; what has been briefly described here is the way in which they have been applied to this problem. For the reader interested in looking further into the philosophy of GAs, some good starting points are [4] and [5]. Introductory reading on the application of GAs can be found in [6] and [7]

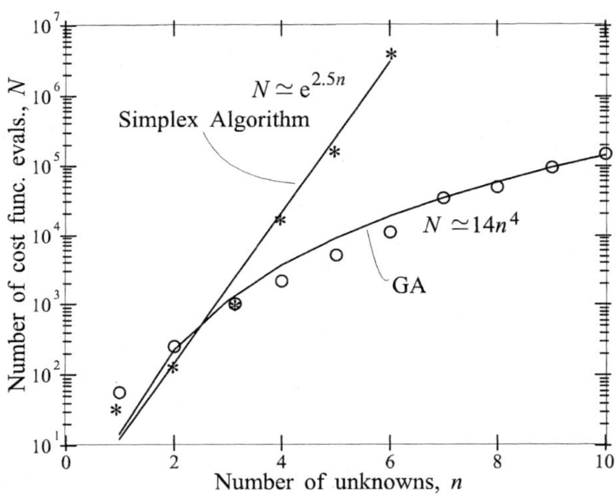

Figure 5 Growth in the number of cost function evaluations with the number of unknowns required for a simplex algorithm and a genetic algorithm

4 Optimised Model for F/A–18

In the previous section, it has been demonstrated that GA procedures can be used to efficiently optimise a mathematical model using simulated data of up to 10 unknown stiffness properties. The real problem involves solving for the stiffness, inertia and damping properties of a complete aircraft; this involves a search in >100 dimensional space.

A detailed structural dynamic finite element model of an F/A-18 is shown in Figure 6. A GVT on an F/A–18 was carried out in Canada (a sample of the data is shown in Figure 7 and the data was supplied to DSTO for this work [8] and the work described here is presented in greater detail in [9].

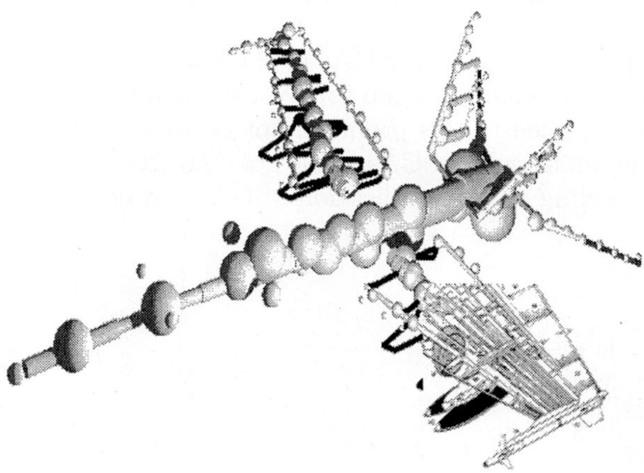

Figure 6 Finite element model (stick model) for structural dynamic analysis of an F/A–18; such a model is a considerable idealisation of the actual structure that is shown on the left wing.

4.1 Model Complexity Issues

As can be seen in Figure 6, there are a great many separate pieces of information required to fully identify even this simple finite–element model. The required pieces of information include beam bending and torsional stiffnesses, inertial properties (linear and rotational), attachment spring stiffnesses and locations of stiffness and inertial elements. When it comes to incorporating the experimental data to address problems with the predictions in the original model, however, it quickly becomes apparent that there is nowhere near

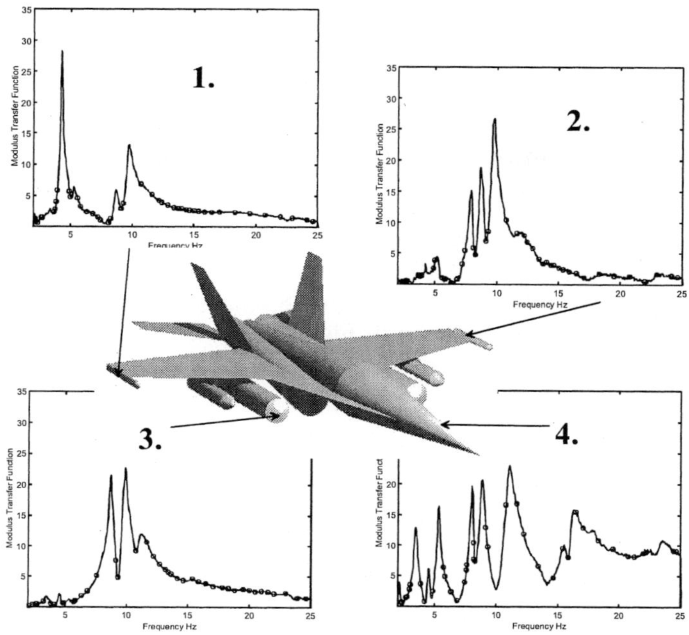

Figure 7 A sample of the Canadian GVT data used in the optimisation process.

the information content in the experimental data that would be required to address the >1000 separate pieces of information needed to make this model. A method of determining models that reflect the complexity of the experimental data is described in detail in [10] and will be briefly described here:

- Run the optimisation procedure — in this case a GA — a number of times such that there are a number of results where the better cost functions (1) are very similar and the model predictions give a satisfactory representation of the experimental data;
- compare the properties found for these results;
- where this comparison shows little variation, assume the property is being determined uniquely;
- where the comparison shows a great deal of variation for similar cost functions, assume that either the property is not required, or that the property and one or more of its neighbours can be combined into one;
- repeat this process until all parameters appear to be defined uniquely *and* the model still gives a satisfactory representation of the data.

This process for dealing with the model complexity issues is clearly very numerically intensive in that many GAs must be solved and the results compared. For such a problem, the computational power of the distributed computing offered by *Beehive*[3] is particularly valuable.

4.2 F/A–18 Model Optimisation Results

Application of the GA optimisation procedures that have been briefly discussed in this paper to the optimisation of the F/A–18 structural dynamic model yields the model as shown in Figure 8 with some sample measurement and model comparisons, as shown in Figure 9.

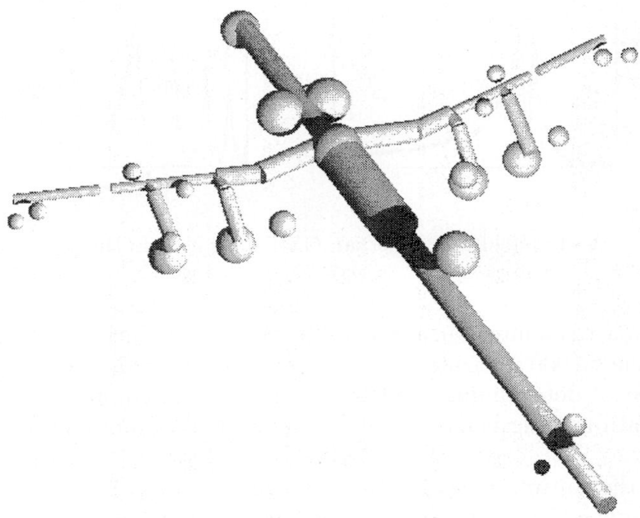

Figure 8 Optimised model for an F/A–18 based on Canadian GVT data.

The results of the F/A-18 model optimisation process shown here, results in a model that gives good agreement with the experimental data for the frequencies of interest for typical flutter calculations. Also, the model is based on the physics of mass/spring/damper systems and can, therefore, be expected to have predictive capabilities for the structural response if changes are made to the system (typically for different stores carried on the wings). Using this structural dynamic model, and an unsteady aerodynamic model (using doublet–lattice modelling [11]), flutter analyses can now be performed in accordance with the airworthiness standards.

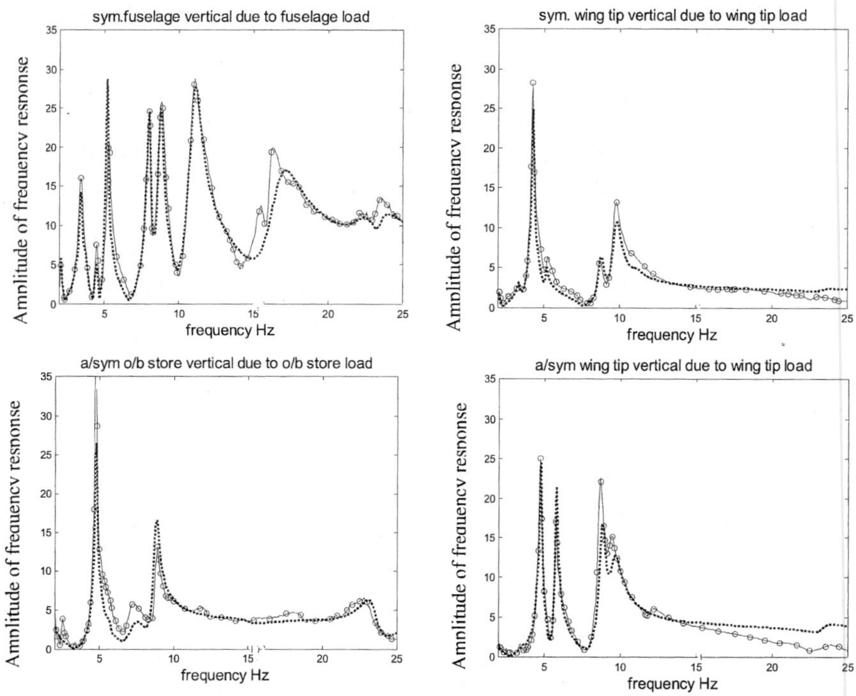

Figure 9 Ground vibration test measurements (solid line) compared with optimised model predictions (dashed line) — the ∘ show the frequencies where the cost function evaluations are made.

5 A Step Further — Modelling Uncertainty

When examining the results of such a process, it becomes apparent that, whilst there is one 'best' model, there may be many 'very good' models - *i.e.* models that have very nearly the same correlation with the experimental data. The question that must be asked is: is there something special about the 'best' model that makes it the only model to be considered, or is it likely that the 'very good' models also have some merit? Whilst the changes from the 'best' to the 'very good' models may not appear to be very significant, in terms of the properties they represent and their effect on the experiment/model comparison, the real object of the exercise is to estimate the airspeed at which an instability will result. From the optimisation exercise carried out, it is not clear what the effects on the stability calculations will be, particularly when the model properties are changed to reflect different aircraft states.

Using both the 'best' and the 'very good' models, with the unsteady aerodynamic modelling, full flutter solutions can be run yielding a range of estimated

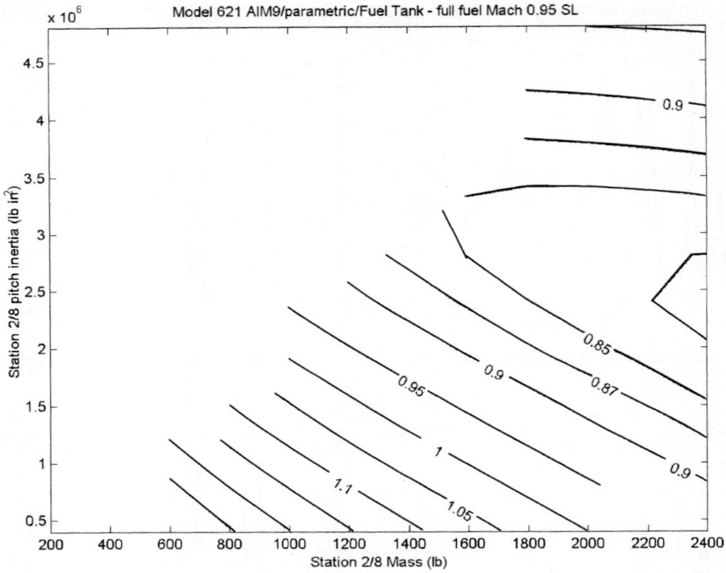

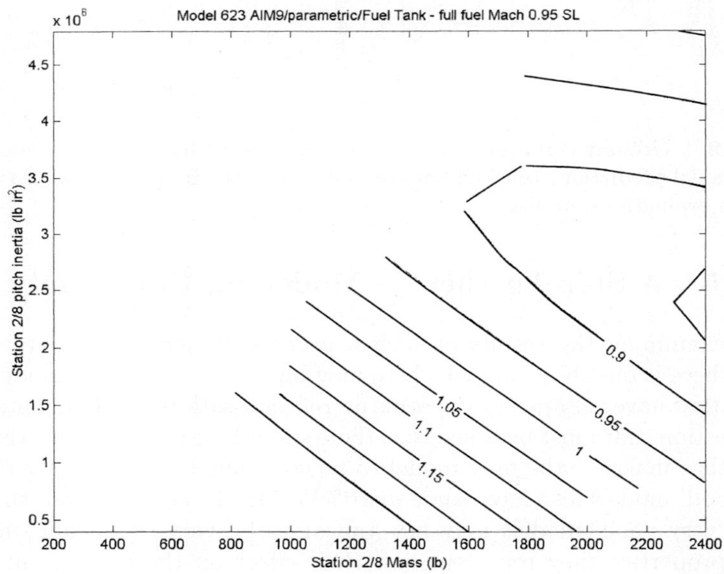

Figure 10 Normalised flutter speed contours for carriage of a parametric store on the outboard pylon of an F/A–18 — the two plots are for results based on independently optimised structural models.

flutter boundaries — *i.e.* the point at which there is a danger of the aircraft breaking up in flight. By such a process, the probable error in the flutter solutions due to uncertainties in the structural model can now be estimated. Examples of two such model results are shown in Figure 10. In this figure, a parametric study of the effects of outboard store load on the flutter speeds are shown where the two models involve independently optimised structural dynamic models that resulted in near–equivalent cost functions. By such a process of comparison of model results, estimates can now be made on the likely errors in flutter speed that may arise due to remaining errors in the structural dynamic model.

6 Conclusion

In this paper, the status of structural dynamic model updating using genetic algorithms has been reviewed. It has been shown that genetic algorithms can be used to create a physically based structural dynamic model to be used for aeroelastic analyses and that this model gives very good agreement with the available experimental data. It has also been discussed how, using such model creation processes and considering models created from independent optimisation runs, estimates can be made as to the errors that might be expected in the final flutter speed calculations.

An important point to remember is that these processes need not be unique to structual dynamic modelling. These techniques could be applied to any situations where there is a significant amount of experimental or historical data available, along with a good understanding of the physical framework of the system under investigation. As well as using these processes to create models that more faithfully reflect reality, the analyst also gains very useful insight into the model parameters that are truly important in the mathematical model making process.

References

[1] J.A. Nelder and R. Mead: "A Simplex Method for Function Minimization.", Comput. J. **7**, 1965, 308—313.

[2] S.A. Dunn: "Modified Genetic Algorithm for the Identification of Aircraft Structures." J. Aircraft, **34**, 2, (1997), 251—253

[3] S. Dunn and S. Peucker: "Genetic Algorithm Optimisation of Mathematical Models Using Distributed Computing.", Proceedings of The Fifteenth International Conference on Industrial & Engineering Applications of Artificial Intelligence & Expert Systems, Cairns, Australia, June 17—20, 2002.

[4] J.H. Holland: "Genetic Algorithms.", Scientific American, July (1992) pp44—50.

[5] S. Forrest: "Genetic Algorithms: Principles of Natural Selection Applied to Computation.", Science, **261**, (1993) pp872—878.

[6] D.E. Goldberg: "Genetic Alorithms in Search Optimization & Machine Learning.", Addison–Wesley Publishing Company, inc. Reading MA, USA (1989)
[7] M. Mitchell: "An Introduction to Genetic Algorithms.", The MIT Press, Cambridge, MA, USA (1996).
[8] M. Dickinson: "CF–18/GBU24 Ground Vibration Test Report." Bombardier Inc. Canadair, Montreal, Canada, (1995).
[9] S.A. Dunn: "Optimised Structural Dynamic Aircraft Finite Element Models Involving Mass, Stiffness and Damping Elements." International Forum on Aeroelasticity and Structural Dynamics, Madrid, Spain, 5—7 June, (2001), 387—396
[10] S.A. Dunn: "Technique for Unique Optimization of Dynamic Finite Element Models." J. Aircraft, **36**, 6, (1999), 919—925
[11] E. Albano and W.P. Rodden: "A Doublet–Lattice Method for Calculating Lift Distributions on Oscillating Surfaces in Subsonic Flows.", AIAA Journal, **7**, 1969, p.2192

Techniques for Optimal Shape Design

Olivier Pironneau[1] and Bijan Mohammadi[2]

[1] Université Paris VI & IUF,
Laboratoire J.-L. Lions, 175rue du Chevaleret, F-75013 Paris.
[2] Université Montpellier II & IUF
Mathematics department. , CC51, F-34095 Montpellier. * * *

Summary

In this paper we apply to a few new problems some of the techniques we have developed over the years for optimal shape design. These include optimality conditions, automatic differentiation and the appropriate descent methods.

Introduction

Optimal Shape Design has developed over the years, from the abstract field of calculus of variations to applications in structure mechanics and fluid mechanics becoming now a valuable tool for the design optimization of airplanes. Weight reduction, stress reinforcement, drag reduction and even noise reduction can be obtained.

The field is said to have started with Hadamard[8] who gave the first formula to evaluate the change due to a boundary modification of the domain of a partial differential equation.

As an offspring of the theory of optimal control by Lions[3], ptimality conditions were studied extensively in the seventies by Murat, Simon, Cea, Pironneau, among others (see [5] for a more complete bibliographical study). Stability and existence was shown to be ill posed in many practically important situation by Tartar[7], leading thereby to the introduction of relaxed problems known now as topological optimization and studied by Allaire[21], Sokolovski[6] and Masmoudi[9], among others.

For fluids the problems are usually well posed and some regularization is necessary but only for stabilizing the algorithms which usually require more regularity on the shapes than given by the partial differential equations. Therefore the difficulties are mostly numerical:

– Complex equations such as turbulence models.

* * * Dedicated to J.L. Lions who passed away in May 2001. His encouragements and beliefs in scientific computing and optimization, his leadership as a scientific consultant at Dassault-Aviation played a great role in the career of both of us.

- Fine grids for acceptable precision.
- Complex shapes, such as airplanes.

In a recent book Mohammadi and the author[4] have proposed some techniques to deal with these difficulties:

- Automatic differentiation to handle difficult equations.
- Mesh adaptation to place the points where they are required.
- Reconstruction of the geometries outside their CAD definition.

In this paper we wish to review these tools and present an application to the optimization of a business jet at the company Dassault-Aviation.

We discuss the various techniques for a generic problem of shape optimization, that of minimizing $J = \int_S f(u, \nabla u)$ where u is solution of a partial differential equation in weak form

$$\min_S \int_S \{f(u, \nabla u) \ : \ \int_\Omega a(u, \nabla u, \hat{u}, \nabla \hat{u}) = 0 \ \ \forall \hat{u} \in H(\Omega)\} \tag{1}$$

in a domain Ω for which S is part of its boundary. We consider an approximation $J_h = \int_{S_h} f(u_h, \nabla u_h)$ where S_h is a polygonal approximation of S, q^i are vertices of a triangulation and $u_h \in H_h$ a finite element approximation of $u \in H$.

1 Automatic Differentiation

Assume we have a computer program to calculate J_h for a given S_h and that it has declared N variables $(x_i)_{1 \le i \le N}$ of which $(x_i)_{1 \le i \le n}$ defines S_h and that it produces m *output variables* (m=1 for an optimization problem , m¿2 for a Pareto or Nash problem).

We want to compute the derivatives of the output variables with respect to $(x_i)_{1 \le i \le n}$. The fundamental idea is that each line of the program can be differentiated exactly and so we can in principle compute simultaneously the differential and the value of each instruction. To obtain the gradient we simply put the differential of the independent variable to the value one before starting the computation.

This approach to automatic differentiation is called the *forward* mode. The other approach, the *reverse* mode is computationally more effective when n is larger than 1, at least in theory but it is less straightforward to use.

1. The reverse mode has a *pre-treatment* of the source code to provide the derivated code. Odyssee [23], Adifor [22], Adic [10] use this approach. This is doable for languages with simple data structures (Fortran 77) and difficult for languages with complex data structures (C,C++).

2. The forward mode generates the derivated code on the fly and use extensively the *Operator overloading* features(C++). Fadbad [11] use this approach; Adol-C [24] has both approaches.

Among the first to use operator overloading is A. Griewank et al. with Adol-C and later C. Bendtsen and O. Stauming with Fadbad. M. Masmoudi and M. Grundman[12] presented us this technique in july 97 and we chose it for its simplicity and also because the evolution of C++ compilers were about to provide us with more efficient numerical tools such as expression templates [13], `valarray<>`, traits etc.

The important operators which appear in a program are the arithmetic operators $\{+,-,*,/,+(unary),-(unary),++,--\}$ the logical operators ! $=,==,<,>$ $,<=,=>$ and the mathematical operators $\{sqrt, log, pow, ...\}$. The implementation of the forward mode done by N. Dicesare[1], `FAD`, is based on operator overloading of all the above and templates so as to be able to define n at execution times. It is known then that the generation of temporary arrays by the compiler could be deadly slow. Therefore a careful optimization using expression template was done. It was found that for problems with up to 50 independent variables the forward mode was not more expensive than the reverse mode, primarily because of the book-keeping and overhead necessary in the later. So depending on the number of variable we are now using our C++ class library FAD[14] or Odyssee.

2 Mesh Adaptivity

2.1 Automatic triangulation

Les us first recall the basis of a Delaunay-Voronoi mesh generator; the triangulation of a square first. Assume that we are given N+4 points the last four of which defines a square containing all the other points. The points are numbered $(q^0, ...,$ $q^{N-1}, q^N, q^{N+1}, q^{N+2}, q^{N+3})$. We maintain a list of triangles. At start it is the two triangles (q^N, q^{N+1}, q^{N+2}), (q^{N+2}, q^{N+3}, q^N).
Then we perform a loop on the points, backward for simplicity:

for i=N-1 down to 0 do

- case 1: there exists a triangle of the list which contains q^i strictly. then replace this triangle by the 3 subtriangles which have q^i for vertex.
- case 2: the point q^i is on the border of the square. Then find the unique triangle which contains q^i and replace it by the 2 subtriangles which have q^i for vertex.
- case 3: There are two triangles which contain q^i (i.e. q^i is on an inner edge. Then each triangles must be replaced by the two subtriangles which have q^i for vertex..

end loop

Notice that to each inner edge of a triangulation we can associate a quadrangle made of the two triangles adjacent to the edge.

A triangulation is said to be "Delaunay" if for each edge the circle circumscribing one triangle does not contain the fourth vertex.

Edge swap If the 4 vertices of a quadrangle associated to an inner edge are not cocyclic then the two configurations obtained by swapping diagonals in the quadrangle, one is Delaunay, the other is not.

algorithm
Loop until nothing changes
Loop on the inner edges E

- Find the 2 triangles T_k,T_l adjacents to E; denote q^1, q^2, q^3 the vertices of T_k and by q^2, q^1, q^4 those of T_l.
- Check the Delaunay criteria. If it fails replace T_k,T_l by the triangles q^3, q^4, q^1 and q^4, q^3, q^2

end edge loop
end loop

Convergence holds because when a configuration becomes Delaunay by an edge swap the minimum angle in the 2 triangles increase.
The complexity of the algorithm is $O(N)$ (See George [15] for more details).

In practice we use the previous algorithm without interior points; all vertices are on the boundary. We assume that the user has input his request on vertex density through the density of points on the boundary. So each vertex has a weight. For boundary vertices it is the average length of the two surrounding boundary edges. Then we perform the following test on each edge of the triangulation:

- the length of an edge is larger than the average weigth of its vertices then we divided the edge by adding a middle point and assigning to it the average of the weights of the two vertices of the edge.

Then the triangulation algorithm is applied again to the new set of points.
And so on till no edge is divided.

2.2 Mesh adaptation to a function

Mesh adaptation is an important asset of unstructured mesh solvers. In [16] an adaptive procedure is presented, based on a change of metric in the Delaunay - Voronoi algorithm.

The idea is that the interpolation error on a mesh is bounded by

$$\|u - u_h\| < C\|\nabla(\nabla u)\|h^2$$

where $\nabla(\nabla u)$ is the Hessian matrix of u. Therefore an attempt to keep $\boldsymbol{h}^T\nabla(\nabla u)\boldsymbol{h}$ constant is likely to work and build an adapted anisotropic mesh. If several functions are specified as for the mesh adaptation, for instance u and v then it is

$$\min\{\boldsymbol{h}^T\nabla(\nabla u)h, h^T\nabla(\nabla v)\boldsymbol{h}\}$$

which will be kept constant and equal to ϵ. More precisely the method is to apply the Delaunay-Voronoi triangulation algorithm with the distance based on these Hessians (so that circles become ellipses). It is however substantially more complex because a function may have a non positive definite Hessian.
To define a metric the Hessian is diagonalized:

$$\nabla(\nabla u) = \begin{pmatrix} \partial^2 u/\partial x^2 & \partial^2 u/\partial x\partial y \\ \partial^2 u/\partial x\partial y & \partial^2 u/\partial y^2 \end{pmatrix} = \mathcal{R}\begin{pmatrix} \lambda_1 & 0 \\ 0 & \lambda_2 \end{pmatrix}\mathcal{R}^{-1},$$

where R is the eigenvectors matrix of $\nabla(\nabla u)$ and λ_i its eigenvalues. The metric is defined by the scalar product $x^T\mathcal{M}y,\ x, y \in R^2$ with $\mathcal{M}$ defined by:

$$\mathcal{M} = \mathcal{R}\begin{pmatrix} \tilde{\lambda}_1 & 0 \\ 0 & \tilde{\lambda}_2 \end{pmatrix}\mathcal{R}^{-1}, \text{ and } \tilde{\lambda}_i = \min(\max(\frac{|\lambda_i|}{\epsilon}, \frac{1}{h_{max}^2}), \frac{1}{h_{min}^2}),$$

with h_{min} and h_{max} being the minimal and maximal edge lengths allowed in the mesh and ϵ the tolerance.
The parameter ϵ is chosen by the user.

3 CAD Free Geometries and Meshes

In engineering the general tendency is to put the CAD system at the backbone of the engineering process, from early design to production. The computational codes are not integrated into the CAD system, therefore reading input files is necessary. There are standard formats such as IGES but they are neither very practical nor completely reliable. Usually one generates a surface triangulation from within the CAD system onto a file. These are too often inadequate for computations so we are faced with the problem of *building a good surface mesh from a bad one*.

This is a standard problem in computer graphics (see Farin[17]) and one solution is to

- Identify in the triangulation the corners and edges which are true geometric singularities.
- Construct a piecewise C^1 surface by Farin's algorithm (the output is a set of triangular Bezier patches).

– Use the temporary Bezier representation to generate a better mesh.

Refinement of the surface mesh follows the same strategy: curvilinear Delaunay-Voronoi mesh and adaptation to a given function or set of functions.

The last step is the generation of the volume mesh; it is done with GHS3D, a software from INRIA written by George-Hecht-Saltel. GHS3D is also capable of anisotropic mesh refinement and adaptation to a set of functions.

3.1 Mesh Displacement

The mesh generator is not coupled with the automatic differentiation library and so the derivative of the interior mesh node positions with respect to the outer mesh node motions is not accessible. Two results allows us to neglect the influence of the inner node motion in the calculation of the derivative of the cost functions with respect to the shape of a PDE domain: the dependency is an order of magnitude smaller[1] and the coupling of mesh refinement with the optimization loops enables inexact gradients[26].

Proposition 1. *For problem (1)* $\nabla_{q^j} J_h = O(h)$ *if* j *corresponds to an inner node.*

This is because everything appears below an integral and integrals are insensitive to piecewise affine continuous change of variable T which map Ω_h onto itself and all vertices onto themselves except vertex q^j, except if q^j is on the boundary. However the theorem invoked here assumes $\nabla \cdot T = 0$ (area or volume preserving) and that is true only up to order h at the discrete level. (see Dicesare et al[1] for more details).

Inexact gradients are dealt with according to Algorithm 3.1 Here $\text{grad}_{zN} J$ is an

Algorithm 1 Steepest descent with mesh refinement and N-inexact gradient

While $h > h_{min}$
. **while**($w = \text{grad}_{zN} J_h(z^m)$, $\|w\| > \epsilon h^\gamma$)
. { **Try to** Find ρ
. $\beta\rho\|w\|^2 < J_h(z^m - \rho w) - J_h(z^m) < -\alpha\rho\|w\|^2$
. **if** success **then**
. Set $z^{m+1} = z^m - \rho w$ and $m := m + 1$
. **else** $N := N + K$;
. }
. $h := h/2$; $N := N(h)$;
. }

inexact gradient of J which becomes exact when $N \to \infty$. Such an algorithm converges under the same hypothesis as the steepest descent with Armijo step size rule (see [26] for details). To apply it to our situation we must ensure that there is always "success" in the test above; the theoretical estimates show that it can be done by a proper choice of ϵ and γ.

Next comes the problem of maintaining the smoothness of the curve/surface when the points are moved in the direction of the gradient of J_h. For a second order problem like (1)a regularization is required to ensure existence of solution(s):

$$J_\epsilon = J + \epsilon \|S\|_{2,S}$$

To give a precise meaning to this statement we assume that

$$S = \{x + \alpha(x) n(x) \ : \ x \in \Sigma\}$$

where Σ is a reference curve/surface near to the solution S^* and which is used to define the local variations of S by a normal displacement of Σ. Then $\|S\|_{2,S} = \int_\Sigma |\alpha''|^2$. It follows that the right scalar product to define the gradient method is

$$< \alpha, \beta >= \int_\Sigma \alpha'' \beta''$$

and so if by automatic differentiation one is able to calculate J'_α one will define the gradient by

$$\int_\Sigma \Delta J_h \Delta\beta = J'_\alpha \beta \quad \forall \beta$$

which is the same as solving a discretization of

$$-\frac{d^4 \mathrm{grad} J}{ds^4} = J'_\alpha \tag{2}$$

where s is the curvilinear coordinate of Σ. This has for effect to smooth J' and so no oscillation will appear (hopefully).

Even though we do not account them in gradJ_h, the final problem to solve is to define the inner mesh node motions. One simple solution is simply to remesh the domain afresh. But that is unnecessarily costly. A better way is to extend J'_α smoothly inside into χ. In fact this operation can be combined with (2) into a single step: solve

$$-\Delta v - \lambda \nabla\nabla \cdot v = 0 \quad v = J'_\alpha \qquad -\Delta\chi = v \quad \frac{\partial \chi}{\partial n}|_\Sigma = 0$$

Indeed the operator above operates from $J'_\alpha \in L^2(\Sigma)$ to $\chi|_\Sigma \in H^2(\Sigma)$ and it is coercive (positive definite after discretization). This so called *smoother* is reinterpreted here as a change in optimization space which is in fact needed by the theory, so it is not a superfluous step.

4 Examples

4.1 The smoother

Consider a symmetric nozzle of length $L > 0$. Incompressible potential flow is modelled by

$$-\Delta\varphi = 0 \text{ in } \Omega, \quad \left.\frac{\partial \varphi}{\partial n}\right|_\Gamma = g_{|\Gamma}, \tag{3}$$

The inflow and outflow velocities can be rescaled so that

$$g_{|\Gamma^1} = -1, \quad g_{|\Gamma^2} = \frac{|\Gamma^1|}{|\Gamma^2|}, \quad g_{|\Gamma^3 \cup \Gamma^4} = 0. \tag{4}$$

where $|\Gamma|$ denotes the length of Γ.

We consider the inverse problem of designing a nozzle that gives a flow as close as possible to a prescribed flow u_d in a subset of Ω, say in a given region $D = [0, L] \times [0, d]$. One way to fulfill this requirement is through the optimization problem.

$$\min_{\Gamma^3} J_0(\Gamma^3) = \int_D \|\nabla\varphi - u_d\|^2, \tag{5}$$

Potential flow in a straight pipe gives u_d constant. So we wish to recover this solution. Figure 1 shows the effect of the regularizing factor and the corresponding smoother.

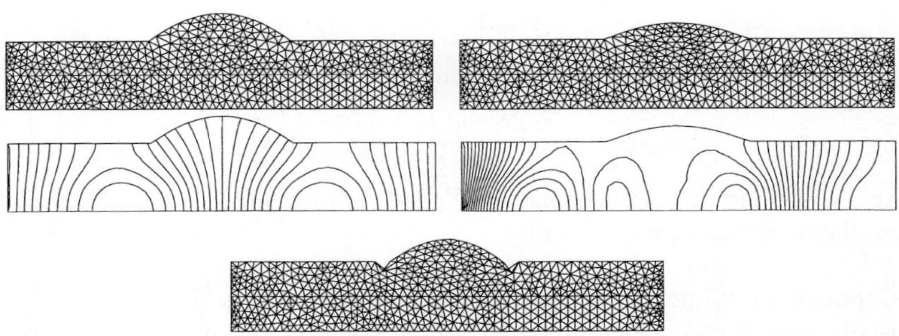

Figure 1 Effect of the smoother (i.e. a good choice for the spaces). Initial shape (left) and after 10 iterations. Without smoother after 4 iterations the mesh is no longer compatible (bottom). The level lines of $|u - u_d|$ for initial guess and the final results are shown; while the first ones are in the interval (0,0.1) the final ones are in (0,0.001). The exact solution is the straight pipe.

4.2 Approximate Gradient

A transonic flow is computed by solving the Euler system of partial differential equation with NSC2KE[4] and the profile is optimized so as to minimize the pressure drag. The state equation is non-linear and the acceleration by approximate gradient is on the number of Newton iterations in the flow solver. The results are shown on Figures 2. It was observed also that most of the variation of a surface integral is due to the change in its normal direction, and so in many cases the adjoint vector part of the gradient is small. However no theory support this fact yet.

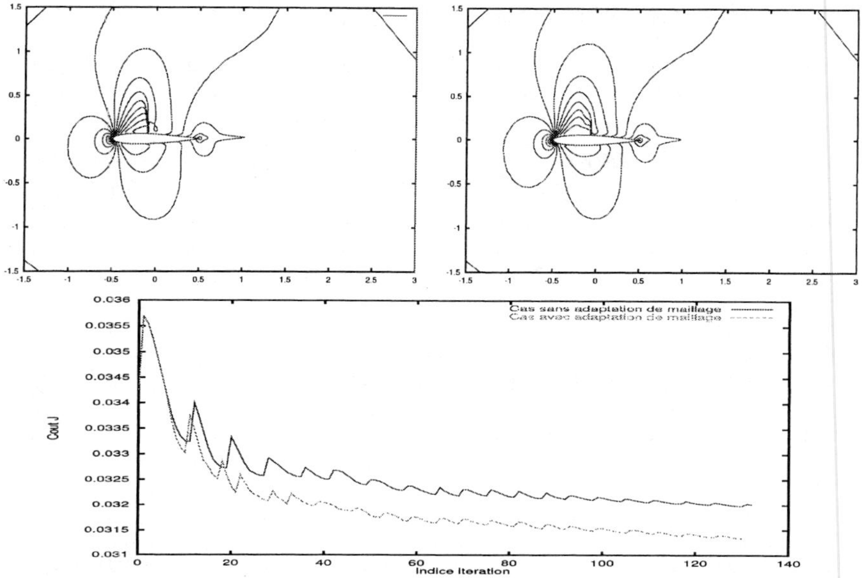

Figure 2 Mach lines for the flow around the airfoil before shape optimization (left) and after. Notice that the shock tends to disappear, an expected result since the drag is a pressure drag. BOTTOM: history of the decrease of the cost function with and without mesh refinement and approximate gradient based on non converged flow solvers. The top curve is without mesh refinement but with control over the iteration number for the flow solver and the curve below is the same with mesh refinement. (Computed by G. Lemarchand [2]).

4.3 CAD Free Mesh Adaptation

A coarse mesh was provided to describe a submarine for which a computation of Navier-Stokes equations with turbulence was desired. A CAD-like Bezier patch based representation of the surface was constructed with Farin's algorithm, including the recognition of corners and sharp edges. Then the Navier-Stokes-like solver was run and the mesh was adapted to $\boldsymbol{u}$. The initial and final mesh are shown on Figure 3.

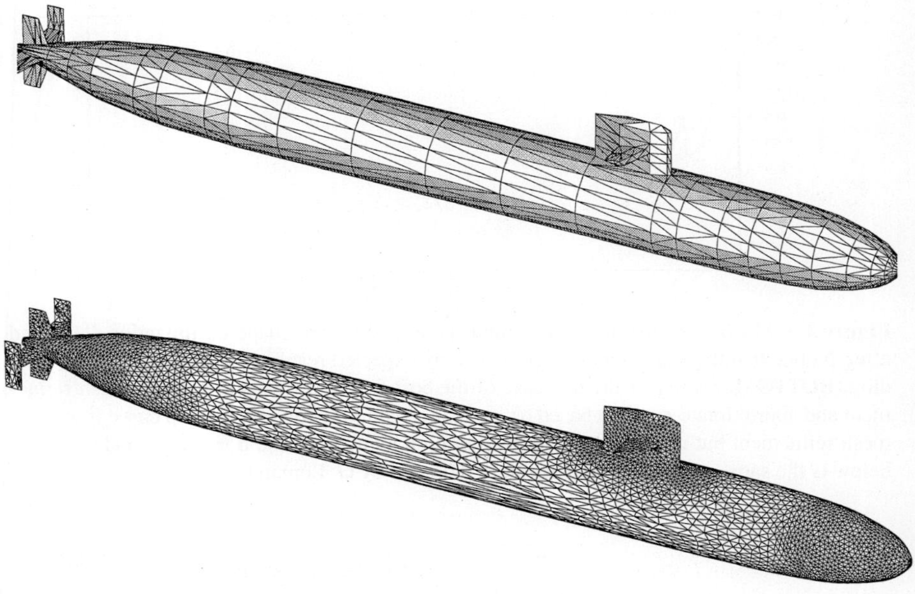

Figure 3 Mesh adaptation from an initial mesh (left) by the solution of Navier-Stokes-like equations around a submarine.

4.4 Optimization of a 3D Business Jet

The company Dassault-Aviation allowed us to test our ideas and implementation on a complete aircraft. Naturally the objective was to minimize the drag subject to various constraints (treated by penalty) such as a given lift and minimum wing thickness. The results are shown on Figure 4. It displays on the top left picture various cross sections of the initial and final shapes; the final shapes are obtained either with an Euler solver or with a Navier-Stokes solver with $k - \epsilon$ turbulence modelling and wall laws. The bottom left picture shows that the changes are both in the position of the wing and in its shape, as shown by the two cross sections displayed. The top right picture displays the decrease in cost function achieved for both cases versus the iteration number. Finally the bottom right plot shows the decrease in drag (lower curves) in the Euler and NS case, and the change in wing thickness and lift, two quantities that were to be kept constant.

5 Perspectives

Optimal Shape Design is still a computer intensive task and the work of specialists. At least in two dimensions it is possible to adapt general PDE solvers like freefem[25] and optimize a shape in minutes, but in 3D it seems not possible in the near future, one reason being that the 3D mesh generators are not in the public domain. There may be a possibility with the fictitious domain method, but here again it is premature to conclude (see Haslinger-Makinen[18] for more details).

Optimizing shapes with commercial softwares which do not make their source code available is even more difficult. Unless the vendor is willing to recompile his code with the automatic differentiation library, only non-differentiable optimization can be used (see Periaux[19] and Schoenauer[20] for instance).

References

[1] N. Di Cesare and O. Pironneau (2000). Shock sensitivity analysis. Computational Fluid Dynamics Journal, vol 9 no 2.

[2] G. Lemarchand, O. Pironneau, E. Polak: Incomplete gradients. Proc Domain Decomposition Method Conf, Lyon 2000. M Garbey ed. (to appear)

[3] J.L. Lions *Contrôle Optimal de systèmes gouvernés par des équations aux dérivées partielles.* Dunod-Gauthier Villars, 1968.

[4] B. Mohammadi, O. Pironneau: *Applied Optimal Shape Design* Oxford University Press, 2001.

[5] O. Pironneau. (1984). Optimal shape design for elliptic systems, Springer-Verlag.

[6] J. Sokolowski and J.P. Zolezio. (1991), Introduction to shape optimization Springer Series in Computational Mathematics.

[7] L. Tartar. (1974). Control problems in the coefficients of PDE, Lecture notes in Economics and Math Systems. A. Bensoussan, Springer Verlag.

[8] J. Hadamard: Leçons sur le calcul des variations. Gauthier-Villars, Paris, (1910).

[9] H. Attouch, M. Cabot, M. Masmoudi, B. Mohammadi, and P. Redont. Coupling dynamic approaches for global minimization, Journal of Convex Analysis, **25**,(2000).

[10] C. Bischof and L. Roh and A.J. Mauer-Oats, ADIC : an extensible automatic differentiation tool for ANSI-C, Software Practice and Experience,27,no 12,pp1427-1456,(1997)

[11] C. Bendtsen and O. Stauning, FADBAD, a flexible C++ package for automatic differentiation using the forward and backward methods, Technical Report IMM-REP-1996-17,(1996)

[12] A. Grundman, Thesis, University Paul Sabatier, Toulouse, (1998).

[13] Todd L. Veldhuizen, Expression templates, C++ Report vol 7, no 5, pp 26–31, june (1995).

[14] P. Aubert, N. dicesare and O. Pironneau, Automatic Differentiation in C++ using Expression Templates and Application to a Flow Control Problem. Accepted in Computer Visualization and Software, 2000.

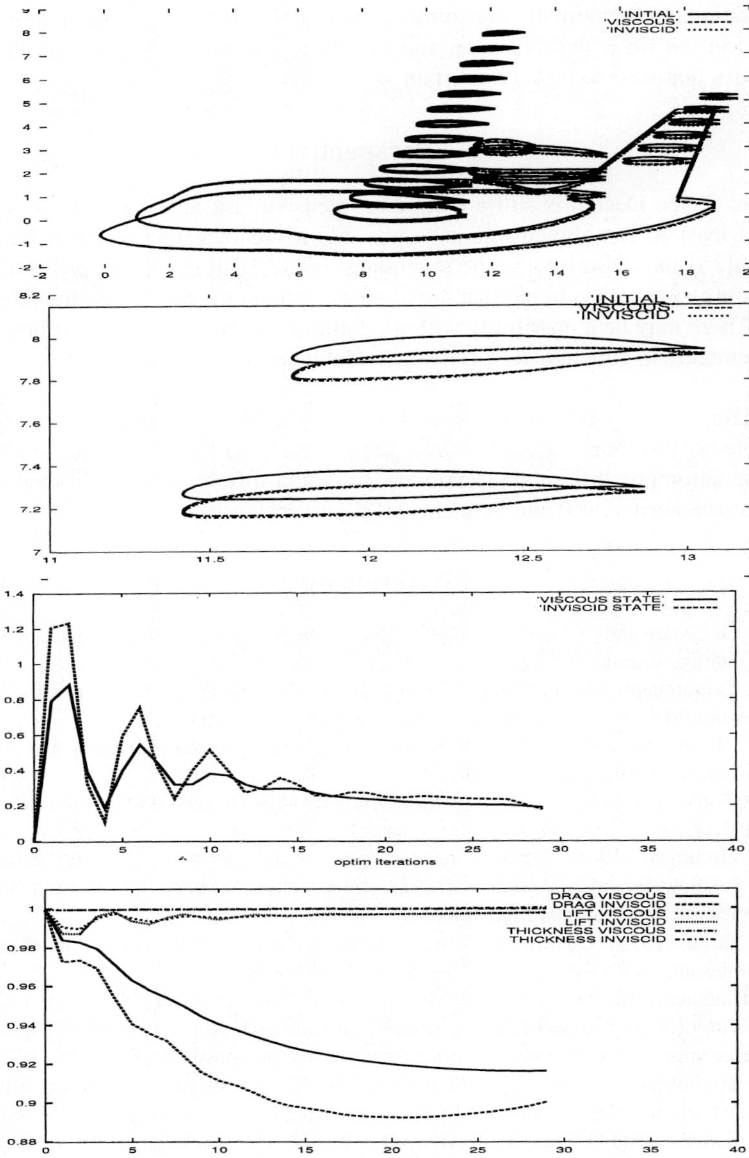

Figure 4 Thanks to the various optimizations the computing time to optimize this business jet is a few hours on a workstation. The third plot is the norm of the gradient of the cost function.

[15] P.L. George, Automatic mesh generation. Applications to finite element method, Wiley, (1991).

[16] F. Hecht and B. Mohammadi. Mesh adaptation by metric control for multi-scale phenomena and turbulence, American Institute of Aeronautics and Astronautics, **97-0859**,(1997).

[17] M. Farin, Algorithms for C.A.D. Wiley, 1987.

[18] J. Haslinger, R. Makinen, Introduction to Shape Optimization, SIAM, (2002).

[19] J. Periaux: Optimization with Genetic Algorithm. These proceedings.

[20] M. Schoenauer and Z. Michalewicz. Evolutionary computation: An introduction. *Control and Cybernetics*, 26(3):307–338, (1997).

[21] Allaire G., Kohn R.V., *Optimal design for minimum weight and compliance in plane stress using extremal microstructures,* Europ. J. Mech. A/Solids 12, 6, pp.839-878 (1993).

[22] C. Bischof, A. Carle, G. Corliss, A. Griewank, and P. Hovland, *ADIFOR : Generating derivative codes from fortran programs,* Scientific Programming, 1(1):11-29, 1992.

[23] N. Rostaing, S. Dalmas, and A. Galligo. *Automatic differentiation in Odyssee,* Tellus, 45a(5):558-568, 1993.

[24] Andreas Griewank, David Juedes, and Jean Utke. *Algorithm 755: ADOL-C : a package for the automatic differentiation of algorithms written in C/ C++,* j-TOMS, 22(2):131-167, June 1996.

[25] Bernardi D., F. Hecht, K. Otsuka, O. Pironneau : freefem+, a finite element software to handle several meshes.Dowloadable from ftp://ftp.ann.jussieu.fr/pub/soft/pironneau/, 1999.

[26] Pironneau O. and Polak E.: Consistent Approximations and Approximate Functions and Gradients In Optimal Control (submitted to SIAM).

Evolutionary Algorithms for Multi-Objective Design Optimization

M. SEFRIOUI[1], E. WHITNEY[2], J. PERIAUX[1], and K. SRINIVAS[2]

[1] Dassault Aviation. 78 Quai Marcel Dassault, 92214 Saint-Cloud, France.
[2] School of Aerospace, Mechanical and Mechatronic Engineering, University of Sydney, NSW 2006, Australia.

Summary

This article presents the general principles of Evolutionary Algorithms (EAs), along with a series of applications in the field of aeronautics. Classical EAs are good enough for problems based on simple mathematical models (i.e. linear models). However, as the applications evolve in complexity, we had to develop new algorithms with better capabilities : among these, we will mostly focus on algorithms combining EAs and Game Theory (hence enabling the algorithm to deal with multi-criteria problems) as well as EAs with a hierarchical structure (which speeds up the convergence by using models of increasing complexity). These concepts are then illustrated via experiments on several applications : minimization of the Radar Cross Section (RCS) around a multi-element airfoil in CEM, reconstruction of a 2D nozzle using multiple CFD models, and a coupled minimization (CEM + CFD) of the drag and RCS for an airfoil. These examples open the way for future applications of EAs in multi-disciplinary design optimization.

1 Genetic Algorithms : Basic Principles

GAs are a population-based semi-stochastic optimization technique. Instead of optimizing a single solution, GAs consider a whole set (or population) of solutions and get them to evolve by applying operators inspired by genetics [1, 2]. To understand the mechanism of GAs, let us consider a minimization problem with a cost index $J = f(x)$, where the parameter is x.

1.1 Representation

There are two major schools of representations : either a binary-coded representation (particularly suited for a combinatorial problem for instance) or a real-coded representation.

Binary coding The idea is to encode x as a finite-length string [3]. The length of the binary string is chosen according to the required accuracy. For a binary string of length $l = 8$ bits, the lower bound x_{min} for the variable x is mapped to 00000000 and the upper bound x_{max} is mapped to 11111111, with

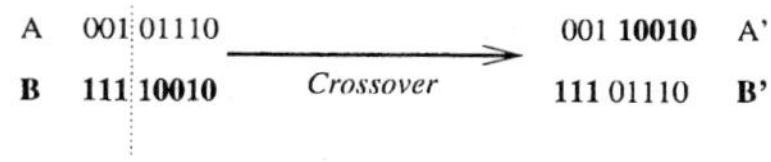

Figure 1. Crossover

a linear mapping in between. Then for any given string, the corresponding value x can be calculated according to: $x = x_{min} + \frac{1}{2^l - 1} \cdot (x_{max} - x_{min})$.

Floating-Point representation A floating-point representation is more natural and need no encoding : the chromosome is a vector of real variables. It is the main representation for Evolution Strategies [2], and is more and more used in the GA community [4].

1.2 Genetic Operators

GA operators are used to create a new population at every generation, by selecting the best solutions, exchanging information between them and mutating them.

Selection The main idea is that selection should depend on the value of the fitness function: the higher the fitness is, the higher the probability is for the individual to be put in the mating pool for the following generation (illustrating the concept of *survival of the fittest*). But it remains a *probability*, which means that is not a deterministic choice: even solutions with a comparatively low fitness may be chosen, and they may sometimes reveal very good properties in the course of events (e.g. if the optimization is trapped in a local minimum). We used a tournament based selection [1].

Crossover Two strings are chosen from the mating pool, and a position along the two strings is picked according to a uniform random law. The paired strings exchange all characters following the cross site.

Mutation It is random alteration of the solution. For binary coding, a mutation means flipping a bit 0 to 1 and vice versa. For real-coded representation, we use a non-uniform mutation [4], where the span of the mutation decreases over time.

1.3 Step by Step Algorithm

(1) *generate* randomly a population of N individuals;
(2) *evaluate the fitness function of each solution;*
(3) select a pair of parents depending on their fitness;

(4) *apply crossover* to form two new children with probability P_c;
(5) *mutate* the two offspring with probability P_m;
(6) repeat steps 3, 4, 5 until a new population have been generated;
(7) go to step 2 until convergence.

2 Coupling GAs and Game Theory for multiple objective optimization

The simplest way to address the problem of multiple objectives optimization is to use a scalar objective, generally obtained through some linear combination of weighted objectives. Such an approach may be of interest in some cases – particularly if the weight of each criterion is known beforehand – but besides its *ad hoc* quality, it has some major drawbacks since there is a loss of information and a need to define the weights associated to each objective. Moreover, the behavior of the algorithm is very sensitive and is biased by the values of these weights [5].

Game Theory is an alternative approach, since it offers a sound theoretical model to deal with conflictual situations [6]. Any *conflict* is considered as a *game*, where each *player* has to consider its own objective.

2.1 Pareto Optimality

Pareto optimality is based upon the principle of *dominance* [7]. For a minimization problem and a Pareto game with n players, a strategy $(\boldsymbol{v_1}^*, .., \boldsymbol{v_n}^*)$ *dominates* a strategy $(\boldsymbol{v_1}, .., \boldsymbol{v_n})$ iff:

$$\begin{cases} \forall i, 1 \leq i \leq n, f_i(\boldsymbol{v_1}^*, .., \boldsymbol{v_n}^*) \leq f_i(\boldsymbol{v_1}, .., \boldsymbol{v_n}) \\ \exists\ i \text{ such as } f_i(\boldsymbol{v_1}^*, .., \boldsymbol{v_n}^*) < f_i(\boldsymbol{v_1}, .., \boldsymbol{v_n}) \end{cases}$$

A strategy is said to be Pareto optimal if it is non-dominated. The Pareto front is the set of all the non dominated solutions [8]. Therefore, a solution belonging to the Pareto Front is optimal in the sense that there is no other design which is better with respect to all design goals. In other words, such a solution can be improved with respect to one goal only by degrading its performance with respect to another.

2.2 Non-dominated Sorting Genetic Algorithm

Non-dominated Sorting Genetic Algorithm (NSGA) has been introduced by [9] and blends a powerful coupling of two techniques to maintain diversity and capture the largest possible front : ranking and sharing.

Pareto ranking is a technique which helps avoid using *directly* the objective functions corresponding to the candidate solutions to perform the selection process. Instead, they are projected in the objective space and the Pareto dominance principle is applied to rank the solutions [1, 10]. The solutions are sorted using Pareto Dominance. The ones in the Pareto Front get a rank of 1. Then, these solutions are removed from the pool of solutions, and the remainder of the pool is iteratively sorted until all the solutions are ranked (the worst a solution, the higher its rank).

Sharing is based upon the notion of competition between similar solutions [1, 11]. In NSGA, all the solutions belonging to the same rank undergo the process of sharing.

These two techniques help distribute the solutions over the entire Pareto front and tend to ensure diversity; a good overview can be found in [8].

2.3 Example : Discontinuous Sinusoidal Problem

Let us consider the following two-objective problem

$$N = 2, x_i \in [0,1]$$
$$f_1 = x_1$$
$$f_2 = (1 + 10x_2)(1 - (\frac{x_1}{1+10x_2})^2 - \frac{x_1}{1+10x_2}\sin(8\pi x_1))$$

This problem has a discontinuous Pareto front with four separate regions. Figure 2 shows the results obtained with a GA coupled with Pareto ranking. It is clear that the GA has managed to converge towards all four regions, and the fact that the solutions are well spread over these regions shows the efficiency of sharing.

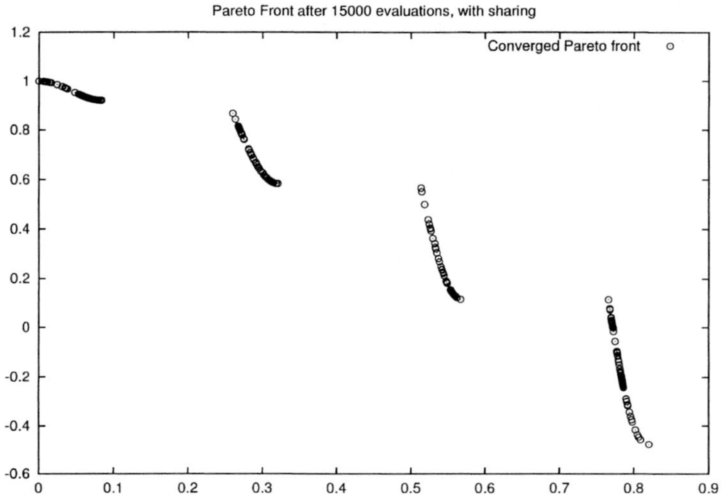

Figure 2. Converged front

2.4 Aerodynamic Application

The objective of this application is to design a two dimensional helicopter aerofoil subject to a two point objective function. The aerofoil is encoded as a geometric mean line and symmetric thickness distribution which is offset about the mean line to produce the desired final shape. Both curves are Beziér curves, with fixed chordwise (x/c) values, with the offsets (y/c) values being the bounded design values. We use 16 parameters : 6 for the mean line and 10 for the thickness distribution. All aerofoils are analyzed using a panel method with 130 panels and at a Reynolds number (Re) of 2×10^6, across a range of flow angles of attack from $\alpha = -4\,\text{deg}$ to $\alpha = +17\,\text{deg}$.

The objective function is to minimize the drag coefficient (c_d) at two separate lift coefficients (c_l):
- $min(c_d)$ @ $c_l = 0.2$
- $min(c_d)$ @ $c_l = 1.0$

Problem constraints are:

$$c_{l_{min}} \leq -0.1.$$
$$-0.005 \leq [c_m \text{ @ } \alpha = 0] \leq +0.005 \text{ (pitching moment).}$$
$$c_{l_{max}} \geq 1.6.$$
$$12\%\tfrac{x}{c} \leq thickness_{max} \leq 13\%\tfrac{x}{c}.$$
$$15\%\tfrac{x}{c} \leq thickness_{position} \leq 35\%\tfrac{x}{c}.$$

All objectives and constraints are defined linearly, scaled by appropriate factors depending on their relative magnitude. Constraints are enforced by adding their value equally to both objectives, such that one solution with equal objective function values to another solution will be dominated by it if it violates any constraints.

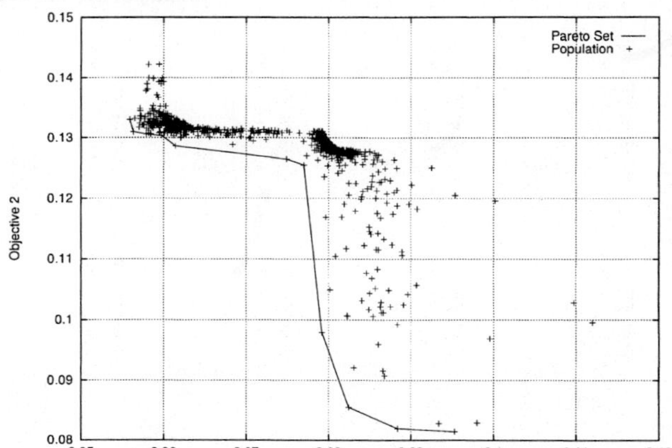

Figure 3. Final position of the front

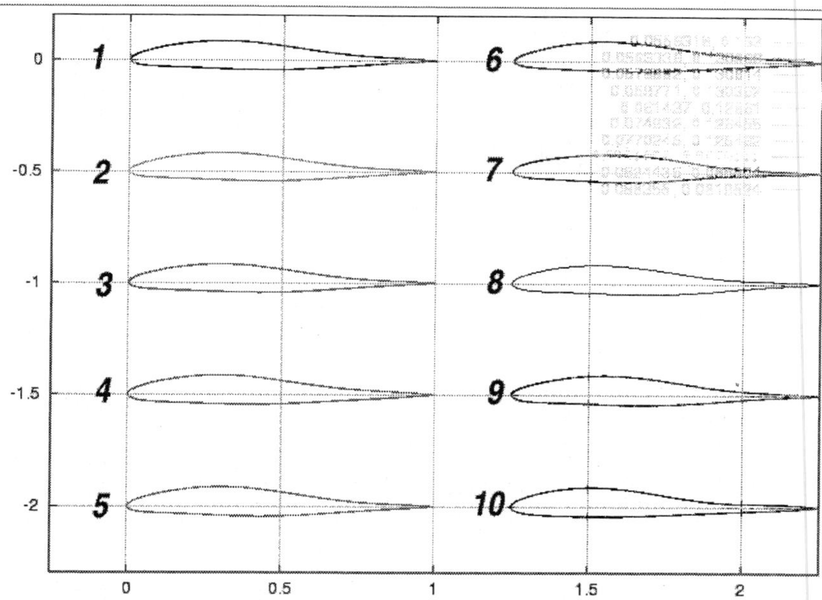

Figure 4. Sampling of the solutions

The example presented was run with 408168 function evaluations occurring in parallel, corresponding to 1200 steps (approx 7 hrs execution time). The final position of the population is shown in Figure 3, where *objective one* is $[c_d @ c_l = 0.2] \times 100$ and *objective two* is $[c_d @ c_l = 1.0] \times 100$. Ten aerofoils sampled at regular intervals from the Pareto set are shown in Figure 4. For example, aerofoil 1 corresponds to 0.056 and 0.133 for the respective values of *objective 1* and *objective 2*, aerofoil 5 to 0.06 and 0.128, and aerofoil 10 to 0.088 and 0.082. The shapes can be seen to vary gradually from one end of the set to the other. All aerofoils demonstrate the classic 'underbelly' or 'S' shape exhibited by aerofoils designed for low pitching moments.

2.5 Electromagnetics : Backscattering Minimization

Let us consider a reflecting obstacle R corresponding to an airfoil. Our aim is to minimize the Radar Cross Section (RCS) of R, which corresponds to the backscattered wave. We suppose that K active elements can be distributed among N possible sites on the surface of the airfoil. The problem consists in finding the optimal distribution of K active elements among N possible sites. Since each element can be either active or passive, a straightforward binary representation for the GA can suit very well the problem (Figure 5).

We consider that the BINACA0012 might be illuminated by either a -45° a $+45^o$ incident waves. The optimization task consists in finding the best distribution in order to cope as well as possible with both cases. The first

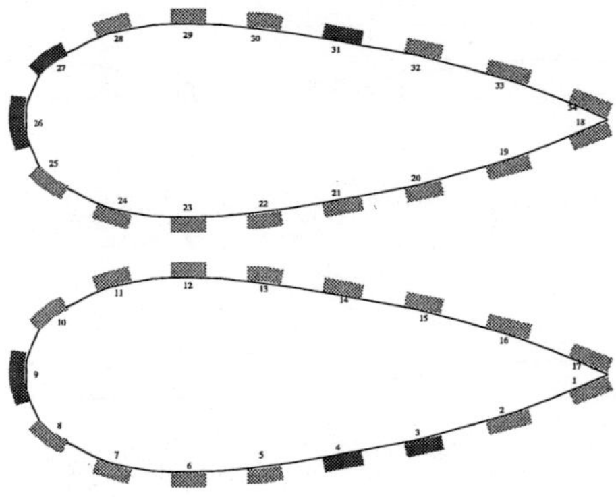

Figure 5. BINACA0012 with active and passive antennas

criteria c^1 is the RCS computed for $\alpha \in [-35^o, -55^o]$. The second criteria c^2 is the RCS computed for $\alpha \in [+35^o, +55^o]$. The problem consists in finding a distribution which is a compromise between both criteria c^1 and c^2.

The Pareto Front obtained after convergence is shown in Figure 6. Since there are 23 solutions on the Pareto Front, we have taken the Pareto solution for which $c^1 = c^2$ as an example. The optimal distribution for that solution is shown in Figure 7.

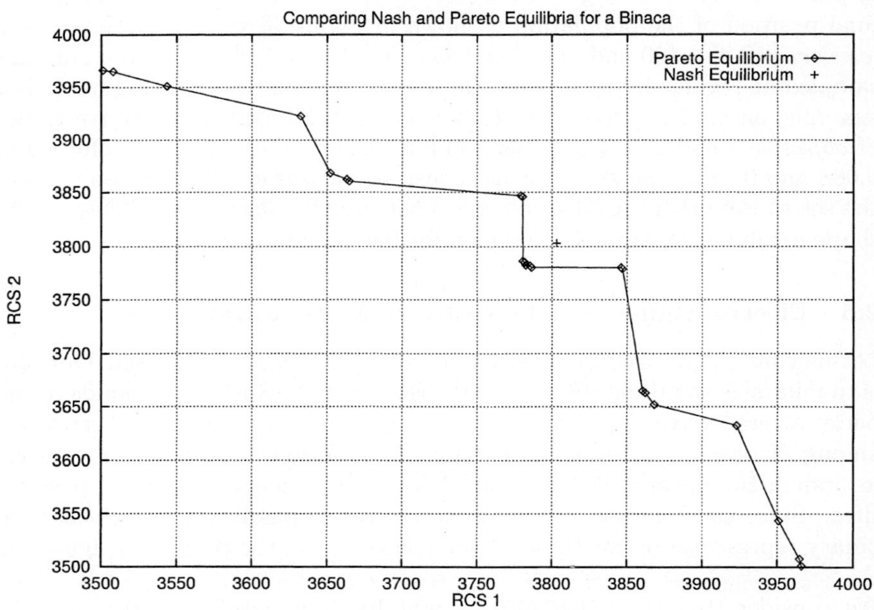

Figure 6. Pareto Front for 2 incident waves at -45° and $+45^o$

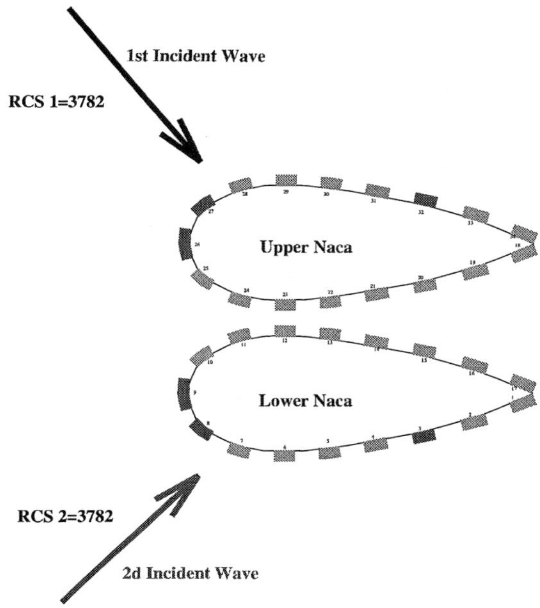

Figure 7. Optimal Distribution, BINACA0012 with $c^1 = c^2$

3 Hierarchical Genetic Algorithms

3.1 Hierarchical Topology

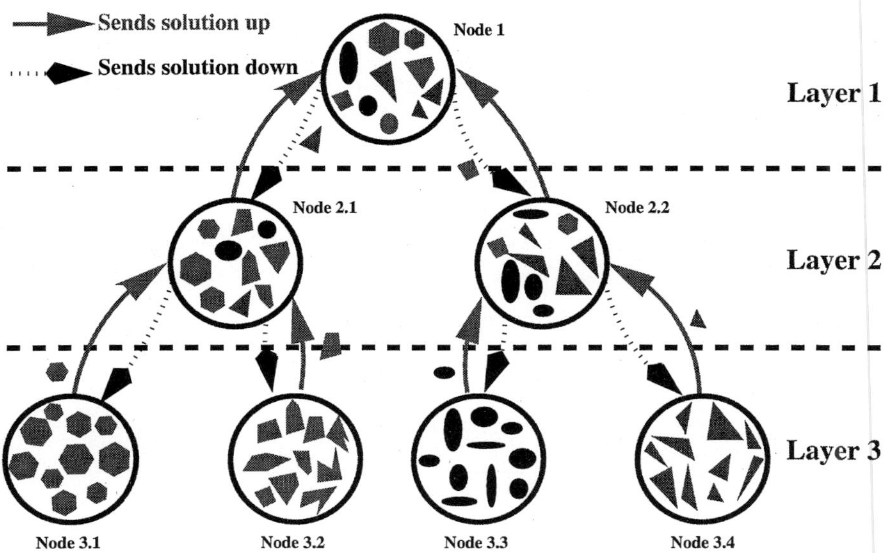

Figure 8. Hierarchical topology

HGAs are a particular approach we developed, based on Parallel GAs [12]. They use a hierarchical topology for the layout of the sub-populations (Figure 8). Solutions go up and down, and the nodes of each layer may have a different purpose, defined by their associated GA:

1. Top layer : refining solutions, by tuning the GA in a way that makes mutation take very small steps.
2. Intermediate layer : compromise between exploitation and exploration.
3. Bottom layer : full-on exploration. That means that the GA can make big leaps in the search space via a large mutation span.

3.2 Multiple Models

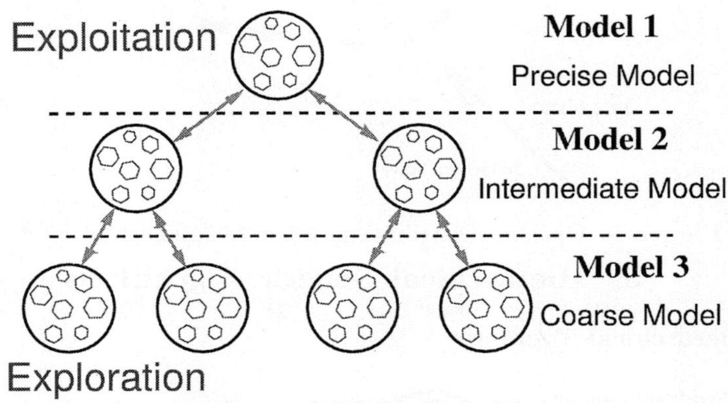

Figure 9. Multiple models

Each GA may have a different fitness function and can be use a different model for a given optimization problem. With this hierarchical structure, the top layer uses a very precise model – meaning a time-consuming solver. But at the same time, the sub-populations of the bottom layer do not need a very precise result, as their main goal is to explore and can use simple models and much fast solvers. Figure 9 shows how that can be done.

3.3 Application : Nozzle reconstruction

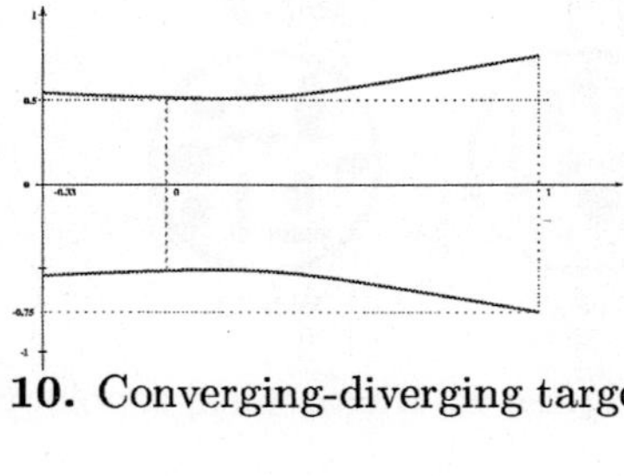

Figure 10. Converging-diverging target nozzle

Problem definition The strategy we just presented is applied to the reconstruction of the shape of a converging-diverging nozzle for a transonic flow involving a shock. To be a bit more explicit, for a given target nozzle (Figure 10), the corresponding Mach number distribution is computed using a quasi-steady one-dimensional approximation for the flow. The equations are solved by a time marching technique using a CUSP scheme with an iterative solver [13, 14]. Modifying the size of the grid changes the quality of the Mach number distribution, but it also greatly changes the CPU time needed by the CFD solver to converge. The CPU time is over 15 times larger for a grid size of 100 compared to a grid size of 25.

Single GA vs Hierarchical GA (inviscid) We first tested the problem with a single population GA. We then ran a HGA using a single model (Model 1) for all layers. Finally, we used a HGA with multiple models : Model 1 for the top layer (grid size 100), Model 2 (grid size 50) for the intermediate layer and Model 3 (grid size 25). Results for HGA with multiple models are shown in Figures 11 and 12.

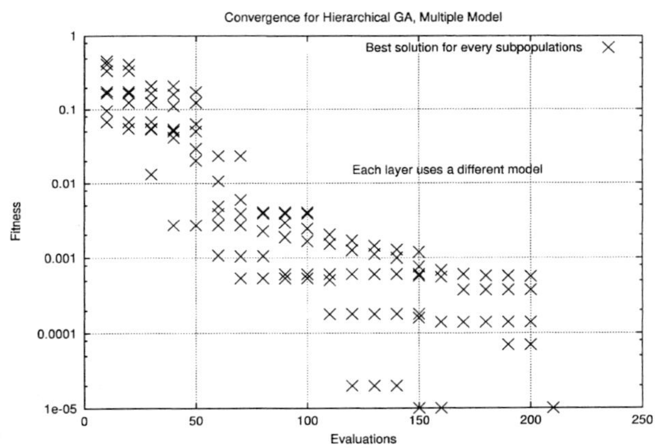

Figure 11. Convergence for Hierarchical GA Multiple Model

Table 1 sums up the performances for traditional GAs, HGAs and HGAs with multiple models. It focuses on the CPU time needed by each of these 3 approaches. It is clear that HGA with multiple models is the best answer to the dilemmas that were considered in the introduction. It is on average 3 times faster that either of the other approaches. For more details, see [15].

Hierarchical GA, Multiple Models and Multiple Solvers We add to the previous experiment a level of complexity by introducing a viscous solver. We mix both viscous and inviscid layers in the following way : we use a viscous solver for the top layer, and we use an inviscid solver for the 2 lower

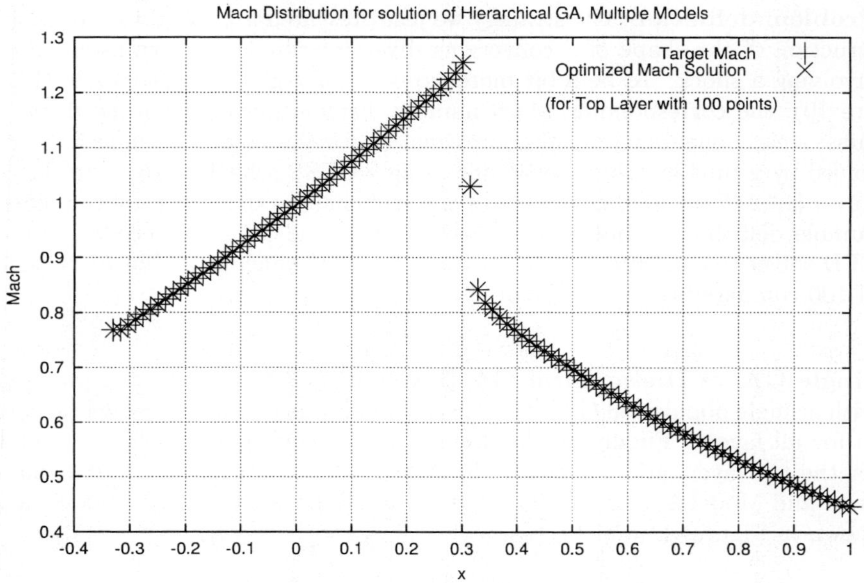

Figure 12. Mach Distribution for Hierarchical GA Multiple Model

Table 1. Inviscid Solver, Hierarchical GA Single GAs

Precision	Single GA Model M_1	HGA Model M_1	HGA M_1,M_2,M_3
10^{-3}	43mn	55mn	15mn
10^{-5}	1h47	1h28	32mn

layers.

For the mixed inviscid/viscous approach, Table 2 sums the results we obtained.

Table 2. Mixed Viscous/Inviscid Solver, Hierarchical GA

Precision	Inviscid M_1,M_2,M_3	Viscous M_1,M_2,M_3	Mixed M_1=Viscous, M_2 M_3=Inviscid
10^{-3}	23mn	48mn	32mn

This example clearly illustrate that GAs handle well approximate models within a hierarchical topology. It shows that using models of different complexity can significantly speed-up an optimization process.

4 Asynchronous Hierarchical Evolution Strategy (AHES)

This part is based on the Hierarchical Topology, but uses an Evolution Strategy instead of a GA, and uses an asynchronous approach to the parallel computations.

4.1 Description

- AHES only generates one candidate solution at a time and only re-incorporates one evaluated individual at a time, rather than an entire population every generation. Solutions can be generated and returned *out of order*. This allows the implementation of asynchronous fitness evaluation.
- AHES uses derandomised mutations : as the algorithm runs, it not only solves the problem, but it also learns *how to solve the problem.* A covariance matrix of the relationship between design parameters is evolved and compensates for possible poor axis scaling in the problem definition.

4.2 Asynchronous Evaluation

Evaluation of candidate solutions is performed in parallel, asynchronously across a network using PVM. The immediate advantages of this are the following:

- there is no waiting time (or *bottleneck*) for individuals to return. As soon as a solution is available, it is incorporated back into the process.
- This method also no longer stipulates an optimum number of processors that should be used. Every additional processor contributes to the solution speed.
- Solver nodes can be added or deleted from the optimization process *during execution.*

4.3 Nozzle Optimization Problem

This method is applied to the nozzle reconstruction already described in the previous section.

- *Single Population:* Single population of size 20, viscous solution over 100 grid points and forced to have synchronous behavior. This is an example of traditional EA optimization.
- *Hierarchical:* Hierarchical model, populations of size 10, viscous solution on all layers, 100-50-25 grid points by layer and forced to have synchronous behavior. This is an example of the current hierarchical approach.

– *Asynchronous:* Hierarchical model, populations of size 10, viscous solution on all layers, 100-50-25 grid points by layer and run ***asynchronously***. This is an asynchronous generalization of "hierarchical".
– *Full-Blown:* Hierarchical model, populations of size 10, viscous solution on the top layer and inviscid on lower layers, 100-50-25 grid points by layer and run ***asynchronously***. This is the current zenith of the method.

The averages and the standard deviations of the execution times and head node evaluations (over the nine node cluster) are:

	Time	Head Evals
Single Population	35 m 27 s ± 12 m 8 s	3383 ± 1137
Hierarchical	16 m 52 s ± 6 m 46 s	470 ± 149
Asynchronous	18 m 4 s ± 4 m 32 s	495 ± 123
Full-Blown	12 m 6 s ± 3 m 58 s	411 ± 112

5 Conclusion and Future

A series of numerical experiments has shown in this paper the benefit of coupling Evolutionary Algorithms with Game Theory (cooperative Pareto or competitive Nash games in particular) as well as Hierarchy for the solutions of multi-objective or multi-disciplinary CFD or CEM optimization problems.

Furthermore the efficiency of a new Asynchronous Hierarchical Parallel EA running on a distributed parallel environment like a cluster of PCs has been also verified for the capture of a global solution, or a Pareto set of solutions of multi-objective CFD problems.

The combined Games-EAs approach investigated here is a preliminary step towards distributed optimization for modern multi-disciplinary design under conflict or with cooperation. This interactive approach will require the set-up of real or virtual integration platforms with new programming languages specifically designed for distributed computing and the seamless treatment of multi-physics interfaces.

Acknowledgments

We would like to thank Dr H. Q. Chen, from the NUAA and Dr S. Dunn from DSTO for their collaboration on some parts of this work.

References

[1] D.E. Goldberg. *Genetic Algorithms in Search, Optimization, and Machine Learning.* Addison-Wesley, Reading, Mass., 1989.

[2] T. Bäck, D. B. Fogel, and H.-P. Schwefel. *Handbook of Evolutionary Computation.* Oxford University Press, New York, 1997.

[3] J. H. Holland. *Adaptation in Natural and Artificial Systems.* University of Michigan Press, Ann Arbor, 1975.

[4] Z. Michalewicz. *Genetic algorithms + data structures = evolution programs.* Artificial Intelligence. Springer-Verlag, New York, 1992.

[5] J. T. Richardson, M. R. Palmer, G. Liepins, and M. Hilliard. Some guidelines for genetic algorithms with penalty functions. In Schaffer [16], pages 191–195.

[6] J. von Neumann. Zur theorie der gesellschaftsspiele. *Math .Ann.*, 100:295–320, 1928.

[7] V. Pareto. *Cours d'Economie Politique.* Rouge, Lausanne, Switzerland, 1896.

[8] C. M. Fonseca and P. J. Fleming. An overview of evolutionary algorithms in multiobjective optimisation. In *Evolutionary Computation*, volume 3, pages 1–16, 1995.

[9] N. Srinivas and K. Deb. Multiobjective optimisation using non-dominated sorting in genetic algorithms. In *Evolutionary Computation 2 (3)*, pages 221–248, 1995.

[10] K. Deb and D. E. Goldberg. An investigation of niches and species formation in genetic function optimization. In Schaffer [16].

[11] J. Horn, N. Nafpliotis, and D. Goldberg. A niched pareto genetic algorithm for multiobjective optimization. In *Proceedings First IIIE Conference on Evolutionary Computation Symposium on the theory of Computing*, 1994.

[12] M. Sefrioui. *Algorithmes Evolutionnaires pour le calcul scientifique. Application l'electromagnetisme et la mcanique des fluides numriques.* PhD thesis, University Pierre et Marie Curie, Paris, Avril 1998.

[13] A. Jameson. Analysis and design of numerical schemes for gas dynamics, 2: Artificial diffusion and discrete shock structure. *Int. J. Comput. Fluid Dynamics*, 5:1–38, 1995.

[14] K. Srinivas. Computation of cascade flows by a modified cusp scheme. *Computational Fluid Dynamics Journal*, 8(2):285–295, 1999.

[15] M. Sefrioui and J. Periaux. A hierarchical genetic algorithm using multiple models for optimization. In *Parallel Problem Solving from Nature – PPSN 6*, Paris, September 2000. Springer.

[16] D. Schaffer, editor. *Proceedings of the 3d International Conference on Genetic Algorithms*, Los Altos, CA, 1989. Morgan Kaufmann.

Some Remarks on Fluid-Structure Interaction Problems in Case of Rigid Body plus Small Perturbations

CÉLINE GRANDMONT[1] and YVON MADAY[2]

[1] CEREMADE, Université de Paris Dauphine, place du Maréchal de Lattre de Tassigny, 75775 Paris cedex 16, France,
grandmont@ceremade.dauphine.fr,
WWW home page: http://www.ceremade.dauphine.fr/~grandmon/
[2] Université Pierre et Marie Curie, Laboratoire Jacques Louis Lions, 4, place Jussieu, 75252, Paris Cedex 05 France

Summary

For many problems of fluid–structure interaction, the stucture evolves freely in the fluid without being attached to any part of its boundary. The resulting motion of the structure can then be quite important but is mainly composed of a large rigid body motion to which is superposed a small elastic perturbation. Due to its size, this perturbation can be modeled by the linearized elasticity equation. The resulting problem leads to nonlinear partial differential equations that can be coupled to the Navier–Stokes equations and numerically tackled by the use of modal expensions. The aim of this paper is to derive this formulation under clear assumptions.

1 Introduction

Fluid–structure interaction problems have been the subject of quite intensive investigations both from the numerical and from the theoretical point of view. Despite this intensive research, a global existence result for the system of partial differential equations in the general case is still not under control; similarly there are not so many numerical simulations and all of them have limitations. From the theoretical point of view, many papers deal with motion of balls in infinite fluids ([17], [18]) most of the papers deal with rigid body motions and the time of existence of solutions is most often quite small (see [1], [2], [3], [11], [14], [15], [19], ...). In the 2D case and for a bounded domain a global existence result is proven in [16]. For results of existence of solution for fluid- elastic problem we refer to ([4], [7], [8]) in the non-stationary case and [10] for the stationary case. In order to increase the generality of the approach we consider in this paper the modelization of a quite often encountered situation where the structure evolves freely in a bounded fluid without having any fixed or clamped part. The motion can then be quite large without involving large deformations. In this case the motion can be seen as the superposition of a large rigid body motion and a small deformation. This deformation is naturally well appreciated in a reference frame that moves with the center of mass of each body, and the equation that mimics the derivation of the linear elasticity equations can be derived. The geometrically nonlinear model that we propose follows the lines presented by B. Fraeijs de Veubeyke in [9], we derive it in a rigourous way that, we hope, will allow to clarify its application. This type of approach has also been used in finite element structural analysis (see e.g. [6]).

Thanks to this decomposition we are currently working on the extension of the ideas presented in [18], [19] that deal with large rigid body motions to this more generic situation. This will be the subject of a forthcoming paper. In what follows, we present the modelization of the problem under consideration. The resulting problem is nevetheless nonlinear, actually highly nonlinear, since the model of rigid body plus small perturbations is already nonlinear, the nonlinearity of the fluid (Navier–Stokes equations) has to be added and on top of this the interaction itself (involving the domain of the fluid as an unknown).

We show that this problem possesses *a priori* stability properties that allow us to be confident in its well posedness, which is still a subject of research. The results quoted here extend those in two dimensions presented in [13].

2 Decomposition of the L^2-vector fields

2.1 Decomposition of the deformations

In this subsection we follow the same lines as in [9]. Let us assume that the elastic body is represented by a reference configuration denoted by $\mathcal{B}$ which is a regular bounded open subset of $\mathbb{R}^3$ with a Lispchitz-continuous boundary. The center of mass is denoted by $\mathbf{G}$ and is assumed to be the origin of $\mathbb{R}^3$. The behaviour of the body is represented by the knowledge of the deformation $\mathbf{\Phi}$ that is a vector field attached to any point $\boldsymbol{\xi}$ of $\mathcal{B}$ and depends on time. This deformation $\mathbf{\Phi}(\boldsymbol{\xi}, t)$ represents the position of a running point $\boldsymbol{\xi}$ at time t.

In this section we want to extract from $\mathbf{\Phi}$ the main information that is the rigid body part. In order to simplify the notation, the time dependancy that is not relevant here is dropped out, it will appear again in the following sections. We thus want to decompose $\mathbf{\Phi}$ as $\mathbf{\Phi}(\boldsymbol{\xi}) = \boldsymbol{\tau} + \mathcal{R}\overrightarrow{\mathbf{G}\boldsymbol{\xi}} + \mathcal{R}\mathbf{d}(\boldsymbol{\xi})$ where $\boldsymbol{\tau}$ represents a translation vector, $\mathcal{R}$ represents a rotation operator and $\mathbf{d}$ represents the remaining part of the deformation. This decomposition is certainly not unique but we want that "$\boldsymbol{\tau} + \mathcal{R}$" represents the main part of the motion, in the sense that $\mathbf{d}$ is small. We thus want to characterize the decomposition by having

$$\mathcal{J}(\boldsymbol{\tau}, \mathcal{R}) = \int_{\mathcal{B}} \mathbf{d}^2 = \int_{\mathcal{B}} \left(\mathbf{\Phi}(\boldsymbol{\xi}) - \boldsymbol{\tau} - \mathcal{R}\overrightarrow{\mathbf{G}\boldsymbol{\xi}}\right)^2 \tag{1}$$

as small as possible.

It is an easy matter to realize that

$$\int_{\mathcal{B}} \left(\mathbf{\Phi} - \boldsymbol{\tau} - \mathcal{R}\overrightarrow{\mathbf{G}\xi}\right)^2 = \int_{\mathcal{B}} \left(\mathbf{\Phi} - \boldsymbol{\tau}\right)^2 - 2\int_{\mathcal{B}} \left(\mathbf{\Phi} \cdot \mathcal{R}\overrightarrow{\mathbf{G}\xi}\right) + \int_{\mathcal{B}} \|\overrightarrow{\mathbf{G}\xi}\|^2 \quad (2)$$

by definition of the center of mass ($\int_{\mathcal{B}} \overrightarrow{\mathbf{G}\xi} = 0$). Thus it appears that $\mathcal{J}(\boldsymbol{\tau}, \mathcal{R}) = \mathcal{J}_1(\boldsymbol{\tau}) + \mathcal{J}_2(\mathcal{R}) + C$ (where C is a generic constant) so that $\mathcal{J}$ admits at least one minimum. Indeed $\mathcal{J}_1$ is continuous and strictly convex coercive while $\mathcal{J}_2$ is continuous over the set of all rotation which is a compact set. In addition the minimizing translation appears immediately as being the average of $\mathbf{\Phi}$, $\boldsymbol{\tau}_* = \frac{1}{|\mathcal{B}|}\int_{\mathcal{B}} \mathbf{\Phi}$. This implies that the corresponding $\mathbf{d}$ satisfies $\int_{\mathcal{B}} \mathbf{d} = 0$. The definition of the rotation is slightly more complex and we follow here the nonsingular definition known as the Rodrigues-Hamilton representation (see [9]) where the rotation is represented as $\mathcal{R} = \mathrm{I}_3 + 2\beta\mathbf{B} + 2\mathbf{B}^2$ where $\mathbf{B}$ is the antisymetric matrix representing the operator $\mathbf{b}\wedge$ with $\mathbf{b} \in \mathbb{R}^3$. This decomposition of the rotation is possible with $\beta^2 + \|\mathbf{b}\|^2 = 1$, note that $\mathbf{b}$ is oriented along the axis of rotation and $\beta = \pm\cos\frac{\theta}{2}$ where θ is the angle through which the body has rotated. With this representation, $\mathcal{J}_2$ can be written

$$\mathcal{J}_2(\mathcal{R}) = -2\int_{\mathcal{B}} \mathbf{\Phi} \cdot \overrightarrow{\mathbf{G}\xi} + 4\beta\int_{\mathcal{B}} \mathbf{b} \cdot (\mathbf{\Phi} \wedge \overrightarrow{\mathbf{G}\xi}) - 4\int_{\mathcal{B}} \mathbf{\Phi} \cdot \left(\mathbf{b} \wedge (\mathbf{b} \wedge \overrightarrow{\mathbf{G}\xi})\right) \quad (3)$$

the last quadratic part can be further prepared as follows

$$\int_{\mathcal{B}} \mathbf{\Phi} \cdot \left(\mathbf{b} \wedge (\mathbf{b} \wedge \overrightarrow{\mathbf{G}\xi})\right) = -\mathbf{b}^t \Big(\int_{\mathcal{B}} \overrightarrow{\mathbf{G}\xi} \cdot \mathbf{\Phi}\ \mathrm{I}_3\Big)\mathbf{b} + \frac{1}{2}\mathbf{b}^t \Big(\int_{\mathcal{B}} \mathbf{\Phi} \otimes \overrightarrow{\mathbf{G}\xi} + \int_{\mathcal{B}} \overrightarrow{\mathbf{G}\xi} \otimes \mathbf{\Phi}\Big)\mathbf{b} \quad (4)$$

so that, up to the constant term $-2\int_{\mathcal{B}} \mathbf{\Phi} \cdot \overrightarrow{\mathbf{G}\xi}$, $\mathcal{J}_2(\mathcal{R})$ appears as a quadratic function in β and $\mathbf{b}$. This expression has to be minimized under the constraint $\beta^2 + \|\mathbf{b}\|^2 = 1$. The corresponding necessary optimality conditions read

$$2\mathbf{b}\int_{\mathcal{B}} \mathbf{\Phi} \wedge \overrightarrow{\mathbf{G}\xi} - \lambda\beta = 0, \quad (5)$$

$$4\int_{\mathcal{B}} (\mathbf{\Phi} \cdot \overrightarrow{\mathbf{G}\xi})\mathbf{b} - 2\Big(\int_{\mathcal{B}} \mathbf{\Phi} \otimes \overrightarrow{\mathbf{G}\xi} + \int_{\mathcal{B}} \overrightarrow{\mathbf{G}\xi} \otimes \mathbf{\Phi}\Big)\mathbf{b} - \lambda\mathbf{b} = 0. \quad (6)$$

where λ is the Lagrange multiplier associated with the normalization constraint. This is an eigenvalue problem more readable as

$$\mathcal{A}(\mathbf{\Phi})\begin{pmatrix}\mathbf{b}\\ \beta\end{pmatrix} := \begin{pmatrix} M & m \\ m^t & O\end{pmatrix}\begin{pmatrix}\mathbf{b}\\ \beta\end{pmatrix} = \frac{\lambda}{2}\begin{pmatrix}\mathbf{b}\\ \beta\end{pmatrix} \quad (7)$$

where $M = 2\int_{\mathcal{B}} (\mathbf{\Phi} \cdot \overrightarrow{\mathbf{G}\xi})\mathrm{I}_3 - \int_{\mathcal{B}}\left(\mathbf{\Phi} \otimes \overrightarrow{\mathbf{G}\xi} + \overrightarrow{\mathbf{G}\xi} \otimes \mathbf{\Phi}\right)$ and $m = \int_{\mathcal{B}} \mathbf{\Phi} \wedge \overrightarrow{\mathbf{G}\xi}$. Let $(\mathbf{b}_*, \beta_*, \lambda_*)$ be a solution of the eigenvalue problem (7) and $\mathcal{R}_*$ be the associated rotation, an easy calculation then leads to

$$\mathcal{J}_2(\mathcal{R}_*) = -2\int_{\mathcal{B}} \mathbf{\Phi} \cdot \overrightarrow{\mathbf{G}\xi} + \lambda_* \quad (8)$$

which means that the rotation $\mathcal{R}_*$ will effectively correspond to a minimum of $\mathcal{J}_2$ and will be unique if the minimal eigenvalue of problem (7) is simple. This gives a condition on $\mathbf{\Phi}$ which says that $\mathcal{A}(\mathbf{\Phi})$ has its minimal eigenvalue simple. The problem comes now to determine the conditions under which problem (7) has a unique minimal eigenvalue. In order to be relevant, this condition has to be expressed in terms of $\mathbf{d}$.

In what follows, we assume that $\mathcal{R}_*$ and $(\mathbf{b}_*, \beta_*, \lambda_*)$ correspond to a minimal eigenvalue of problem (7). We note that any rotation $\mathcal{R}$ can be written as

$$\mathcal{R} = \widetilde{\mathcal{R}}\mathcal{R}_* \tag{9}$$

where $\widetilde{\mathcal{R}}$ is also a rotation. Let us write $\mathcal{J}_2$ in terms of $\widetilde{\mathcal{R}}$, this leads to

$$\widetilde{\mathcal{J}_2}(\widetilde{\mathcal{R}}) = -2\int_{\mathcal{B}}\left(\mathbf{\Phi}\cdot\widetilde{\mathcal{R}}\mathcal{R}_*\overrightarrow{\mathbf{G}\xi}\right) \tag{10}$$

that we try to minimize. It is already known that $\widetilde{\mathcal{R}} = \mathrm{I}_3$ is a solution to this problem, we want to find now a sufficient condition so that this solution is unique. The optimality conditions for this problem, in terms of $(\tilde{\beta}, \tilde{\mathbf{b}})$ (where $(\tilde{\beta}, \tilde{\mathbf{b}})$ are associated to $\widetilde{\mathcal{R}}$) leads to the eigenvalue problem

$$\begin{pmatrix} \tilde{M} & \tilde{m} \\ \tilde{m}^t & O \end{pmatrix}\begin{pmatrix} \tilde{\mathbf{b}} \\ \tilde{\beta} \end{pmatrix} = \frac{\tilde{\lambda}}{2}\begin{pmatrix} \tilde{\mathbf{b}} \\ \tilde{\beta} \end{pmatrix} \tag{11}$$

where now $\tilde{M} = 2\int_{\mathcal{B}}(\mathbf{\Phi}\cdot\mathcal{R}_*\overrightarrow{\mathbf{G}\xi})\mathrm{I}_3 - \int_{\mathcal{B}}\left(\mathbf{\Phi}\otimes\mathcal{R}_*\overrightarrow{\mathbf{G}\xi} + \mathcal{R}_*\overrightarrow{\mathbf{G}\xi}\otimes\mathbf{\Phi}\right)$ and similarly $\tilde{m} = \int_{\mathcal{B}}\mathbf{\Phi}\wedge\mathcal{R}_*\overrightarrow{\mathbf{G}\xi}$. Let $(\tilde{\mathbf{b}}_*, \tilde{\beta}_*, \tilde{\lambda}_*)$ be a solution of the eigenvalue problem (11) and $\widetilde{\mathcal{R}}_*$ be the associated rotation, an easy calculation then leads to

$$\widetilde{\mathcal{J}_2}(\widetilde{\mathcal{R}}_*) = -2\int_{\mathcal{B}}\mathbf{\Phi}\cdot\overrightarrow{\mathbf{G}\xi} + \lambda_* + \tilde{\lambda}_* = \mathcal{J}_2(\mathcal{R}_*) + \tilde{\lambda}_* \tag{12}$$

Thus $\tilde{\lambda}_*$ is positive. By construction, $\tilde{\lambda}_* = 0$ is a solution to this problem, it corresponds to $\widetilde{\mathcal{R}}_* = \mathrm{I}_3$, $\tilde{\beta}_* = 1$ and $\tilde{\mathbf{b}}_* = 0$. It results that $\tilde{m} = \int_{\mathcal{B}}\mathbf{\Phi}\wedge\mathcal{R}_*\overrightarrow{\mathbf{G}\xi} = 0$, which can also be written $\int_{\mathcal{B}}\mathbf{d}\wedge\overrightarrow{\mathbf{G}\xi} = 0$, where $\mathbf{d} = \mathcal{R}_*^t\left(\mathbf{\Phi} - \boldsymbol{\tau}_*\right) - \overrightarrow{\mathbf{G}\xi}$. The simplicity of this trivial eigenvalue is then equivalent to the fact that $\tilde{M}$ is non singular. This matrix is symmetric non negative (since 0 is the smallest eigenvalue of (11)) and an easy calculation gives from the decomposition $\mathbf{\Phi} = \boldsymbol{\tau}_* + \mathcal{R}_*\overrightarrow{\mathbf{G}\xi} + \mathcal{R}_*\mathbf{d}$ (which defines $\mathbf{d}$)

$$\tilde{M} = 2\left(\int_{\mathcal{B}}(\overrightarrow{\mathbf{G}\xi}+\mathbf{d})\cdot\overrightarrow{\mathbf{G}\xi}\right)\mathrm{I}_3 - \mathcal{R}_*\left(\int_{\mathcal{B}}(\overrightarrow{\mathbf{G}\xi}+\mathbf{d})\otimes\overrightarrow{\mathbf{G}\xi}+\overrightarrow{\mathbf{G}\xi}\otimes(\overrightarrow{\mathbf{G}\xi}+\mathbf{d})\right)\mathcal{R}_*^t \tag{13}$$

and $\tilde{M} = \mathcal{R}_*\mathcal{M}(\mathbf{d})\mathcal{R}_*^t$ where

$$\mathcal{M}(\mathbf{d}) = 2\left(\int_{\mathcal{B}}(\overrightarrow{\mathbf{G}\xi}+\mathbf{d})\cdot\overrightarrow{\mathbf{G}\xi}\right)\mathrm{I}_3 - \left(\int_{\mathcal{B}}(\overrightarrow{\mathbf{G}\xi}+\mathbf{d})\otimes\overrightarrow{\mathbf{G}\xi}+\overrightarrow{\mathbf{G}\xi}\otimes(\overrightarrow{\mathbf{G}\xi}+\mathbf{d})\right) \tag{14}$$

It is readily checked that, for $\mathbf{d} = 0$, $\mathcal{M}(\mathbf{d})$ is equal to $2\mathcal{I}$ where $\mathcal{I}$ is the rotation inertia matrix of $\mathcal{B}$, so that, for $\mathbf{d}$ small enough, $\tilde{M}$ is positive definite. This leads to define the sets

$$\mathbf{X}_0 = \left\{ \mathbf{\Phi} \in L^2(\mathcal{B})^3; \quad \mathcal{A}(\mathbf{\Phi}) \text{ has a simple minimal eigenvalue} \right\} \tag{15}$$

$$\mathcal{E}_0 = \left\{ \mathbf{d} \in L^2(\mathcal{B})^3; \quad \int_{\mathcal{B}} \mathbf{d} = 0, \quad \int_{\mathcal{B}} \mathbf{d} \wedge \overrightarrow{\mathbf{G}\xi} = 0 \right\} \tag{16}$$

$$\mathbf{Y}_0 = \left\{ \mathbf{d} \in \mathcal{E}_0, \quad \mathcal{M}(\mathbf{d}) \text{ is positive definite} \right\} \tag{17}$$

$$\mathbf{Z}_0 = \mathbb{R}^3 \times \mathcal{O}_3 \times \mathbf{Y}_0 \tag{18}$$

where $\mathcal{O}_3$ is the set of all rotations in $\mathbb{R}^3$. We have just proven that to any $\mathbf{\Phi} \in \mathbf{X}_0$ the minimization of $\mathcal{J}$ allows us to associate a unique triplet $(\boldsymbol{\tau}, \mathcal{R}, \mathbf{d})$ in $\mathbf{Z}_0$. Reciprocally, let $(\boldsymbol{\tau}, \mathcal{R}, \mathbf{d})$ in $\mathbf{Z}_0$ be given, the previous discussion over the equivalence between $\tilde{M}$ and $\mathcal{M}(\mathbf{d})$ allows us to state that $\mathbf{\Phi} = \boldsymbol{\tau} + \mathcal{R}\overrightarrow{\mathbf{G}\xi} + \mathcal{R}\mathbf{d}$ belongs to $\mathbf{X}_0$. We thus have established the following

Proposition 1. *The mapping H from $\mathbf{Z}_0$ onto $\mathbf{X}_0$ defined by*

$$H(\boldsymbol{\tau}, \mathcal{R}, \mathbf{d}) = \boldsymbol{\tau} + \mathcal{R}\overrightarrow{\mathbf{G}\xi} + \mathcal{R}\mathbf{d} \tag{19}$$

is a continuous one to one mapping. In addition, H is continuous from $\mathbb{R}^3 \times \mathcal{O}_3 \times (\mathbf{Y}_0 \cap H^s(\mathcal{B})^3)$ onto $\mathbf{X}_0 \cap H^s(\mathcal{B})^3$ for any real number $s \geq 0$.

We are going to prove now that this mapping is a $\mathcal{C}^1$-diffeomorphism.

2.2 Decomposition of the velocities

The decomposition of L^2–vector fields that we have introduced in the previous subsection is well suited for the deformations of the structure but not for the velocities that are the derivatives with respect to time of the deformations. Indeed, it can be established that the derivative of H satisfies

$$\mathrm{D}H(\boldsymbol{\tau}, \mathcal{R}, \mathbf{d}).(\overline{\boldsymbol{\tau}}, \overline{\mathcal{R}}, \overline{\mathbf{d}}) = \overline{\boldsymbol{\tau}} + \overline{\Theta} \wedge \mathcal{R}(\overrightarrow{\mathbf{G}\xi} + \mathbf{d}) + \mathcal{R}\overline{\mathbf{d}} \tag{20}$$

with

$$\overline{\Theta} = -2\overline{\beta}\mathbf{d} + 2\beta\overline{\mathbf{d}} + 2\mathbf{d} \wedge \overline{\mathbf{d}}. \tag{21}$$

We are now in position to prove that $\mathrm{D}H(\boldsymbol{\tau}, \mathcal{R}, \mathbf{d})$ is an isomorphism. Starting from the direct orthogonal decomposition with obvious notations

$$[L^2(\mathcal{B})]^3 = <\boldsymbol{\tau}> \perp <\Theta \wedge \overrightarrow{\mathbf{G}\xi}> \perp <\tilde{\mathbf{d}}> \tag{22}$$

for any $\boldsymbol{\tau}$ and Θ in $\mathbb{R}^3$ and any $\tilde{\mathbf{d}} \in \mathcal{E}_0$, it is an easy matter to get

$$[L^2(\mathcal{B})]^3 = <\boldsymbol{\tau}> \perp <\Theta \wedge \mathcal{R}\overrightarrow{\mathbf{G}\xi}> \perp <\mathcal{R}\tilde{\mathbf{d}}> \tag{23}$$

for any rotation $\mathcal{R}$. We remark now that

$$\int_{\mathcal{B}} (\Theta \wedge (\mathcal{R}\overrightarrow{\mathbf{G}\boldsymbol{\xi}} + \mathbf{d})) \cdot (\Theta \wedge \mathcal{R}\overrightarrow{\mathbf{G}\boldsymbol{\xi}}) = \Theta^t \tilde{M} \Theta$$

with $\tilde{M}$ introduced in (13) which means that this scalar product is non zero (even strictly positive) for any $\Theta \neq 0$ whenever $\mathbf{d} \in \mathcal{E}_0$. This proves the decomposition

$$[L^2(\mathcal{B})]^3 = < \boldsymbol{\tau} > \oplus < \Theta \wedge \mathcal{R}(\overrightarrow{\mathbf{G}\boldsymbol{\xi}} + \mathbf{d}) > \oplus < \mathcal{R}\tilde{\mathbf{d}} > \tag{24}$$

hence the isomorphism of $\mathrm{D}H(\boldsymbol{\tau}, \mathcal{R}, \mathbf{d})$ from $\mathbb{R}^3 \times \mathcal{O}^3 \times \mathcal{E}_0$ onto the set $[L^2(\mathcal{B})]^3$. Next, we remark that

$$\int_{\mathcal{B}} (\overline{\Theta} \wedge \mathcal{R}(\overrightarrow{\mathbf{G}\boldsymbol{\xi}} + \mathbf{d})) \wedge \mathcal{R}\overrightarrow{\mathbf{G}\boldsymbol{\xi}} = \Big(\int_{\mathcal{B}} (\overrightarrow{\mathbf{G}\boldsymbol{\xi}} + \mathbf{d}) \cdot \overrightarrow{\mathbf{G}\boldsymbol{\xi}}\Big)\overline{\Theta} - \int_{\mathcal{B}} (\mathcal{R}(\overrightarrow{\mathbf{G}\boldsymbol{\xi}} + \mathbf{d}) \otimes \mathcal{R}\overrightarrow{\mathbf{G}\boldsymbol{\xi}})\overline{\Theta}$$

Let us set

$$B = \Big(\int_{\mathcal{B}} (\overrightarrow{\mathbf{G}\boldsymbol{\xi}} + \mathbf{d}) \cdot \overrightarrow{\mathbf{G}\boldsymbol{\xi}}\Big)\mathrm{I}_3 - \int_{\mathcal{B}} \mathcal{R}(\overrightarrow{\mathbf{G}\boldsymbol{\xi}} + \mathbf{d}) \otimes \mathcal{R}\overrightarrow{\mathbf{G}\boldsymbol{\xi}}$$

we remark easily that $B + B^t = \tilde{M}$, hence, for $\mathbf{d} \in \mathcal{E}_0$, it is invertible. For any $\mathbf{v}$ in $[L^2(\mathcal{B})]^3$, the decomposition

$$\mathbf{v} = \overline{\boldsymbol{\tau}} + \overline{\Theta} \wedge \mathcal{R}(\overrightarrow{\mathbf{G}\boldsymbol{\xi}} + \mathbf{d}) + \mathcal{R}\overline{\mathbf{d}}$$

suggested by (24) is thus achieved in the following way

$$\begin{cases} \overline{\boldsymbol{\tau}} = \displaystyle\int_{\mathcal{B}} \mathbf{v} \\ \overline{\Theta} = B^{-1}\Big(\int_{\mathcal{B}} \mathbf{v} \wedge \mathcal{R}\overrightarrow{\mathbf{G}\boldsymbol{\xi}}\Big) \\ \overline{\mathbf{d}} = \mathcal{R}^{-1}\Big(\mathbf{v} - \overline{\boldsymbol{\tau}} - \overline{\Theta} \wedge \mathcal{R}\overrightarrow{\mathbf{G}\boldsymbol{\xi}}\Big) \end{cases} \tag{25}$$

and $\overline{\mathcal{R}}$ is recovered from (21). This allows us to state that $\mathrm{D}H(\boldsymbol{\tau}, \mathcal{R}, \mathbf{d})$ is actually a bicontinuous mapping.

3 Modelization and analysis of the structure

3.1 Derivation of the equations for the stucture

From now on we introduce back the dependancy in time of the displacement. Let $T > 0$ be given, for any $t \in [0, T]$, we derive from Proposition 1 the existence of a unique element of $\mathbf{Z}_0$ such that $\boldsymbol{\Phi}(\boldsymbol{\xi}, t) = \boldsymbol{\tau}(t) + \mathcal{R}(t)\overrightarrow{\mathbf{G}\boldsymbol{\xi}} + \mathcal{R}(t)\mathbf{d}(\boldsymbol{\xi}, t)$. Let $\mathbf{u} = \boldsymbol{\Phi} - \mathrm{I}_3$ be the associated displacement. We now recall the main steps of the modelling leading to the equations satisfied by $\mathbf{u}$ in case the material is assumed to be a Saint Venant-Kirchhoff hyperelastic media. For such a media, the stored energy function takes the form

$$W(\mathbf{u}) = \check{W}(\mathbf{E}) = \frac{\lambda}{2}\Big[\mathrm{tr}(\mathbf{E})\Big]^2 + \mu \mathrm{tr}\Big[\mathbf{E}^2\Big]$$

where $\mathbf{E}(\mathbf{u}) = (\nabla \mathbf{u} + \nabla \mathbf{u}^t + \nabla \mathbf{u}^t \nabla \mathbf{u})/2$ and λ and μ are the Lamé constants. It is immediate to note that $\mathbf{E}(\mathbf{u}) = \mathbf{E}(\mathbf{d})$. Taking now into account the assumption that $\mathbf{d}$ is small, we linearize $\mathbf{E}(\mathbf{d})$ at the first order and replace it by $\overline{\varepsilon}(\mathbf{d}) = (\nabla \mathbf{d} + \nabla \mathbf{d}^t)$, this leads, as is more detailled in [13] to the following weak formulation for the equations modeling the displacement $\mathbf{u}$ (associated to the perturbation $\mathbf{d}$)

$$\begin{cases} \forall\ \mathbf{v} \text{ regular enough, with } \overline{\mathbf{d}} \text{ uniquely defined by (25)} \\ \rho_{\mathcal{B}} \int_{\mathcal{B}} (\partial_{tt}\mathbf{u}) \cdot \mathbf{v} + \int_{\mathcal{B}} \overline{\sigma}(\mathbf{d}) : \overline{\varepsilon}(\overline{\mathbf{d}}) = \int_{\mathcal{B}} \mathbf{f} \cdot \mathbf{v} + \int_{\partial \mathcal{B}} \mathbf{g} \cdot \mathbf{v}, \end{cases} \tag{26}$$

where $\overline{\sigma}(\mathbf{d}) = \lambda \mathrm{tr}(\overline{\varepsilon}(\mathbf{d}))\mathrm{I}_2 + 2\mu\overline{\varepsilon}(\mathbf{d})$, $\rho_{\mathcal{B}}$ is the density of the body, $\mathbf{f}$ is the prescribed body force and $\mathbf{g}$ is the exterior force applied to the boundary of $\mathcal{B}$.

Remark 1. It is important to note that this equation is nonlinear since $\mathbf{d}$ depends nonlinearly on $\mathbf{u}$ and moreover that the test function $\overline{\mathbf{d}}$ is attached to $\mathbf{v}$ also through a nonlinear way involving the data of $\mathbf{d}$.

3.2 Discretization of the stucture and existence results

Following [13], we are able to prove that this problem, associated with proper initial conditions has a unique solution at least as long as the associated $\mathbf{d}$ belongs to $\mathbf{Y}_0$. We refer to [12] for more details about the proof. The main lines are first to obtain appropriate classical *a priori* energy estimates by taking $\mathbf{v} = \partial_t \mathbf{u} = \dot{\tau} + \omega \wedge \mathcal{R}(\boldsymbol{G\xi} + \mathbf{d}) + R\partial_t \mathbf{d}$. The next line consists in defining an appropriate finite dimensional version of (26). The choice of the first eigenfunctions of the operator of linearized elasticity ($-\mathrm{div}(\overline{\sigma})$) comes actually in mind for at least two reasons. First since this is a positive self-adjoint operator from L^2 into L^2 with bounded compact inverse from $\mathcal{E}_0$ into $\mathcal{E}_0$, so that its eigenfunctions form a basis of $\mathcal{E}_0$, second because this is the basis that can actually be used for simulations and is known as the representation of the structure following the modal synthesis approach. It is then natural to express system (26) in terms of $(\boldsymbol{\tau}, \mathcal{R}, \mathbf{d})$. The equations are

$$\left\{\begin{aligned}
& m\ddot{\boldsymbol{\tau}} = \int_{\mathcal{B}} \mathbf{f}, \\
& \dot{\mathcal{R}} = \boldsymbol{\omega} \wedge \mathcal{R}, \\
& \mathcal{R}\mathcal{I}\mathcal{R}^t\dot{\boldsymbol{\omega}} - \int_{\mathcal{B}} (\boldsymbol{\omega} \wedge \mathcal{R}(\boldsymbol{G\xi} + \mathbf{d}))(\boldsymbol{\omega} \cdot \mathcal{R}(\boldsymbol{G\xi} + \mathbf{d})) \\
& \qquad +2\int_{\mathcal{B}} \Big(\partial_t \mathbf{d} \cdot (\boldsymbol{G\xi} + \mathbf{d})\Big)\boldsymbol{\omega} - 2\int_{\mathcal{B}} \Big(\mathcal{R}\partial_t \mathbf{d} \otimes \mathcal{R}(\boldsymbol{G\xi} + \mathbf{d})\Big) \cdot \boldsymbol{\omega} \\
& \qquad\qquad - \int_{\mathcal{B}} \mathcal{R}(\partial_{tt}\mathbf{d} \wedge \mathbf{d}) = - \int_{\mathcal{B}} \mathbf{f} \wedge \mathcal{R}(\boldsymbol{G\xi} + \mathbf{d})), \\
& \int_{\mathcal{B}} \partial_{tt}\mathbf{d} \cdot \bar{\mathbf{d}} + 2\int_{\mathcal{B}} \mathcal{R}(\partial_t \mathbf{d} \wedge \bar{\mathbf{d}}) \cdot \boldsymbol{\omega} + \int_{\mathcal{B}} \Big(\mathcal{R}(\boldsymbol{G\xi} + \mathbf{d}) \otimes \mathcal{R}\bar{\mathbf{d}}\boldsymbol{\omega}) \cdot \boldsymbol{\omega} \\
& \qquad -\|\boldsymbol{\omega}\|^2 \int_{\mathcal{B}} \bar{\mathbf{d}} \cdot (\boldsymbol{G\xi} + \mathbf{d}) + \mathcal{R}\int_{\mathcal{B}} (\mathbf{d} \wedge \bar{\mathbf{d}}) \cdot \dot{\boldsymbol{\omega}} + \int_{\mathcal{B}} \overline{\sigma}(\mathbf{d}) : \overline{\varepsilon}(\bar{\mathbf{d}}) \\
& \qquad\qquad = \int_{\mathcal{B}} \mathbf{f} \cdot \mathcal{R}\bar{\mathbf{d}} \qquad\qquad \forall \bar{\mathbf{d}} \in \mathcal{E}_0 \cap H^1(\mathcal{B})^3,
\end{aligned}\right. \tag{27}$$

where m is the mass of the body $\mathcal{B}$. The last equation corresponds to the equations of linearized elasticity but written in a moving frame, so additional terms such as the Coriolis term appear. Those equations are obtained from (26) by taking successively particular test functions following the velocity decomposition. First we take $\mathbf{v}$ constant in $\mathbb{R}^3$, then $\mathbf{v} = \Theta \wedge \mathcal{R}(\boldsymbol{G\xi} + \mathbf{d}), \Theta \in \mathbb{R}^3$, and finally $\mathbf{v} = \mathcal{R}\bar{\mathbf{d}}, \bar{\mathbf{d}} \in \mathcal{E}_0 \cap (H^1(\mathcal{B}))^3$.

Considering a finite number of modes to describe the elastic part of the structure consists in taking $\mathbf{d}(\boldsymbol{\xi}, t) = \sum_{i=1}^N \alpha_i(t)\phi_i(\boldsymbol{\xi})$ and $\bar{\mathbf{d}}(\boldsymbol{\xi}, t) = \sum_{i=1}^N \beta_i(t)\phi_i(\boldsymbol{\xi})$ where ϕ_i are well chosen modal functions.

4 Definition and preliminary analysis of the fluid-structure interaction problem

In this section we briefly present the equations that can describe a coupled fluid–structure interaction when the structure is modeled by the previous equations and the fluid is assumed to be newtonian viscous incompressible with a behaviour that is described by the Navier–Stokes equations. We also derive for this model, at least formally, an energy estimate following the same lines as in [5].

We suppose that the fluid fills, at time $t = 0$, a smooth domain of $\mathbb{R}^d, d = 2, 3$. In this cavity denoted by Ω, we consider, to simplify, a single deformable body $\mathcal{B}$. We denote by $\mathcal{B}(t)$, the position (and the shape) of the body at time t and by $\Omega(t)$ the fluid domain at time t so that it is defined as $\Omega(t) = \Omega \setminus \mathcal{B}(t)$. The Navier–Stokes equations are then used to model the fluid behaviour in $\Omega(t)$ which is an unknown domain depending on time and on the displacement of the structure.

The equations for the fluid part are

$$\begin{cases} \partial_t \mathbf{w} + (\mathbf{w}.\nabla)\mathbf{w} - \nu \Delta \mathbf{w} + \nabla p = \mathbf{f} & \text{in } \Omega(t) \\ \text{div } \mathbf{w} = 0 & \text{in } \Omega(t) \\ \mathbf{w}(.,t) = 0 & \text{on } \partial\Omega \\ \mathbf{w}(.,0) = \mathbf{w}_0 & \text{in } \Omega(0), \end{cases} \tag{28}$$

where $\mathbf{w}$ denotes the fluid velocity, p its pressure, ν its viscosity. The applied exterior force $\mathbf{f}$ is given together with the initial velocity $\mathbf{w}_0$. Note that this system is not complete since no boundary condition is proposed yet on the interface $\partial\Omega(t) \setminus \partial\Omega$.

For the structure part we recall the following weak formulation of the equations satisfied by the displacement $\mathbf{u}$ and by the associated perturbation $\mathbf{d}$:

$$\begin{cases} \forall \mathbf{v} \text{ sufficiently smooth, } \bar{\mathbf{d}} \text{ defined uniquely by (25),} \\ \displaystyle\int_{\mathcal{B}} (\partial_{tt}\mathbf{u}) \cdot \mathbf{v} + \int_{\mathcal{B}} \overline{\sigma}(\mathbf{d}) : \overline{\epsilon}(\bar{\mathbf{d}}) = \int_{\mathcal{B}} \mathbf{h} \cdot \mathbf{v} + \int_{\partial\mathcal{B}} (\mathbf{g} + T_f) \cdot \mathbf{v}, \end{cases} \tag{29}$$

where $\mathbf{h}$ and $\mathbf{g}$ are external, respectively body and surface forces applied on the structure. This system has to be completed by initial data: $\mathbf{u}(\cdot,0) = \mathbf{u}_0$ together with $\partial_t \mathbf{u}(\cdot,0) = \mathbf{u}_1$. The first coupling condition at the interface between the fluid and the structure is

$$\mathbf{w}(\boldsymbol{\xi} + \mathbf{u}(\boldsymbol{\xi},t),t)) = \partial_t \mathbf{u}(\boldsymbol{\xi},t) \quad \text{on } \partial\mathcal{B}, \tag{30}$$

which states the equality of the velocities: since the fluid is viscous and therefore sticks to the boundaries. Moreover the forcing term T_f appearing in the right hand side of the structure equations corresponds to the fluid force and can be defined by:

$$\int_{\partial\mathcal{B}} T_f \cdot \mathbf{v} = \int_{\partial\mathcal{B}(t)} (p \cdot \mathbf{n} - \nu \frac{\partial \mathbf{w}}{\partial \mathbf{n}}) \cdot \mathbf{v} \circ \chi^{-1}, \forall \mathbf{v}, \tag{31}$$

where $\chi(t,\boldsymbol{\xi}) = \boldsymbol{\xi} + \mathbf{u}(\boldsymbol{\xi},t), \forall \boldsymbol{\xi} \in \mathcal{B}$, $\mathbf{n}$ denotes the outer unit normal of $\mathcal{B}(t)$, and $\mathbf{v} \circ \chi^{-1}$ is to be taken in the sense $\mathbf{v}(\chi^{-1}(\boldsymbol{\xi},t),t)$.

Remark 2. We have to point out that — contrarily to what is the general case — the pressure in this context is not determined up to an additive constant. Actually the level of pressure (its average) is the Lagrange mutiplier associated with the constraint that the solid $\mathcal{B}(t)$ has a constant volume. Note also that this invariance of the global volume is not explicitly stated but is a consequence of the incompressibility of the fluid together with the equality of the fluid and structure velocities expressed in (30) **.**

Taking into account these two coupling conditions one can write a global weak formulation of the equations describing the behaviour for our coupled system:

$$\left\{\begin{array}{l}\displaystyle\int_{\Omega(t)} \partial_t \mathbf{w}\cdot\phi + \int_{\Omega(t)} (\mathbf{w}.\nabla)\mathbf{w}\cdot\phi + \nu\int_{\Omega(t)} \nabla\mathbf{w}\cdot\nabla\phi \\ \displaystyle\qquad + \int_{\mathcal{B}} (\partial_{tt}\mathbf{u})\cdot\mathbf{v} + \int_{\mathcal{B}} \overline{\sigma}(\mathbf{d}) : \overline{\epsilon}(\bar{\mathbf{d}}) \\ \displaystyle\qquad\quad = \int_{\Omega(t)} \mathbf{f}\cdot\phi \int_{\mathcal{B}} \mathbf{h}\cdot\mathbf{v} + \int_{\partial\mathcal{B}} \mathbf{g}\cdot\mathbf{v}, \\ \forall(\phi,\mathbf{v}) \text{ such as } \operatorname{div}\phi = 0 \text{ in } \Omega(t), \phi(\chi(\boldsymbol{\xi},t),t) = \mathbf{v}(\boldsymbol{\xi},t) \text{ for } \boldsymbol{\xi}\in\partial\mathcal{B}, \\ \phi = 0 \text{ on } \partial\Omega \text{ and where } \bar{\mathbf{d}} \text{ is associated to } \mathbf{v} \text{ and is defined by (25).}\end{array}\right. \tag{32}$$

Note that the space of test functions heavily depends on the solution.

We remark that $(\mathbf{w}, \partial_t\mathbf{u})$ is a couple of admissible test functions. Indeed, div $\mathbf{v} = 0$ in $\Omega(t)$, $\mathbf{v} = 0$ on $\partial\Omega$. Moreover thanks to the coupling condition (30) and the definition of χ, $\mathbf{w}(\chi(\boldsymbol{\xi},t),t) = \partial_t\mathbf{u}(\boldsymbol{\xi},t)$ for $\boldsymbol{\xi}\in\partial\mathcal{B}$. Futhermore we remark that the $\bar{\mathbf{d}}$ function associated to $\partial_t\mathbf{u}$ is $\partial_t\mathbf{d}$, since $\partial_t\mathbf{u} = \dot{\tau} + \omega\wedge\mathcal{R}(\overrightarrow{\boldsymbol{G\xi}} + \mathbf{d}) + \mathcal{R}\partial_t\mathbf{d}$.

Taking $(\mathbf{w}, \partial_t\mathbf{u})$ as test function, one next obtains, at least formally:

$$\begin{aligned}&\int_{\Omega(t)} \partial_t\mathbf{w}\cdot\mathbf{w} + \int_{\Omega(t)} (\mathbf{w}.\nabla)\mathbf{w}\cdot\mathbf{w} + \nu\int_{\Omega(t)} \nabla\mathbf{w}\cdot\nabla\mathbf{w} \\ &\quad + \frac{1}{2}\frac{d}{dt}\int_{\mathcal{B}} \|\partial_t\mathbf{u}\|^2 + \frac{1}{2}\frac{d}{dt}\int_{\mathcal{B}} \overline{\sigma}(\mathbf{d}) : \overline{\epsilon}(\mathbf{d}) \\ &\qquad = \int_{\Omega(t)} \mathbf{f}\cdot\mathbf{w}\int_{\mathcal{B}} \mathbf{h}\cdot\partial_t\mathbf{u} + \int_{\partial\mathcal{B}} \mathbf{g}\cdot\partial_t\mathbf{u}\end{aligned}$$

Now taking into account the fluid incompressibility and integrating by parts (with the choice that was made on $\mathbf{n}$) we have

$$\int_{\Omega(t)} (\mathbf{w}.\nabla)\mathbf{w}\cdot\mathbf{w} = -\frac{1}{2}\int_{\partial\mathcal{B}(t)} \|\mathbf{w}\|^2\mathbf{w}\cdot\mathbf{n}$$

Recalling that

$$\frac{d}{dt}\int_{\Omega(t)} k = \int_{\Omega(t)} \partial_t k - \int_{\partial\mathcal{B}(t)} k\mathbf{w}\cdot\mathbf{n},$$

this leads to

$$\int_{\Omega(t)} \partial_t\mathbf{w}\cdot\mathbf{w} + \int_{\Omega(t)} (\mathbf{w}.\nabla)\mathbf{w}\cdot\mathbf{w} = \frac{1}{2}\frac{d}{dt}\int_{\Omega(t)} \|\mathbf{w}\|^2,$$

which is the derivative with respect to the time of the kinetic energy of the fluid. This yields

$$\begin{aligned}&\frac{1}{2}\frac{d}{dt}\int_{\Omega(t)} \|\mathbf{w}\|^2 + \nu\int_{\Omega(t)} \|\nabla\mathbf{w}\|^2 \\ &\quad + \frac{1}{2}\frac{d}{dt}\int_{\mathcal{B}} \|\partial_t\mathbf{u}\|^2 + \frac{1}{2}\frac{d}{dt}\int_{\mathcal{B}} \overline{\sigma}(\mathbf{d}) : \overline{\epsilon}(\mathbf{d}) \\ &\qquad = \int_{\Omega(t)} \mathbf{f}\cdot\mathbf{w}\int_{\mathcal{B}} \mathbf{h}\cdot\partial_t\mathbf{u} + \int_{\partial\mathcal{B}} \mathbf{g}\cdot\partial_t\mathbf{u}\end{aligned}$$

Energy estimates can then be derived in a standard way. One can obtain the same results with a finite number of modal functions to describe the elastic part of the structure (i.e. $\mathbf{d}(\boldsymbol{\xi}, t) = \sum_{i=1}^{N} \alpha_i(t)\phi_i(\boldsymbol{\xi})$). The only thing one will have to change is the definition of the space of test function which now has to be taken such as $\bar{\mathbf{d}}(\boldsymbol{\xi}, t) = \sum_{i=1}^{N} \beta_i(t)\phi_i(\boldsymbol{\xi})$.

References

[1] C. Conca, J. San Martin and M. Tucsnak: "Existence of solutions for the equations modelling the motion of a rigid body in a viscous fluid", Comm. Partial Differential Equations **25**, no. 5-6, 1019–1042, 2000

[2] B. Desjardins, M. J. Esteban: "Existence of weak solutions for the motion of rigid bodies in a viscous fluid", Arch. Ration. Mech. Anal. **146**, no. 1, 59–71, 1999

[3] B. Desjardins, M. J. Esteban: "On weak solutions for fluid-rigid structure interaction: compressible and incompressible models", Comm. Partial Differential Equations **25**, no. 7-8, 1399–1413, 2000

[4] B. Desjardins, M. J. Esteban, C. Grandmont, P. Le Tallec: "Weak solutions for a fluid–structure interaction model", Rev. Mat. Complut. **14** , no. 2, 523–538, 2001

[5] D. Errate, M. Esteban, Y. Maday: "Couplage fluide structure, un modèle simplifié", Note aux C.R.A.S., Paris, t.318, Série I, 1994

[6] C. A. Felippa, "A Systematic Approach to the Element-Independent Corotational Dynamics of Finite Elements", Report CU-CAS-00-03, January 2000, revised May 2000.

[7] F. Flori and P. Orenga: "On a nonlinear fluid–structure interaction problem defined on a domain depending on time", k Nonlinear Anal. **38**, no. 5, Ser. B: Real World Appl., 549–569, 1999

[8] F. Flori and P. Orenga: "Fluid–structure interaction: Analysis of a 3D compressible model" Ann. Inst. H. Poincaré Anal. Non Linéaire, **17**, no. 6, 753–777, 2000

[9] B. Fraeijs de Veubeke: "The dynamics of Flexible Bodies", Int. J. Engng SCi., **14**, 895–913, 1976

[10] C. Grandmont: "Existence for a three-dimensional steady state fluid-structure interaction problem", J. Math. Fluid Mech. **4**, no. 1, 76–94, 2002

[11] C. Grandmont and Y. Maday: "Existence for an unsteady fluid–structure interaction problem", M2AN Math. Model. Numer. Anal. **34**, no. 3, 609–636, 2000

[12] C. Grandmont and Y. Maday: "in preparation" 2002.

[13] C. Grandmont, Y. Maday and P. Métier: "Existence of a solution for a unsteady elasticity problem in large displacement and small perturbation", C. R. Math. Acad. Sci. Paris **334**, 2002, no. 6, 521–526

[14] M. D. Gunzburger, C. Lee and G. Seregin: "Global existence of weak solutions for incompressible viscous flow around a moving rigid body in three dimensions", J. Math. Fluid Mech. **2**, no. 3, 219–266, 2000

[15] K.-H. Hoffmann and V. N. Starovoitov: "On a motion of a solid body in a viscous fluid. two dimensional case", Adv. Math. Sci. Appl. **9**, no. 2, 633–648, 1999

[16] J. San Martín, V. Starovoitov and M. Tucsnak: "Global weak solutions for the two-dimensional motion of several rigid bodies in an incompressible viscous fluid.", Arch. Ration. Mech. Anal. **161**, 2002, no. 2, 113–147

[17] D. Serre: "Chute libre d'un solide dans un fluide visqueux incompressible. Existence", Japan J. Appl. Math., Japan J. Appl. Math. **4**, no. 1, 99–110, 1987

[18] T. Takahashi and M. Tucsnak: "Global strong solutions for the two dimensional motion of a rigid body in a viscous fluid", preprint

[19] T. Takahashi, "Existence of strong solutions for the equations modelling the motion of a rigid–fluid sustem in a bounded domain", preprint

An Investigation of the Least-Squares Finite Element Method in Electromagnetism

JOHN KOT

CSIRO Radiophysics Laboratory, PO Box 76, Epping 1710, Australia,
John.Kot@csiro.au,
WWW home page: http://www.csiro.au

Summary

The least-squares FEM applied directly to Maxwell's equations is investigated as an alternative to the Galerkin FEM. The least-squares FEM is introduced via a simple model problem. Determinacy and ellipticity of the Maxwell system is investigated using Clifford algebra. Some simple results for a practical component are presented.

1 Introduction

Finite element method (FEMs) may be classified according to the underlying variational principle, such as Rayleigh-Ritz, Galerkin, or Least-Squares. The Galerkin FEM is now widely used in engineering electromagnetics; however, the application of the method was not without problems. The first attempts to use node-based finite elements for electromagnetic wave propagation problems applied Galerkin's method to second-order systems derived from Maxwell's equations; but the method was plagued with non-physical solutions. These were usually referred to as "spurious modes" since the problems arose in eigenvalue problems for modal propagation in waveguides. These non-physical solutions typically violated one of "divergence" equations in Maxwell's equations of electromagnetism. These problems were overcome by the use of "edge" elements [1, 2], based upon Whitney's work on Geometric Integration Theory [3].

The edge element approach recognised that the language of differential forms was a better framework for electromagnetism than the traditional vector calculus. Using "Whitney forms" as finite elements lead naturally to the correct continuity between elements and overcame many of the problems associated with the nodal elements. However the crucial operator in differential form methods in electromagnetism is the *duality* operator that maps the *flux forms* D and B to the *work forms* E and H. This operator is called the *Hodge star* operator. While this duality has a natural representation on dual geometrical meshes, the correct representation of the Hodge star operator on single unstructured meshes via Galerkin's method has been a major thrust of the work on edge element based FEMs [4].

An alternative approach that promises to avoid the problems of spurious modes, but that uses traditional node-based elements and single unstructured meshes is the *least squares* finite element method (LSFEM). The LSFEM is a simple method that leads to positive-definite matrix equations, even for non-self-adjoint problems, and so may in principle be applied directly to Maxwell's system of first-order equations for electromagnetism. However, the standard literature cautions against least-squares methods on two grounds:

1. Least-squares methods lead to ill-conditioned systems compared to Galerkin methods
2. Least-squares methods can give rise to non-physical solutions unless additional conditions are imposed upon the system.

Moreover, many engineering textbooks on electromagnetism state that Maxwell's equations are an over-determined system, with the "divergence" equations being in some way dependent upon the "curl" equations. At this point the case for the LS-FEM does not look strong; but the promising results obtained by proponents of this method in fluid dynamics and electromagnetism [5] have inspired the present investigation.

In the following we shall compare and contrast the least-squares and the Galerkin approaches, and look at the application to a prototypical Maxwell-type system. In doing this we shall substantially follow the approach in [5], but we introduce a different analysis for Maxwell-type systems that we feel is simpler and may be more useful especially for problems involving radiation into unbounded regions. The main results may be summarised as:

- The LS and the Galerkin FEM, when properly applied, lead to matrix equations with the same order of condition number.
- Spurious solutions arise through attempts to solve systems that are not strictly equivalent to the original problem.
- Maxwell's equations for monochromatic fields form a properly-determined elliptic system.

2 A Model Problem in 1D

In this section, we will consider the application of Galerkin's method and the LS-FEM to a model first-order problem in one dimension:

$$u_e'(x) = f(x) \tag{1}$$

$$u_e(0) = 0. \tag{2}$$

We aim to find the "best" approximation $u(x)$ to the exact solution $u_e(x)$ of (1) in a space $\mathbb{V} = \text{lin. span}\{\phi_i(x) \mid i = 1..N\}$ where the basis functions $\phi_i(x)$ possess sufficient derivatives and satisfy the boundary condition (2). Our unknowns are the coefficients a_i in the expansion:

$$u(x) = \sum_{i=1}^{N} a_i \phi_i(x). \tag{3}$$

The equation residual is

$$r_u(x) = u'(x) - f(x) \tag{4}$$

and we let (u, v) denote the usual L_2 inner product.

2.1 Galerkin's Method

In Galerkin's method we make the equation residual (4) orthogonal to the dual space $\mathbb{V}^*$ by setting to zero all the inner products (r_u, ϕ_i). This leads to N linear equations in the N unknowns a_i. Galerkin's method can be stated as the variational problem: find $u \in \mathbb{V}$ such that

$$(u' - f, v) = 0 \; \forall \, v \in \mathbb{V}. \tag{5}$$

2.2 Least Squares Method

In the least squares method we minimize the square of the L_2 norm of r_u over the space $\mathbb{V}$. The least squares method can be stated as the minimization problem: find $u \in \mathbb{V}$ such that

$$\|r_u\|^2 = (u' - f, u' - f) \leq \|r_v\|^2 \; \forall \, v \in \mathbb{V}. \tag{6}$$

Defining a quadratic form $F(u) = \|r_u\|^2$, a necessary condition for u to satisfy (6) is that

$$\lim_{t \to 0} \frac{d}{dt} F(u + tv) = 0 \; \forall \, v \in \mathbb{V}. \tag{7}$$

Evaluating the limit leads to the variational formulation of the least-squares problem: find $u \in \mathbb{V}$ such that

$$(u' - f, v') = 0 \; \forall \, v \in \mathbb{V}. \tag{8}$$

2.3 A Numerical Example (from [5])

Let $\mathbb{V}$ be the usual space of linear finite elements ("chapeau" functions) on $[0, 1]$, with a uniform mesh spacing h, and let

$$f(x) = 20e^{-20(1-x)}.$$

The Galerkin FEM gives rise to a matrix:

$$\frac{1}{2h}\begin{bmatrix} 0 & 1\,0 & 0 \cdots 0 \\ -1 & 0\,1 & 0 & 0 \\ 0 & -1\,0 & 1 & 0 \\ \vdots & & \ddots & \vdots \\ 0 & 0 & -1 & 0\,1 \\ 0 & \cdots 0 & 0 & -1\,1 \end{bmatrix}$$

We recognise the first few rows of the matrix as being the central difference formula for the first derivative, and so we expect trouble numerically with this linear system.

The LSFEM gives rise to the matrix:

$$\frac{1}{h^2}\begin{bmatrix} 2 & -1 & 0 & 0 \cdots & 0 \\ -1 & 2 & -1 & 0 & 0 \\ 0 & -1 & 2 & -1 & 0 \\ \vdots & & \ddots & & \vdots \\ 0 & & 0 & -1 & 2\; -1 \\ 0 & \cdots & 0 & 0\; -1 & 1 \end{bmatrix}$$

Clearly this linear system is related to the central difference formula for the second derivative, which should not lead us into trouble. Thus the LSFEM applied to our first-order model problem gives rise to a system that looks like the standard Galerkin method for some second order system. This raises the question, do we not then require one extra boundary condition to give a properly-posed problem? We return to this point below. Let us look at the numerical solution to our systems. Figure 1 shows the solutions for the Galerkin and LS FEM. Clearly the Galerkin FEM is a bad choice of method for this problem. For this simple problem it is easy to devise a Petrov-Galerkin method (where we project onto a different space from $\mathbb{V}^*$) that works well, but in general choosing this space for the Petrov-Galerkin method will be problem specific.

2.4 The Euler-Lagrange equation

Taking the variational formulation of the least-squares problem and integrating by parts, we arrive at the Euler-Lagrange equation for the least-squares formulation:

$$u'' - f' = 0 \text{ for } x \in [0,1] \tag{9}$$

$$u' - f = 0 \text{ at } x = 1 \tag{10}$$

$$u = 0 \text{ at } x = 0 \tag{11}$$

The LSFEM for our first-order model problem (1 - 2) is equivalent to a Galerkin FEM for the second-order problem (9 - 11), with the extra boundary condition being the original equation as a natural boundary condition at $x = 1$ (10). Hence the

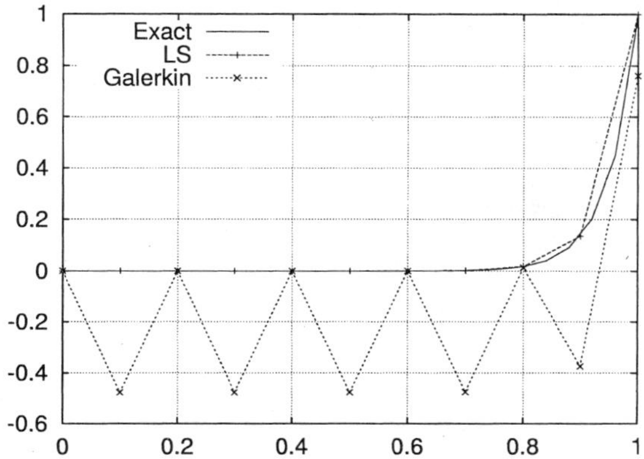

Figure 1 Numerical solutions for the model problem in 1-D

LSFEM applied to our model problem results in a symmetric, positive-definite matrix equation with optimal rate of convergence, and with condition number $\mathcal{O}(h^{-2})$ [6].

Neither of the criticisms of the LSFEM is valid here. Provided the method is applied to the underlying first-order equation, the result is a well-posed problem with convergance and conditioning identical to that obtained from applying the Galerkin FEM to a self-adjoint, second-order equation. The extra boundary condition needed arises as a natural boundary condition and so presents no extra complication for the FEM.

3 The LSFEM for a general first-order system

We consider a general first-order differential operator in $\mathbb{R}^N$:

$$Au = \sum_{i=1}^{N} A_i \frac{du}{dx_i} + A_0 u \tag{12}$$

where $u : \mathbb{R}^N \to \mathbb{R}^P$ and A_i are $M \times P$ real matrices. Examining the operator under a change of variables we are lead to the characteristic equation

$$\det\left(\sum_{i=1}^{N} \alpha_i A_i\right) = 0. \tag{13}$$

For an elliptic operator, the characteristic equation (13) has no real roots. Our aim is to solve the first-order linear boundary-value problem

$$Au = f \text{ for } x \in \Omega \tag{14}$$

$$Bu = 0 \text{ for } x \in \Gamma = \partial\Omega. \tag{15}$$

Let (u, v) denote the usual inner product for functions with support in Ω, and define the equation residual as

$$r_v = Av - f. \tag{16}$$

Choosing a space of approximating functions $\mathbb{V}$ the variational formulation of the least-squares problem is: find $u \in \mathbb{V}$ such that:

$$(Au - f, Av) = 0 \; \forall \, v \in \mathbb{V}. \tag{17}$$

Integrating (17) by parts we arrive at the Euler-Lagrange equations for the least-squares method:

$$A^*Au - A^*f = 0 \text{ for } x \in \Omega \tag{18}$$

$$Au - f = 0 \text{ for } x \in \Gamma = \partial\Omega \tag{19}$$

$$Bu = 0 \text{ for } x \in \Gamma. \tag{20}$$

Again we see that the LSFEM method for (15) is equivalent to a Galerkin method for a second-order, self-adjoint problem. The "extra" boundary condition required to render the second-order system well-posed is provided by the original first-order equation appearing as a natural boundary condition (19).

4 Maxwell's Equations and the "Div-Curl" System

As a path to Maxwell's equations, Jiang considers the "div-curl" system [5]. Here we follow a similar path but presented in the framework of Clifford algebra as this leads to the result more succinctly and we hope shall lead to useful results especially for problems involving open boundaries and radiation. A very brief introduction to Clifford algebra is given in the appendix.

4.1 The div-curl system

We consider the following linear first-order boundary-value problem:

$$\nabla \cdot u = \rho \quad \text{for } x \in \Omega \tag{21}$$

$$\nabla \times u = f \quad \text{for } x \in \Omega \tag{22}$$

$$n \cdot u = 0 \quad \text{for } x \in \Gamma_1 \tag{23}$$

$$n \times u = 0 \quad \text{for } x \in \Gamma_2 \tag{24}$$

where $\Gamma = \Gamma_1 \bigcup \Gamma_2 = \partial\Omega$. The system involves four scalar equations in three unknowns and so appears overdetermined in the same way that Maxwell's equations are commonly described. Let us re-express the system in terms of the Clifford algebra Cl_3. For a general Clifford number in Cl_3

$$A = \langle \alpha | a | b | \beta \rangle$$

where α is the scalar part, a the vector part, b the bivector part, and β the pseudoscalar part, let

$$\mathcal{Q}(A) = \alpha + a$$

and

$$\tilde{\mathcal{Q}}(A) = b + \beta.$$

Then our div-curl system becomes

$$Du = \omega \quad \text{for } x \in \Omega \tag{25}$$
$$\mathcal{Q}(nu) = 0 \quad \text{for } x \in \Gamma_1 \tag{26}$$
$$\tilde{\mathcal{Q}}(nu) = 0 \quad \text{for } x \in \Gamma_2 \tag{27}$$

where $\omega = -\rho + f\sigma$ (the operation of σ is explained in the appendix). To study the determinacy and ellipticity of the system we introduce a dummy scalar potential into the problem. Let

$$\hat{u} = \theta + u$$

where $\theta \in \Lambda^0$, *i.e.* a scalar function on Ω. Our system becomes

$$D\hat{u} = \omega \quad \text{for } x \in \Omega \tag{28}$$
$$\mathcal{Q}(n\hat{u}) = 0 \quad \text{for } x \in \Gamma_1 \tag{29}$$
$$\tilde{\mathcal{Q}}(n\hat{u}) = 0 \quad \text{for } x \in \Gamma_2. \tag{30}$$

Formally, the system has not changed. However, now there are four equations in four scalar unknowns, and the system is easy to classify. By examining the vector part of (28) we see immediately that the potential θ is monogenic in Ω, and hence harmonic. It is then easy to show from the boundary condition (29) that θ is identically zero in Ω, *i.e.* introducing the dummy potential has not changed our system, and the two systems are completely equivalent. From (13) we find that the characteristic equation is

$$\alpha_1^2 + \alpha_2^2 + \alpha_3^2 = 0.$$

The characteristic equation has no real roots. Hence our original div-curl system is elliptic and properly determined.

4.2 The monochromatic Maxwell's equations

The monochromatic form of Maxwell's equations in simple media is

$$\nabla \times E - i\omega\mu H = 0 \tag{31}$$
$$\nabla \times H + i\omega\epsilon E = 0 \tag{32}$$
$$\nabla \cdot H = 0 \tag{33}$$
$$\nabla \cdot E = \frac{\rho}{\epsilon}. \tag{34}$$

This system comprises a pair of div-curl systems coupled by the zero-order terms, hence the principle part of the system retains the div-curl structure and hence by the same argument is properly determined and elliptic.

5 A Practical Example

The LSFEM has been applied to the analysis of a Rotman lens beam-forming network [7]. Figure 2 shows a photograph of the Rotman lens. The lens was analysed by solving the two-dimensional Maxwell equations in the lens interior, using first-order triangular elements. A mesh size of approximately one-tenth of a wavelength was needed to obtain a convergent solution. Figures 3 and 4 show the computed solution for the real and imaginary parts of the principle component of the electric field inside the lens with the central beam-port excited by a 50Ω microstrip transmission line.

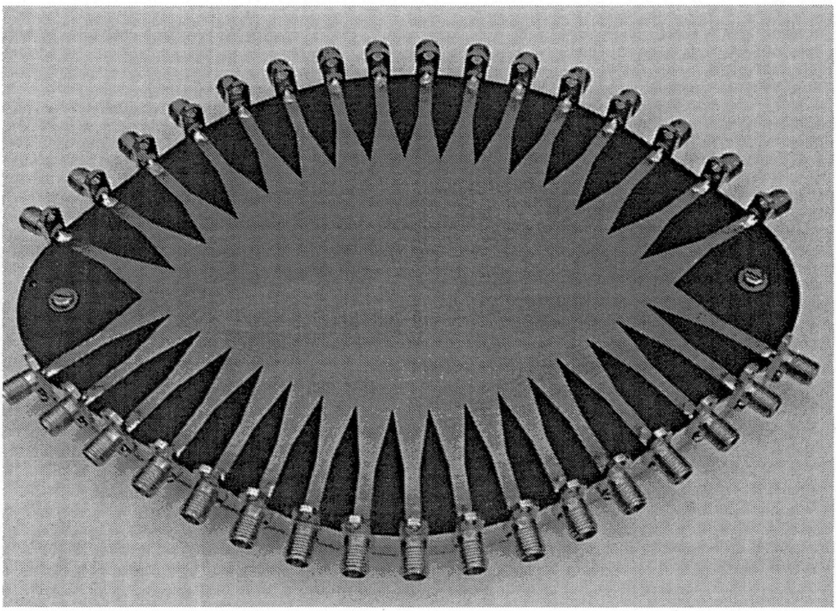

Figure 2 Rotman lens beam-forming network

6 Conclusion

In starting from Maxwell's equations in their original, first-order form, the LSFEM avoids many of the pitfalls that have arisen in trying to translate Maxwell's system into "equivalent" second-order equations. The LSFEM does not require the construction of special finite element spaces, and leads to symmetric, positive definite matrices. It also treats electric and magnetic fields on an equal footing, which for

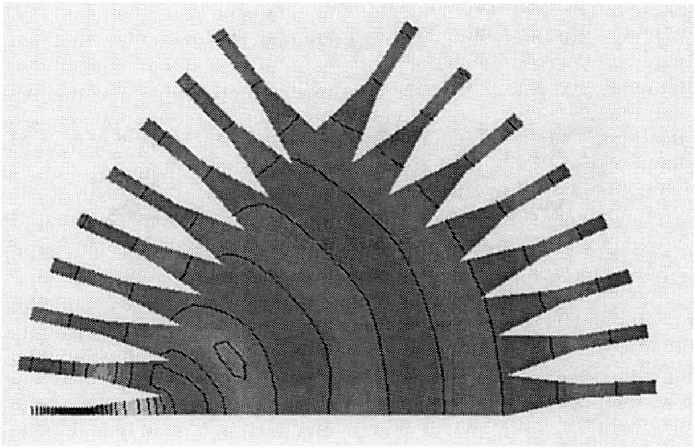

Figure 3 Real part of E_z with central beam port excited

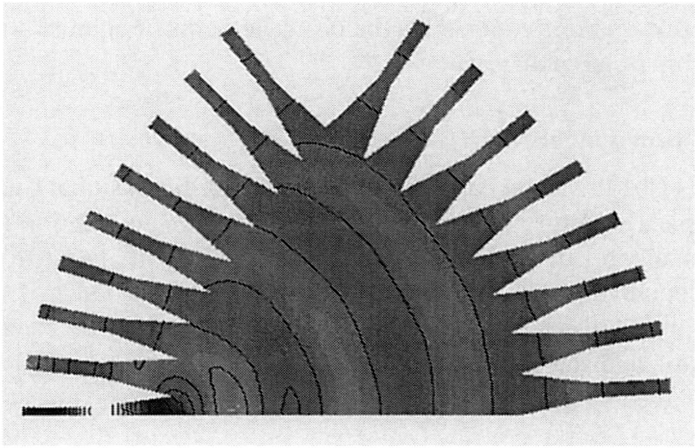

Figure 4 Imaginary part of E_z with central beam port excited

some applications is an advantage. Finally, when coupled with the methods of Clifford algebra, the least-squares method appears to offer a powerful tool for analysis. Against this, the LSFEM has a cost in efficiency in the need to deal with a larger system of equations and a larger number of unknowns. For microwave engineering, the attraction of the FEM is its power to analyse complex, composite structures. As the degree of integration of microwave systems increases, there is a growing need for accurate modelling of closely-packed metal and dielectric structures with complex geometrical shapes, as found in typical microwave integrated circuits and their packaging. The LSFEM's promise of freedom from spurious, non-physical solutions coupled with the "nice" numerical properties of the resulting matrix equations make it a very attractive method for the analysis of such structures.

References

[1] A. Bossavit. "Whitney forms: a class of finite elements for three-dimensional computations in electromagnetism". IEE PROCEEDINGS, Vol. 135, Pt. A, (**8**), Nov. 1988, pp 493-500.

[2] J. C. Nedelec. "Mixed Finite Elements in R^3". Numer. Math. (**35**), 1980, pp. 315-341.

[3] H. Whitney. Geometric Integration Theory, Princeton University Press, 1957.

[4] A. Bossavit. "The Hodge Operator: Considered Central in Computational Electromagnetics", PIERS 2000, Boston, USA, 2000.

[5] B. Jiang. The Least-Squares Finite Element Method, Springer-Verlag, 1998.

[6] S. C. Brenner, L. R. Scott. The Mathematical Theory of Finite Element Methods, Springer-Verlag, 1994.

[7] N. Yuan, J. S. Kot, A. J. Parfitt. "Analysis of Rotman lenses using a hybrid least-squares FEM / transfinite element method", IEE Proc Microwaves Antennas Propag, vol. 148 (**3**), 2001, pp 193-198.

A A Brief Introduction to Clifford Algebra

In this appendix we briefly introduce the basic elements of algebra and analysis for Clifford algebra in three dimensions.

A.1 The Clifford algebra Cl_3

Let $\{e_1, e_2, e_3\}$ be the usual cartesian basis for three-dimensional Euclidean space. The vector space structure of $\mathbb{R}^3$ tells us how to add vectors and to multiply them by scalars. An algebra defines a product of two vectors, and in a Clifford algebra we define an associative product between vectors where the product of vectors e_i and e_j is the new quantity $e_i e_j$, and the product of three vectors e_i, e_j, and e_k is the new quantity $e_i e_j e_k$, and so on, subject to the rules:

$$e_1^2 = e_2^2 = e_3^2 = -1 \tag{35}$$

$$e_i e_j = -e_j e_i \text{ for } i \neq j. \tag{36}$$

We can see immediately that the only quantities we can make are scalars, vectors, bivectors like $e_1 e_2$, and trivectors like $e_1 e_2 e_3$. A product like $e_i e_j e_k e_l$ must contain at least one repetition of subscripts, and so collapses to a unit of lower grade, by (35). Moreover, from (36) we can see that the space of bivectors has dimension 3, and the space of trivectors has dimension 1 (hence the alternate name "pseudoscalars"). The spaces of scalars, vectors, bivectors, and pseudoscalars are said to be of grade 0, grade 1, grade 2, and grade 3, and the spaces themselves are usually denoted Λ^0, Λ^1, Λ^2, and Λ^3, respectively.

Defining vectors $u = u_1 e_1 + u_2 e_2 + u_3 e_3$ and $v = v_1 e_1 + v_2 e_2 + v_3 e_3$, and the pseudoscalar unit $\sigma = e_1 e_2 e_3$ we calculate the Clifford product between vectors as

$$\begin{aligned} uv &= u_1 v_1 e_1 e_1 + u_1 v_2 e_1 e_2 + \dots \\ &= -(u_1 v_1 + u_2 v_2 + u_3 v_3) + (u_1 v_2 - u_2 v_1) e_1 e_2 + \dots \\ &= -u \cdot v + u \times v \sigma. \end{aligned}$$

The Clifford product between vectors gives rise to a scalar part and a bivector part. A general Clifford number has the structure:

$$\begin{array}{lll} u = u_0 + & \text{scalar part} & \in \Lambda^0 \\ \quad u_1 e_1 + u_2 e_2 + u_3 e_3 + & \text{vector part} & \in \Lambda^1 \\ \quad u_4 e_2 e_3 + u_5 e_3 e_1 + u_6 e_1 e_2 + & \text{bivector part} & \in \Lambda^2 \\ \quad u_7 e_1 e_2 e_3 & \text{pseudoscalar part} & \in \Lambda^3. \end{array}$$

A.2 Clifford analysis in Cl_3

The Dirac operator D in Cl_3 is defined as

$$D = e_1 \frac{\partial}{\partial x_1} + e_2 \frac{\partial}{\partial x_2} + e_3 \frac{\partial}{\partial x_3}. \tag{37}$$

For a vector u

$$Du = -\nabla \cdot u + \nabla \times u\sigma \tag{38}$$

The scalar part is the negative of the divergence, called by Maxwell the "convergence" of the vector field u. A function that satisfies $Du = 0$ is said to be *monogenic*, a generalization of the concept of a *complex analytic* function. Finally we note that D is the square root of the negative Laplacian, a positively-determined operator:

$$D^2 = -\Delta.$$

This result is usually attributed to Dirac although it was known to Maxwell via Quaternion analysis and attributed by Maxwell to his colleague Taite. Hence a monogenic function is automatically harmonic.

Fluids, Finite Elements and Multi-physics

A. N. STOKES, N. G. BARTON, X.-L. LUO, Z. ZHU
CSIRO Mathematical and Information Sciences
Private Bag 10, Rosebank MDC, Clayton 3169, Australia

Summary

We describe various applications of the finite element package, *Fastflo*, developed by CSIRO. *Fastflo* has been successfully used for diverse industrial applications including magnetohydrodynamics, geological fluid dynamics, and flows in porous media with complex thermal effects, chemistry and strain. The capabilities of *Fastflo* will be highlighted, especially those that relate to the operator-splitting method in fluid dynamics. We summarise the application of *Fastflo* to turbulent flow, non-Newtonian flow and combustion. In the area of multi-physics, we describe work on chromatography and geophysics. Finally we describe an algorithm for fluid-structure interaction problems, based in this instance on laminar flow around the structure and time-dependent linear elasticity within the structure. The moving mesh capability of *Fastflo* is an important feature of this algorithm.

1 Introduction

About twelve years ago, we started a research group in Computational Fluid Dynamics, which had to address and solve industrial problems arising in an Australian context. We saw a need for a strong multi-physics capability with the requirements

- the ability to handle complex geometries specified by CAD *etc.*,
- the ability to specify different but coupled physics on different regions,
- the ability to handle moving meshes or grids,
- a language that would allow us to specify such problems at a macro level.

The need for complex geometries pointed us towards finite elements. We needed a robust, fast and flexible algorithm for the underlying Navier-Stokes solution and, courtesy of connections with Prof Roland Glowinski, we built a core algorithm based on operator-splitting.

To enable the inclusion of multi-physics, we developed a language to specify the steps in operator-splitting, together with:

- the ability to write down systems of PDEs for coupled problems,
- a systematic expression of natural boundary conditions, to allow coupling of different regions in which different physical models applied,

- an expression of the transformations needed for general moving meshes.

The language was developed, and the resulting finite element environment was called *Fastflo*. It was highly tensor based, and allowed the entire definition of the physics and algorithm to be separated from the definition of geometry, and to be made available for user creativity or modification at a command macro level. At the same time, the user was liberated from the need to develop code for routine finite element operations like assembly, solvers and graphics. This paper sketches the development that occurred and highlights some current work.

2 The Operator-Splitting Algorithm

The core algorithm implemented, which guided the development of the language, was that used by Glowinski and co-workers [3,5,6] for the solution of the Navier-Stokes equations for incompressible fluid flow. These can be expressed, in a tensor notation with a summation convention, as:

$$\rho\dot{v}_i + \rho v_j v_{i,j} = -p_{,i} + (\eta(v_{i,j} + v_{j,i}))_{,j} + f_i \tag{1}$$

where p is the pressure, $\mathbf{v}$ the velocity, η the viscosity, ρ the density and $\mathbf{f}$ the body force per unit mass, and an equation expressing conservation of mass:

$$v_{i,i} = 0 \tag{2}$$

The theta-splitting algorithm of Glowinski [5] involved three fractional steps within each timestep. We use a notation in which the values of the state variables v and p are superfixed at the end of each step, so v^3 is the value of v at the end of the first complete timestep. The convention for writing increments is that $\delta v^n = v^{n+1} - v^n$, $\delta p^n = p^{n+1} - p^n$ Then the successive substeps are:

1. $$\rho\frac{\delta v_i^{3n}}{\delta t^{3n}} + \delta p_i^{3n} - \alpha(\eta\delta v_{i,j}^{3n})_{,j} = -\rho v_j^{3n} v_{i,j}^{3n} - p_i^{3n} + (\eta(v_{j,i}^{3n} + v_{i,j}^{3n}))_{,j} + f_i \tag{3}$$

$$\delta v_{i,i}^{3n} = -v_{i,i}^{3n}$$

2. $$\rho\frac{\delta v_i^{3n+1}}{\delta t^{3n+1}} + \rho v_j^{3n+1}\delta v_{i,j}^{3n+1} - \alpha(\eta\delta v_{i,j}^{3n})_{,j} = -\rho v_j^{3n+1} v_{i,j}^{3n+1} - p_i^{3n+1} + (\eta(v_{j,i}^{3n+1} + v_{i,j}^{3n+1}))_{,j} + f_i \tag{4}$$

3. As for substep 1.

The ratio of time subintervals δt^{3n}: δt^{3n+1}: δt^{3n+2} *is* θ*:1-2*θ*:*θ and we use Glowinski's value $\theta = 1-1/\sqrt{2} \approx 0.292893$. This has some optimal properties in terms of time accuracy. Then we choose $\alpha = \theta(1-2\theta)/(1-\theta)$. The first and third fractional steps describe a quasi-Stokes problem (linear, with pressure computed to satisfy incompressibility); the second fractional step involves the nonlinear convective derivative, but omits the complication of incompressibility. Special algorithms are used for the quasi-Stokes and convective fractional steps. The algorithm was also tuned for use of iterative solvers [15,18], storage of matrix factors, use of pre-conditioners and optimal use of memory. Our overall experience is that the solver is robust and accurate, as well as fast in that large timesteps can be taken.

3 Survey of Fluids Applications in *Fastflo*

3.1 Three-Dimensional Turbulent Flow

We coded three standard two-equation turbulence models using *Fastflo* - linear k-ε [8], nonlinear k-ε [17] and renormalization group [19]. For 3D calculations, this means that there are six degrees of freedom on each node: three velocity components, pressure, the turbulent kinetic energy k and the viscous dissipation ε. Our calculations were typically based on computation of the mean flow using the basic operator-splitting algorithm, together with computation of the turbulence variables separately at the conclusion of each timestep. Where required, special attention was given to implementation of wall boundary conditions for turbulent flow.

These algorithms were successfully applied to several test cases, especially those associated with benchmark tests arranged by the World User Association for Applied Computational Fluid Dynamics. The 1994 test involved turbulent flow in a narrowing cylindrical U-bend. Luo's [9] successful results based on implemention of the nonlinear k-ε algorithm [17] in *Fastflo* showed good behaviour for separation at the U-bend, as well as length of the separation bubble. The 1996 test, sponsored by Daimler-Benz, required computation of 3D turbulent flow around a model automobile. It was carried out on a mesh with 96,000 nodes, and converged satisfactorily in about 30-50 timesteps in 128 Mb RAM on a modest workstation computer. The results [10] were comparable with the best that were presented at the meeting. Figure **1** shows the particle pathlines.

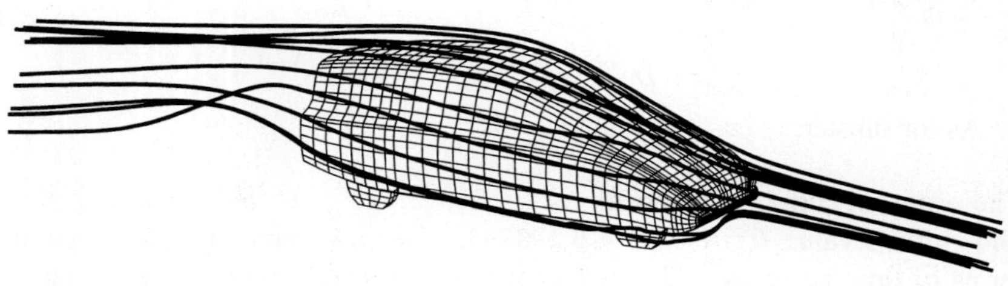

Figure 1 Particle pathlines for turbulent flow around a model motor vehicle, from the 1996 WUA- CFD benchmark study.

3.2 Non-Newtonian Flows

A convenient test problem for algorithms of viscoelastic flow is the falling sphere in a tube. In 1988, the problem of a sphere of radius a falling in a cylindrical tube of radius $2a$ filled with a viscoelastic fluid described by the Upper-Convected Maxwell (UCM) model was proposed as a benchmark problem at the *5th International Workshop on Numerical Methods in Non-Newtonian Flows,* and has been widely studied since. For this particular problem, the Weissenberg number is defined as

$$Wi = \lambda U / a$$

where λ is the relaxation time in the UCM model and U is the speed of the falling sphere. Although significant progress has been made, it remains a challenge to find a method that will converge beyond $Wi = 2$. Apart from theoretical interest, an important practical aspect of the problem is to understand the drag reduction mechanism of polymer solutions. Calculations have predicted a significant drag reduction for the UCM fluid in the sphere-in-tube problem, as compared to Newtonian fluid.

This problem was solved using a decoupled transient algorithm with a Streamline Upwind Petrov-Galerkin (SUPG) technique. The algorithm incorporated a pre-conditioned conjugate gradient (PCG) method for solving viscoelastic flows. The high efficiency of the decoupled operator-splitting algorithm enabled problems with more than 10^5 DOF, on a mesh of 17,000 nodes, to be solved on a 133 MHz DEC workstation in about 7 hours. Good convergence was obtained for Wi up to 2.8. The calculation used an unstructured triangular mesh. The results presented in Figure 2, computed for $Wi = 2.5$, show that the velocity and pressure fields have a significant downstream shift, as compared to the upstream-downstream symmetry in the pure viscous creeping flow case. The extra stress field is characterised by large gradients and thin boundary layers adjacent to the sphere surface.

3.3 Swirling Flows with Combustion

In this study, various combustion models were coupled to a swirling turbulent axial flow. The purpose was to compare numerical and experimental results, in particular to determine whether the combustion model or the turbulence model was the more significant factor. A hybrid method with SIMPLE-type algorithm and artificial compressibility was used to solve for the velocity and pressure. For turbulence and combustion effects, the following models were examined: an RNG-based k-ε turbulence model; a conventional k-ε turbulence model; a mixed-is-burnt combustion model; an eddy dissipation combustion model: a P_1 spherical harmonics radiation model

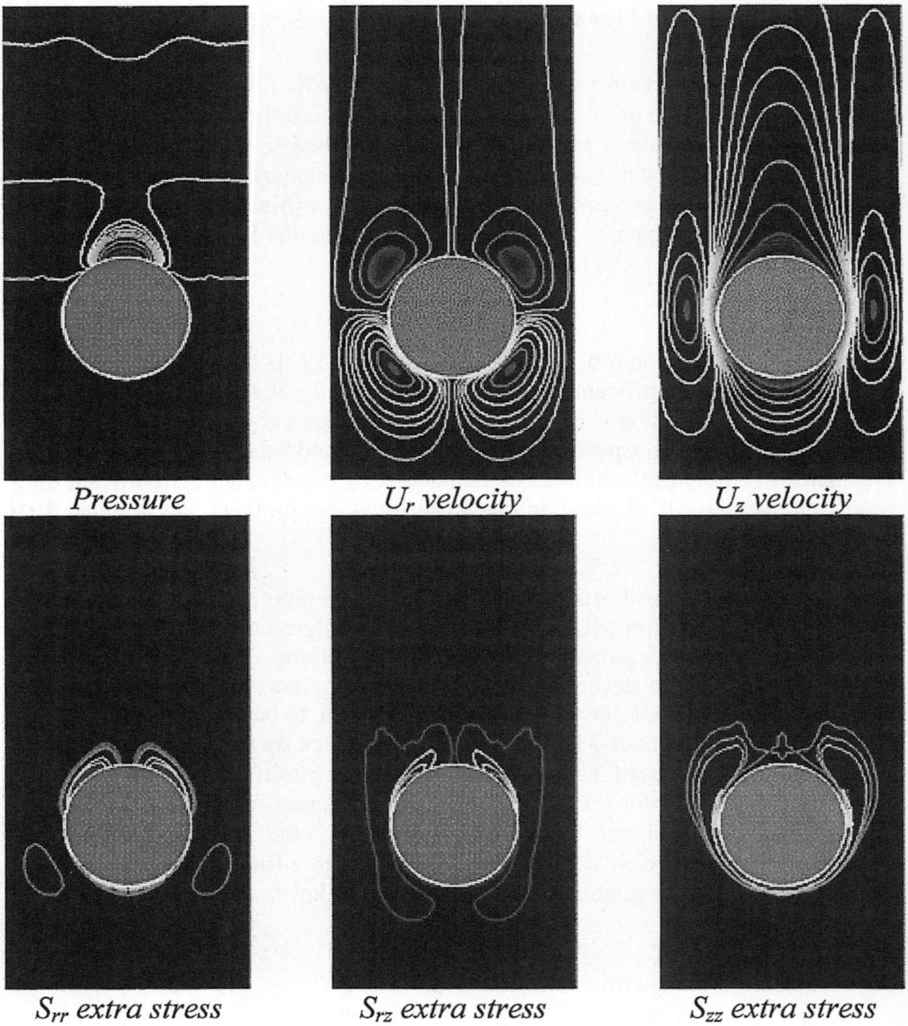

Figure 2 Results for the UCM sphere-in-tube benchmark viscoelastic problem.

As the combusting flow has a low Mach number, there are two approaches for eliminating the acoustic wave effect. One approach is to rescale the Navier-Stokes momentum equation, the other approach is to apply a perturbation expansion to the momentum equation. These two approaches were both tried for the test case. Results presented here were obtained from the conventional k-ε turbulence model and the eddy dissipation combustion model. As the flow is axi-symmetric, only half of the cross-section needs to be resolved. This cross-section was discretised with 4,028 bi-linear quadrilateral elements, and a converged solution was reached within 500 iterations, which took about 5 hours CPU time on a DEC 3000 Alpha workstation.

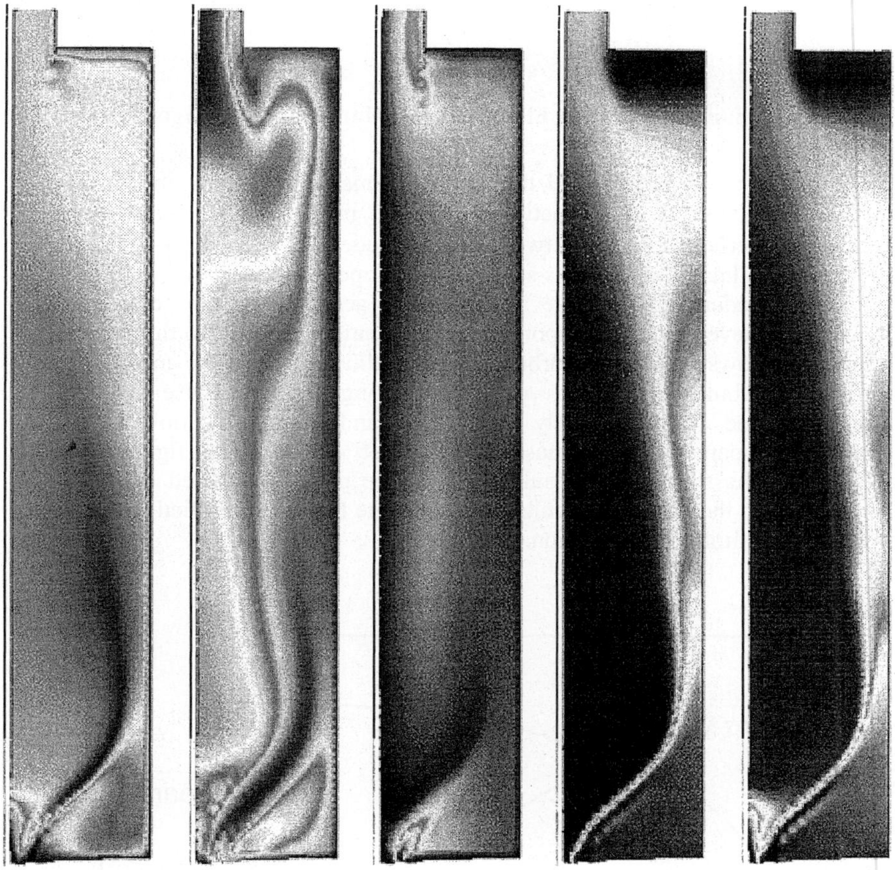

Temperature *Turbulence* *swirl* O_2 CO_2

Figure 3 Results for swirling combusting flows.

An axisymmetric burner is illustrated schematically in Figure 3. At the bottom section, fuel is pumped into the burner through a jet located along the axis, whilst air enters the combustion chamber through an annulus around the fuel jet. At the annulus inlet, the air has a tangential velocity of average value 11.7 m/s, and the average value of the axial velocity is 13.7 m/s. The Reynolds number in this case is Re = 139,081, based on the radius of the burner and the air's average axial velocity. The fuel is natural gas consisting of 91% methane by mass. The walls of the burner are all cooled to 400°C.

The comparison [20] showed that a good match with experimental results could be obtained with a four stage combustion model; others models were much less successful. The turbulence model used had much less effect than the combustion model.

3.4 Aerodynamic Turbulent Flows – the Spalart-Allmaras Algorithm

In recent work, we have coded the Spalart-Allmaras turbulence model [16] in *Fastflo.* This uses a one-equation turbulence model and thus involves less computing overheads than the two-equation models such as the k-ε algorithm. The Spalart-Allmaras model was specially developed for aerodynamic flows with turbulent boundary layers close to designated aerodynamic surfaces, and this *Fastflo* code development is important for applications relevant to this workshop. Figure [4] shows results for turbulent flow. Illustrated are the mean velocity profile, the turbulence variable ν_T and the shearing stress τ. These results, even though obtained with a relatively coarse mesh and a simplified flow solver [1], generally compare well with those of Spalart & Allmaras [16, figure 4]. The noisiness in the shear stress results close to the plate is believed to be due to inaccuracies in the pressure calculation at the plate for this simplified solver. This topic is under further investigation.

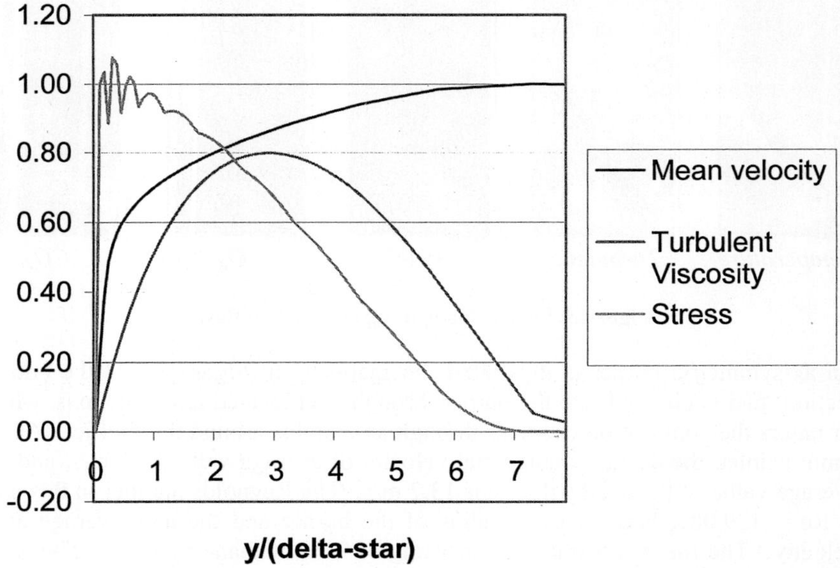

Figure 4 Results for turbulent flow over a flat plate at Reynolds number based on momentum thickness of 1124: mean flow scaled by U_{edge}, ν_T scaled by $0.025U_{\text{edge}}\delta$, τ scaled by τ_{wall}.

4 Complex Physics Applications in *Fastflo*

4.1 Porous Media – Packed Beds

In the past two years, we have carried out substantial *Fastflo* simulations for fluid flow through a chromatography column consisting of a packed bed of very small particles, and supplied by a liquid distribution system. Within the packed bed, the dominant flow is Darcy flow, whereas in the liquid distribution system the flow is laminar. The flow in these two distinct regions has to be computed conjointly, with continuity of flow velocity and pressure at the interface. Luo *et al.* [11] examined the elastic deformation of the packed bed as caused by the pressure force on the matrix of particles. In a second paper, Luo *et al.* [12] looked at the effect of the liquid distribution system, including simulations for dispersion of tracer within the bed. This work provides a basis for design of systems to give satisfactory chromatographic performance. Finally Luo *et al.* [13] used *Fastflo* to simulate the growth of the bed of particles as particles are transported by a slurry and trapped at the surface of the existing chromatographic bed. This work used the operator-splitting solver and several major capabilities of *Fastflo* - moving meshes and definition of different physical models on different sub-regions. The work also incorporated the Phillips model [14] for two-phase flow.

4.2 Multi-scale Geophysical Flows

Modelling of flows and chemistry over geological periods is now an important tool in predicting the location of ore deposits. Apart from the complexities of geometry, rock structural behaviour and chemical reactions, there are the critical issues of greatly varying timescales, extending to geological time, and spatial scales, since both flow and structural behaviour tend to be dominated by the presence of faults, with a scale of metres relative to the geological scale of many kilometres. The problems of scaling are not unlike those of shocks in fluid mechanics, and need to be handled with a careful choice of invariants, which the FEM method supports. Here we show an indicative computation of the flow in a faulted environment in the goldfields of Western Australia. It is of relatively modest complexity, but still includes fluid flow through faults that are subject to opening and closing in response to an overall shearing force. There is a source of fluid in the fault on the left and a sink in the fault on the right. Fluid flow in the rock obeys Darcy's law. Displacement vectors are calculated under the assumptions of linear elasticity, uniform rock properties and a shear applied across the region, and assuming that all the faults are free to either expand or contract. We have also extended the simulations to include the effect of chemical reactions and coupling between the elasticity and fluid flow.

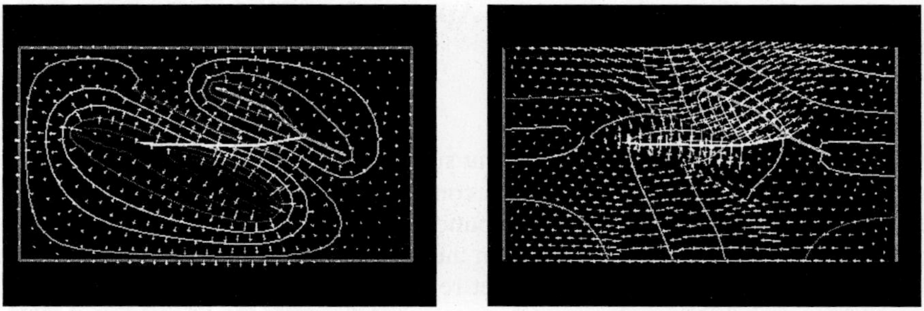

Figure 5 Illustrative calculations for fluid flow and elastic stress-strain in a 2D rock system with faults. Left: pressure contours for fluid flow predicted by Darcy law with a source in the LH fault and sink in the RH fault; right: displacement arrows and contour lines of the trace of the stress tensor.

4.3 Seismic Propagation in Liquid-Saturated Media

Seismic investigation of sedimentary rock formations is an important tool in the search for oil. Much information can be obtained on the geological formations, but it would be very desirable to learn something from the response as to the nature of liquids present in the rock, and particularly something about their density and pressure. Because of this, modelling of elastic wave propagation in a porous medium containing liquid is an active area of research. The basic problem is still to compute solutions of the original model due to Biot [2], whilst more recent work is described in [4]. The equations are:

$$\begin{aligned} \rho u_{tt} + \rho_f w_{tt} &= M\nabla^2 u + F\alpha^2\nabla^2 u + F\alpha\nabla^2 w\,, \\ \rho_f u_{tt} + (\rho_f + \rho_a) w_{tt} + \frac{\eta}{\kappa} w_t &= F\alpha\nabla^2 u + F\nabla^2 w \end{aligned} \tag{5}$$

where u is the displacement of the solid, and w the displacement of liquid relative to solid. The viscosity of liquid is η, the density of phases is denoted by ρ, subscripted f for fluid or a for averaged. M, F and α are derived elastic constants for the two-phase system. A centred difference scheme was applied for timestepping of these equations, which were treated as fully coupled. The iterative solver BiCGSTAB [18] was used.

In the absence of viscosity, there are two possible P-wave solutions [2], one faster than the other. These are shown in the corresponding numerical solutions. However, with even small viscosities, the two waves exchange energy, and eventually only a single wave is seen. The merging seems to happen on a very short distance scale. The following shows the propagation of waves in the absence of viscosity from a spherical subsurface cavity in response to an applied stress, with the time profile of a differentiated Gaussian. The spatial region is 2 km square in cross-section, with radial symmetry. An unstructured triangular mesh with 4000 nodes was used:

These snapshots of the magnitude of the displacement of the liquid relative to the solid clearly show the Biot fast and slow P-waves.

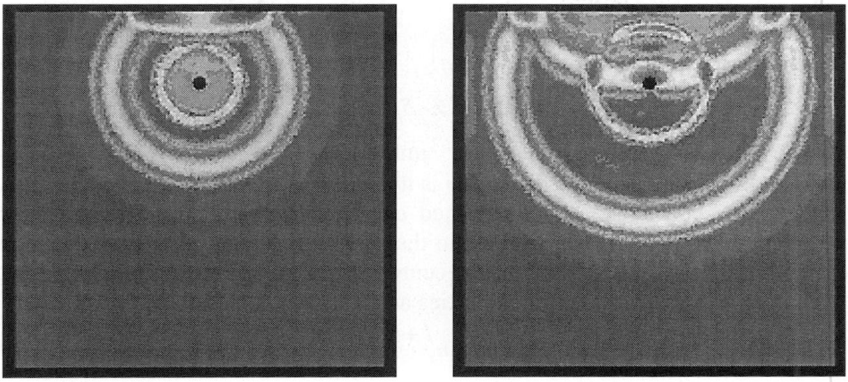

Figure 6 Two snapshots of the magnitude of the liquid displacement relative to the solid – characteristic frequency 25 Hz.

4.4 Recent Work on Fluid-Structure Interaction

In the spirit of this workshop, which considers coupling of fluids, structures and waves, we conclude by presenting recent work on the development of a fluid-structure interaction model in *Fastflo*. The model involves 2D laminar flow of an incompressible fluid around an elastic constriction in a duct. The basic inlet flow oscillates sinusoidally between zero and a maximum, and the outlet boundary condition is zero stress throughout the full cycle. The model is a crude representation of oscillatory flow past a valve, and it provides a basis for further development, especially using a turbulence model for the basic flow. If required, it would also be possible to represent other surrounding fluids, such as two-phase, non-Newtonian, or compressible.

The fluid-structure algorithm was developed especially to exploit design features in *Fastflo*'s finite element methodology. The working variable represents the fluid velocity in the fluid region and the rate of change of elastic displacement (*i.e.* a velocity) in the valve material. The variable is continuous across the interface between the sub-regions. Moreover, the equations are formulated in a way that the natural boundary condition for both sub-regions is that of stress. This natural boundary condition, which arises by integration by parts in the finite element method, is then by default continuous across the interface.

The equations are:

Fluid $$\rho\frac{\partial v_i}{\partial t}+\rho v_j v_{i,j}=-p_{,i}+\left[\eta\left(v_{i,j}+v_{j,i}\right)\right]_{,j} \tag{6}$$

$$v_{j,j}=0 \tag{7}$$

Solid $$\rho\frac{\partial^2 u_i}{\partial t^2}=\left[\sigma_{ij}\right]_{,j} \tag{8}$$

$$\sigma_{ij}=\lambda_1\left(u_{i,j}+u_{j,i}\right)+\lambda_2\delta_{ij}u_{k,k} \tag{9}$$

In these equations, u_i represent the cumulative elastic displacements, v_i the velocity components, ρ is the density, σ is the stress, η the fluid viscosity, λ_1 and λ_2 are the Lamé constants. As mentioned earlier, we use the time derivative of displacement as the working variable in the solid. Denote the time level by l and express the elasticity equations using central differences and a modified Crank-Nicolson formulation. The key properties are

$$v_i=\frac{\partial u_i}{\partial t} \quad \text{so that} \quad u_i^{l+1}=u_i^l+\Delta t.v_i^{l+1} \tag{10}$$

using an implicit representation for the incremental displacement. The elasticity equation therefore becomes

$$\begin{aligned}\rho\left(v_i^{l+1}-v_i^l\right)= & \left[\lambda_1\left(u_{i,j}^l+u_{j,i}^l\right)+\lambda_2\delta_{ij}u_{k,k}^l\right]_{,j} \\ & +\frac{\Delta t}{2}\left[\lambda_1\left(v_{i,j}^{l+1}+v_{j,i}^{l+1}\right)+\lambda_2\delta_{ij}v_{k,k}^{l+1}\right]_{,j}\end{aligned} \tag{11}$$

In the fluid sub-region, the fluid equations are solved using a timestepping solver [1], with an iterative loop within each timestep to ensure that the pressure is such that the continuity equation is satisfied. In the solid sub-region, equation (8) is solved. It is observed that both the working variable (the velocity) and the stress are continuous across the interface. The stress is matched by the boundary integrals arising from integrating the respective momentum equations by parts. These discretised equations were solved at each timestep using a Uzawa algorithm, with 6-noded triangle elements used for velocity, and 3-noded elements for pressure.

For the movement of the mesh, we displace the nodes so that the boundary moves to the correct position, while the interior nodes do not move in such a way as to lead to unreasonably distorted elements. We have found that the best way to do this is to solve a static elasticity problem, in which the "stiffness" of the mesh can be adjusted as required for sensitive parts of the field of computation. For the present study the following is appropriate:

$$\left(\alpha\lambda_1\left(u_{i,j}+u_{j,i}\right)+\alpha\lambda_2\delta_{ij}u_{k,k}\right)_{,j}=0 \tag{12}$$

where u is now the mesh displacement. The boundary displacement for each timestep is set equal to the physical displacement, which was the computed velocity at the boundary multiplied by the timestep. It would be possible to specify only the normal displacement, but that created some difficulties at junctions, and was not necessary for this problem. As the mesh for the computations is moving, there should be an extra advection term in the discretised momentum equation. However, we found this term had negligible effect for our test problem.

Figure 7 shows results obtained using this algorithm for the valve problem. The mesh had 914 6-noded triangles with a total of 1913 nodes. The working fluid was water, density $\rho = 998$ kg.m^{-3}, viscosity $\eta = 1.002e\text{-}3$ kg/(m.s), channel width 0.007 m, inlet flow maximum 0.2 m/s, flow oscillation period 0.08 s. The nominal Reynolds number of the flow based on peak inlet velocity and channel width is 1400. The flow can be visualised by the cumulative displacement of the mesh and the streamfunction ψ, which are shown in Fig at four stages of the inlet flow oscillation. The figures clearly show the elastic displacement of the mesh.

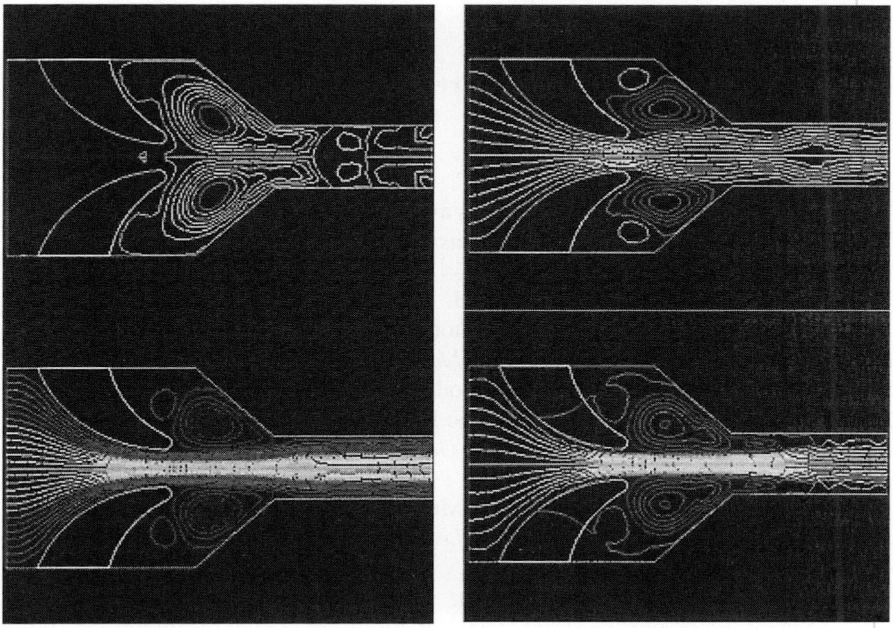

Figure 7 Stream function plots at four stages of the systole/diastole cycle.

5 Conclusion

We have surveyed previous major applications of *Fastflo* and described a fresh application to a fluid-structure interaction simulation. Important capabilities that are required for coupled problems in aerodynamics have been described: these include mesh transformation, the ability to frame and solve separate problems on sub-regions with coupling at the interface, and the use of sophisticated high performance solvers for various fluid models. Some of these features were used in the fluid-structure example.

Our view is that *Fastflo* provides a good framework for coupled problems in the aeronautics context. Attractive developments in the medium term would be to implement the fluid-structure interaction problem using the Spalart-Allmaras algorithm for the basic flow, and then to make calculations for a realistic 3D aeroelasticity problem. Development of a robust compressible flow solver, if required, is feasible. The applicability of *Fastflo* to coupled electromagnetic and flow problems has already been displayed in the context of MHD. We are keen to develop research collaborations on these topics in an aeronautics context. In all cases, these developments are expected to require minimal or no further development of the basic *Fastflo* package, rather algorithmic development within *Fastflo*'s macro command language.

References

[1] N.G. Barton, and A.N. Stokes. "Development of Intermediate-Level Navier-Stokes Solvers in *Fastflo*". CSIRO MIS Technical Report, 2002.

[2] M.A. Biot. "Theory of Elastic Waves in a Fluid Saturated Porous Solid. I. Low Frequency Range". *J. Acoust. Soc. Am*. Volume 28, 1956. pp. 168–178.

[3] E. Dean, R. Glowinski and C.-H. Li. "Application of Operator Splitting Methods to the Numerical Solution of Nonlinear Problems in Continuum Mechanics and Physics". In *Mathematics Applied to Science*, ed. J. Goldstein, S. Rosecrans and G. Sod, 1987.

[4] J. Dvorkin and A. Nur. "Dynamic Poroelasticity: a Unified Model with the Squirt and Biot Mechanisms". *Geophysics*, Volume 58, 1993, pp. 524–533.

[5] R. Glowinski. "Finite Element Methods for the Numerical Simulation of Incompressible Viscous Flow". *Lectures in Applied Mathematics*, Volume 28, 1991.

[6] R. Glowinski and O. Pironneau. "Finite Element Methods for Navier-Stokes Equations". *Anual Revue of Fluid Mechanics*. Volume 24, 1992, pp. 167-204.

[7] *Fastflo* Tutorial Guide. CSIRO Mathematical and Information Sciences, Sydney. 1999.

[8] B.E. Launder and D.B. Spalding. "The Numerical Computation of Turbulent Flows". *Comp. Meth. App. Mech. Engng*., Volume 3, 1974, pp. 269-289.

[9] X.L. Luo. "Operator Splitting Computation of Turbulent Flow in an Axisymmetric Narrowing Bend using Several k-ε Models and Wall Functions". *International Journal of Numerical Methods in Fluids*, Volume 22, 1996, pp. 1189-1205.

[10] X.-L. Luo, A.N. Stokes and N.G. Barton. "Turbulent Flow around a Car Body - Report of *Fastflo* Solutions". *Proc. WUA-CFD Conference*, Freiburg, 1996.

[11] X.-L. Luo, N.G. Barton, K.H. Gebauer and A.N. Stokes. "Finite Element Analysis of a Compressible Packed Chromatographic Bed". *Proc: 6th World Congress of Chem Eng.*, Melbourne, 2001.

[12] X.-L. Luo, N.G. Barton, K.H. Gebauer and A.N. Stokes. "Flow through a Packed Chromatographic Bed supplied by a Liquid Distribution System". *Proc. 6th World Congress of Chem Eng.*,Melbourne, 2001.

[13] X.-L. Luo, N.G. Barton, K.H. Gebauer and A.N. Stokes. "Simulation of Packed Bed Formation in Chromatography Columns". *Proc World Congress on Particle Technologies 4,* Sydney, 2002.

[14] R.J. Phillips, R.C. Armstrong and R.A. Brown. "A Constitutive Equation for Concentrated Suspensions that Accounts for Shear-Induced Particle Migration". *Phys. Fluids* Volume A4, 1992, pp. 30-40.

[15] Y. Saad and M. Schultz. "GMRES: A generalised Minimal Residual Algorithm for Solving Nonsymmetric Linear Systems". *SIAM Journal of Scientific and Statistical Computing* Volume 7, 1986, pp. 856-869.

[16] P. R. Spalart and S. R. Allmaras. "A One-Equation Turbulence Model for Aerodynamic Flows". *La Recherche Aerospatiale*, Volume 1, 1994, pp. 5-21.

[17] C.G. Speziale. "On Non-Linear k-ε and k-l Models of Turbulence". *Journal of Fluid Mechanics*, Volume 178, 1987, pp. 459-475.

[18] H. van der Vorst. "Bi-CGSTAB, A Fast and Smoothly Converging Variant of Bi-CG for the Solution of Nonsymmetric Linear Systems". *SIAM Journal of Scientific and Statistical Computing* Volume 13, 1989, pp. 631-644.

[19] V. Yakhot and S.O. Orszag. "Renormalization Group Methods in Turbulence". *Journal of Scientific Computing*, Volume 3, 1986, p. 1.

[20] Z. Zhu and A.N. Stokes. "Computing Diffusion Flames with Multistage Combustion Models". *Proc. 13th Australasian Fluid Mechanics Conference*, Melbourne, 1998, pp. 1043-1046.

Status and Future Challenges of CFD in a Coupled Simulation Environment for Aircraft Design

F. CHALOT, T. FANION, M. MALLET, M. RAVACHOL and G. ROGE
Dassault Aviation
78 quai Dassault
92214 Saint Cloud - France

Summary

The state of the art of Computational Fluid Dynamics and the axis of improvements are described. The issue of flutter prediction is addressed first : the use of linearized Euler solvers for transonic flutter is explained. Recent advances in optimum aerodynamic shape design are presented next , the results demonstrate the applicability of optimization based on the Euler equations and open the way to multidisciplinary optimum design. Finally, the use of Large Eddy Simulation for accurate turbulent flow simulation is illustrated.

1 Introduction

Computational Fluid Dynamics has reached a high level of maturity and has a considerable impact on aircraft design. This has been foreseen for some time [9] and described in a number of publications for example in [3].

The presentation will illustrate the present capability of CFD and also identify the areas where development efforts are made. Coupling with structural analysis for the prediction of flutter will be presented first : this is an example of a multi - physics simulation for a critical design issue for aircraft. Issues related to design optimization will be discussed next, a lot of progress is made in this area to increase design turn around capabilities. Finally, we will address the question of the accuracy of CFD simulations: this will be illustrated by the problem of turbulence modeling.

2 Fluid Structure Interaction

Multi physics simulation is an area of active development and fluid / structure interaction is typical example. For aircraft design, the main objective is to predict flutter. Flutter is the resonant interaction between unsteady aerodynamic loads and the elastic deformation of the aircraft structure. Flutter is a dangerous phenomena that can lead to the rapid loss of an aircraft. State of the art techniques rely on simplified linear numerical models for aerodynamics. This approach is fast and well calibrated (we emphasize the high number of configurations that must be analysed). However, when nonlinear flow behavior is present, these methods cannot be applied. Emerging techniques consider more complex flow models.

Two approaches can be considered. The first one relies on the coupling of non-linear unsteady CFD and unsteady structural dynamics codes. Significant effort has been devoted to the development of this approach. Critical numerical ingredients have been identified, most notably the need to use efficient staggered schemes and to fulfil the geometric conservation law criteria. This is described for example in [8]. This approach has been implemented and demonstrated in our computational environment SOUPLE. A typical result is presented in [3]. This approach is very general. It is however CPU intensive and this restricts its present application to validation study and to the analysis of the most critical and complex configurations.

An alternate approach is based on the solution of the linearized Euler equations in the frequency domain. The procedure includes the following steps. First a structural displacement basis is considered. For example a modal basis can be selected. The Generalized Aerodynamic Force matrix *GAF(f)* is computed next. For a given frequency, *GAF(f)(i,j)* is the aerodynamic pressure force associated to nodal displacement *i* when the structure oscillates along mode *j* at frequency *f*. Matrix *GAF* is obtained by solving the linearized Euler equations for each mode and a number of frequencies over a selected frequency range. Typically, the number of modes considered is between 4 and 10 and the number of frequencies around 10. Between two frequencies, linear interpolation is performed. This step requires the solution of (number of modes)X(number of frequencies) complex linear systems. This is the most CPU intensive step of the procedure, however its cost is still low.

Once matrix *GAF* is computed, the complete dynamic behaviour of the coupled system can be explored easily. The classical approach uses the 'p-k' method which is based on a loop over the velocity. Starting at zero velocity, for each velocity increment, the frequency and damping associated to each mode is computed. Flutter speed is reached as soon as negative damping is obtained.

The linearized Euler solver was developed on the basis of our industrial Euler code EUGENIE. It is an unstructured solver. A modified Lax Wendroff numerical scheme is selected. A 1st order Steger Warming version can also be used. Automatic differentiation was used to contribute to the generation of the code for the linearized operator. A software tool called *Odyssée* developed by INRIA was used. Significant efforts have been devoted to achieve a high level of efficiency. The GMRES linear solver was selected and incomplete LDU preconditionning is performed. Parallel implementation on distributed memory architecture has also been programmed. A detailed description can be found in [7].

The method was validated using the well known AGARD 445.6 wing test case. The experimental investigation was performed at NASA Langley transonic tunnel.

The wing has an aspect ratio of 4, a taper ratio of 0.6, the ¼ cord sweep angle is 45 degrees. The wing profile in streamwise direction is a NACA65A004. Data is available for a wide Mach number range from 0.338 to 1.141. The first four vibration modes are illustrated on Figure 1, the mesh used for the aerodynamic computations is available on Figure 2. A representative flutter diagram is presented on Figure 3. For three modes, the frequency and damping are obtained as a function of velocity. Flutter occurs when negative damping is predicted. The transonic dip phenomena is illustrated in Figure 4 : the flutter speed index is plotted as a function of the Mach number. A nonlinear behaviour can be observed through the transonic regime. We can verify that the linearized Euler based method can predict this phenomena correctly.

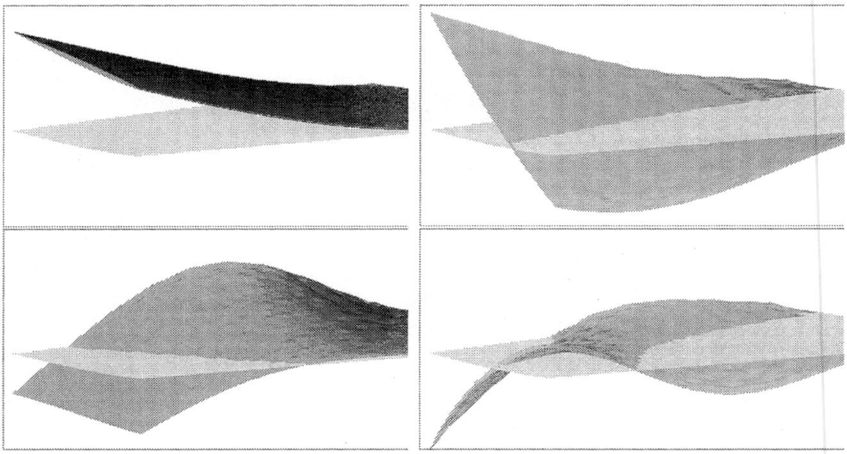

Figure 1 Vibration modes 1 to 4 - AGARD wing

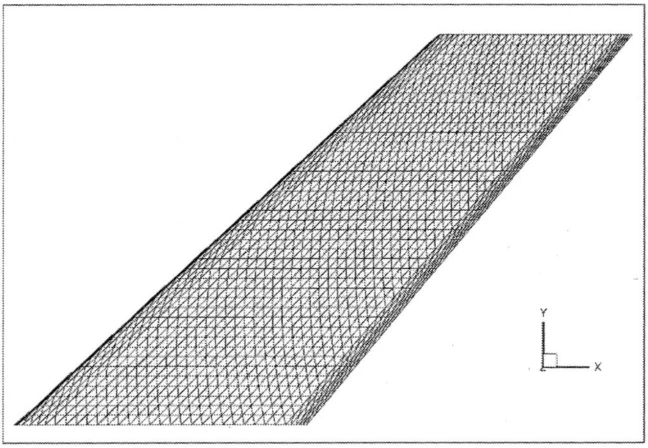

Figure 2 Mesh for Euler computation - AGARD wing

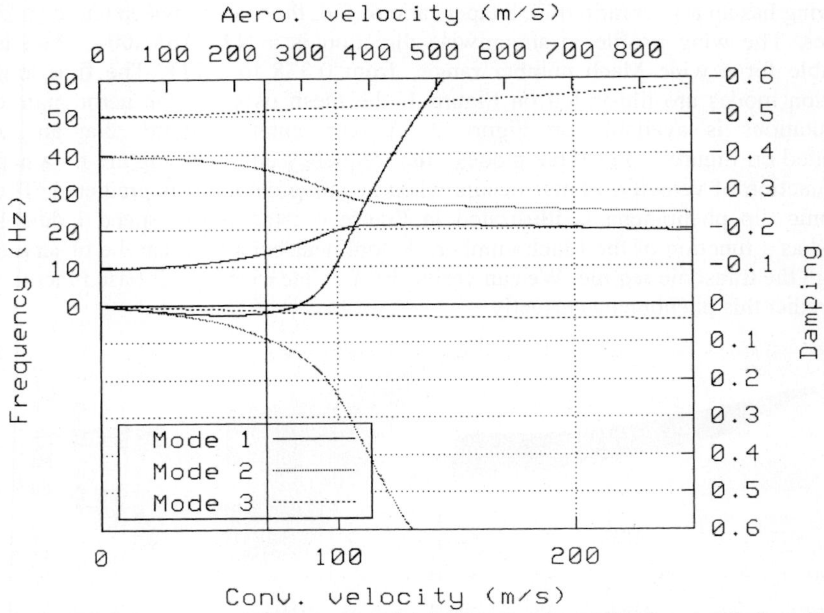

Figure 3 Flutter diagram (frequency and damping vs velocity for three modes) - AGARD wing

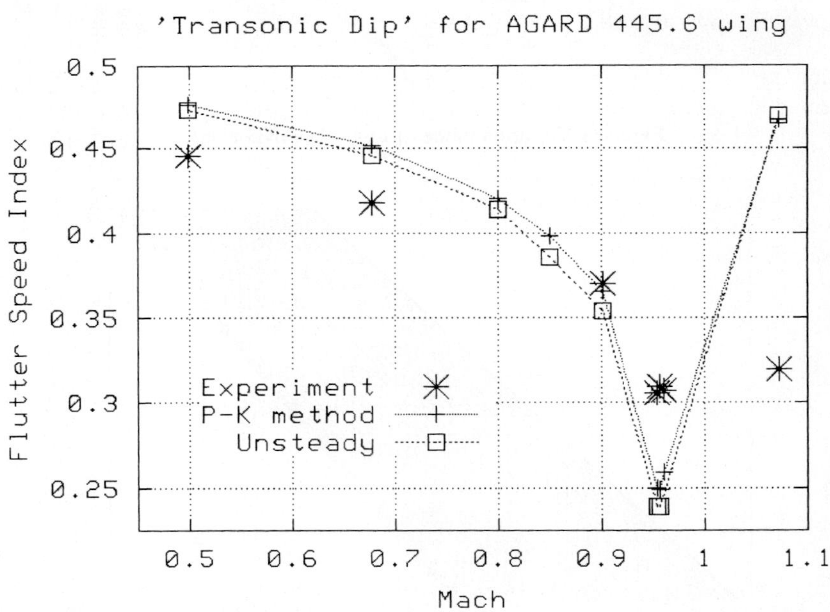

Figure 4 Transonic dip - Flutter speed vs Mach number - AGARD wing

3 Optimum Shape Design

Shape optimization is a very promising tool to improve and accelerate future aircraft design. Optimum shape design can be considered at two levels. At the early design stage, simplified models yield sufficient precision and strong multidisciplinary coupling is performed. When detailed design is reached, more accurate models are required for each discipline and multidisciplinary design is more difficult. One bottleneck that has delayed the impact of multidisciplinary optimization is the high cost of aerodynamic optimization. Considerable effort has been devoted to make progress in this field [6].

The general formulation of a shape optimization problem has been introduced by many authors ([14], [11]) and the key role played by the adjoint equation to obtain the design sensitivities in order to compute efficiently the gradient components of the cost function has been described. The adjoint equations themselves are formed as a system of partial differential equations. These can be discretized and solved in the same manner as the original flow equations. This approach is often called a continuous sensitivity analysis. We have selected a different approach, called a discrete sensitivity analysis which consists in applying the control theory to the discrete equations. The software to iteratively solve the adjoint equation was developed with the help of the *Odyssee* automatic code differentiation tool. The original code is the same 3D parallel unstructured Euler code EUGENIE mentioned in section 2.

A typical example of the design capability that has been reached is illustrated for the design of the transonic wing of a business jet. A complete configuration is considered (wing, fuselage, nacelle, pylon, empennage). The configuration is presented in Figure 5. An inverse design problem is considered: the objective is to reach a target pressure over the wing. The target pressure is defined to reduce the wave drag associated to the shock. The complete aircraft lift is constrained to be equal to a given value. CAD variables are used to define the shape parameters : this includes location, tangent and curvature parameters at control points. 96 shape variables are considered. The result of the optimization process is illustrated in Figure 6. The shock has been considerably reduced. The shape modification is illustrated for one wing section on Figure 7. The cost of the optimization process is only 6 times the cost of the solution of the Euler equations in this configuration. This is extremely efficient if we consider that 15 design iterations have been performed and 96 design variables are updated.

One issue associated to optimum design is non convex optimization where gradient based methods cannot be used alone. In this case, stochastic optimization should be used. See for example [13].

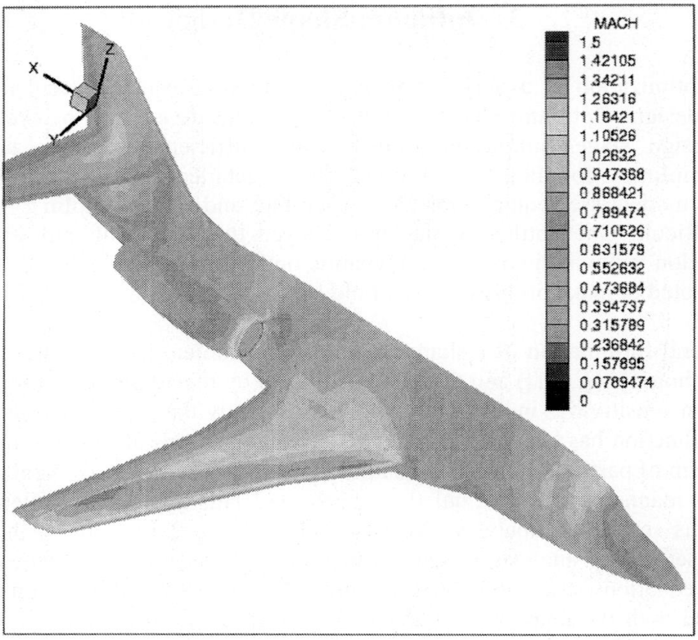

Figure 5 Optimum shape design - Falcon configuration

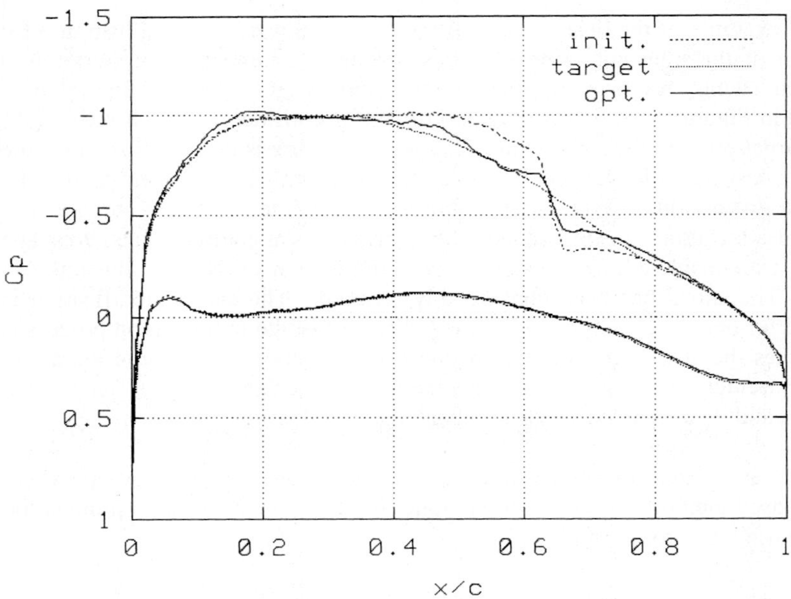

Figure 6 Optimum shape design - Pressure along wing section

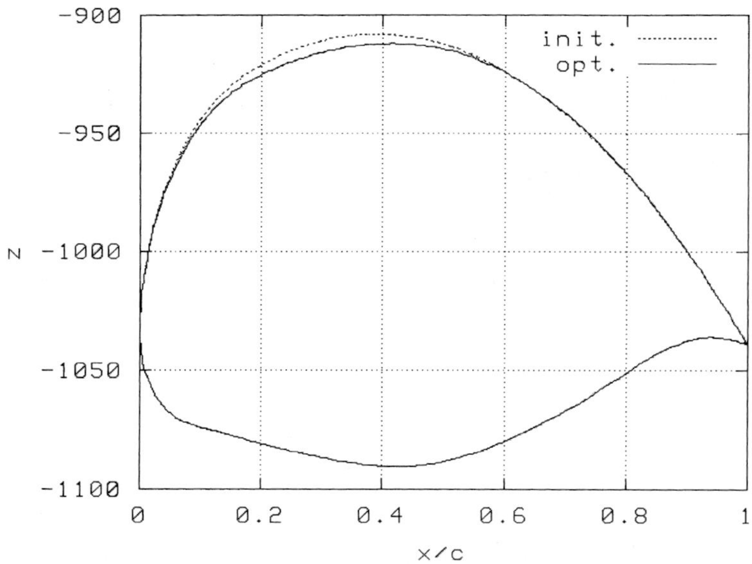

Figure 7 Optimum shape design - Initial and optimized shapes

4 Turbulence Modeling

The impact of CFD for aircraft design is still limited for many applications by a lack of accuracy. This problem can often be traced to inadequate turbulence modeling for complex flow configurations. Examples of such situations include flows with large separations like high angle of attack flows over delta wings or air inlets, flows configurations with active or passive flow control devices, afterbody flows.

One way to improve turbulence models is to consider advanced Reynolds averaged models. Another approach is based on the Large Eddy Simulation model. This approach will not replace simulation based on the solution of the Reynolds Average Navier Stokes equations with models like the (k,ε) model [15]. However it can be used for special challenging applications and for validation.

The work presented in [2] describes how our industrial unstructured NS code *AETHER* was modified to gain an LES capability (see [1] and [10] for a description of the numerical features of *AETHER*). Modifications includes improvement of the time stepping scheme, modification of the boundary condition treatment and implementation of a selective Smagorinsky model [5]. LES was then applied to the study of the mixing enhancement of a compressible mixing layer. This study was presented in [4]. Mixing enhancement is obtained by a small jet blowing perpendicular to the plane mixing layer. We present below the result of an extension of this study : the influence of jet pulsation was studied numerically. The instantaneous temperature field in a plane perpendicular to the mixing layer are presented in Figure 8 for both cases (continuous and pulsed jet). A comparison of the average velocity profile downstream of the injection device is available on Figure 9. Improved mixing is obtained with a pulsed jet at this location not far from the control jet; further downstream similar mixing is obtained. Overall, the pulsed jet leads to a similar mixing as the continuous jet. The main positive aspect of pulsed jet is that it requires a reduced mass flow. This example illustrates the application of LES to the study of flow control.

Further developments are in progress to enhance LES application including the use of wall functions and implicit schemes. The value of VLES or DES models will also be assessed.

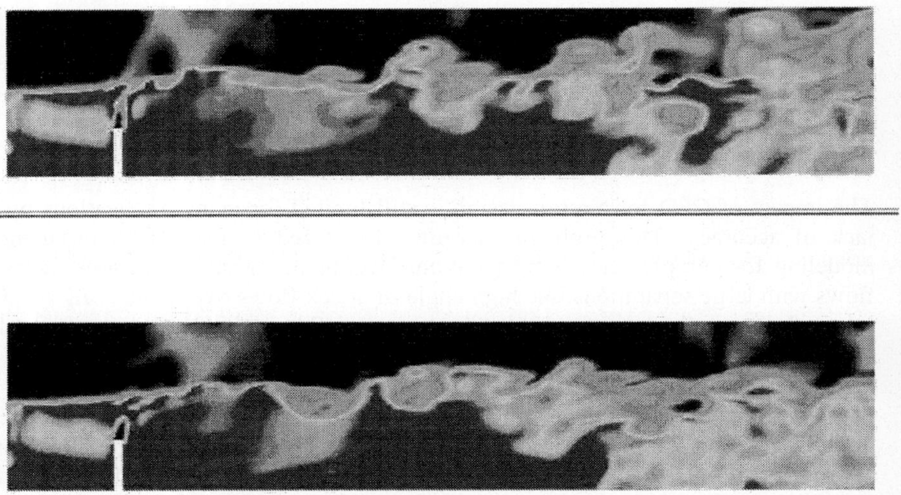

Figure 8 LES simulation of plane shear layer with excitation - Temperature isolines in a plane perpendicular to the mixing layer. Top : continuous blowing, bottom : pulsed blowing.

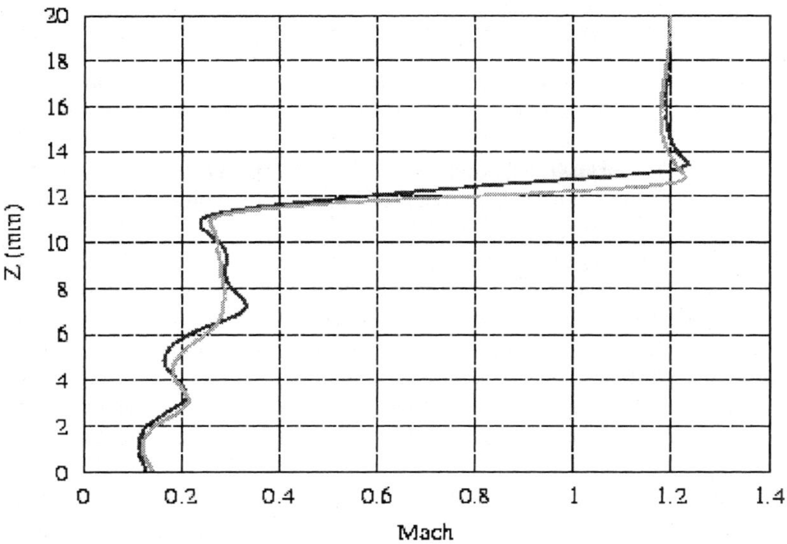

Figure 9 LES simulation of plane shear layer with excitation - Average Mach number profiles 23 mm downstream of the injection jet. Green line : continuous jet, red line: pulsed jet.

5 Conclusions

Recent key advances in simulation have been illustrated via 3 examples: the use of accurate aerodynamics simulation for transonic flutter analysis, the capability of efficient aerodynamic shape optimization for low drag transonic wing design, and the application of accurate LES turbulent modeling for flow control studies.

Multiphysics simulation is already performed in industry as demonstrated by our 1st example. Detailed buffet analysis could have been another example. Progress towards multiphysics optimization is made possible by advances described in our 2nd example.

Acoustics is an important area where multiphysics analysis is needed. Significant work is in progress in this field, part of it brings together CFD and CAA (see [12] for elements on aeroacoustics).

The design of advanced projects like a supersonic business jet will require a high level multiphysics design capability to achieve a quiet and efficient shape. This represents a real challenge in particular to the numerical simulation technology community.

References

[1] F. Chalot, M. Mallet and M. Ravachol, "A Comprehensive Finite Element Navier - Stokes Solver for Low and High-Speed Aircraft Design". AIAA 94-0814.

[2] F. Chalot, B. Marquez,.M. Ravachol, F. Ducros, F. Nicoud, Th. Poinsot, "A consistent Finite Element Approach to Large Eddy Simulation", 29th Fluid Dynamics conference, AIAA 98-2652.

[3] F. Chalot, Q.V. Dinh, M. Mallet, M. Ravachol, G. Rogé, Ph. Rostand, B. Stoufflet, "CFD for aircraft design: recent developments and applications", 4th ECCOMAS Computational Fluid Dynamics Conference, Wiley (1998).

[4] F. Chalot, B. Marquez,.M. Ravachol, F. Ducros, T. Poinsot, "Large Eddy Simulation of a Compressible Mixing Layer : Study of the mixing enhancement", 14th CFD conference, AIAA 99-3358.

[5] E. David, "Modélisation de écoulements compressibles et hypersoniques : une approche instationnaire", Ph.D thesis, INPG, 1994.

[6] Q.V.Dinh, G. Rogé, C. Sevin and B. Stoufflet, "Shape optimization in Computational Fluid Dynamics", European Journal of Finite Elements, Vol 5, 1996, pp 569-594.

[7] Th. Fanion, Ph. D. Thesis, "Etude de la simulation numérique des phéno mènes d'aéroélasticité dynamique. Application au problème du flottement des avions". University of Paris Dauphine, 2001.

[8] C. Farhat and M. Lesoinne, "On the accuracy, stability and performance of the solution of three-dimensional nonlinear transcient aeroelastic problems by partitioned procedures". AIAA 96-1388.

[9] "Grand Challenges: high performance computing and communications. The FY 1992 US research and development program". Report by the committee on Sciences. Office of Science and Technology Policy.

[10] T.J.R. Hughes et al., "A New Finite Element Formulation For Computa tional Fluid Dynamics", Comp. Meth. in Applied Mech. and Eng., North-Holland, 1986.

[11] A. Jameson, "Aerodynamic Design via Control Theory", Journal of Scientific Computing, 3:233-260, 1988.

[12] S. Lemaire and N. Heron, "An overview of Aeroacoustics at Dassault Aviation", Proceeding of the Swing workshop. DLR.

[13] B. Mantel, J. Periaux, M. Sefrioui, B. Stoufflet, J.A. Desideri, S. Lanteri and N. Marco, "Evolutionary Computational Methods for Complex Design in Aerodynamics". AIAA 98-0222.

[14] O. Pironneau, "Optimal Shape Design for Elliptic Systems", Springer - Verlag, New-York, 1984.

[15] M. Ravachol, "Turbulence modeling". Computational Science for the 21th century, Wiley, 1997, pp. 380-391.

Challenges in Development of mini and micro Air Vehicles (mAVs and μAVs)

K.C. WONG
School of Aerospace, Mechanical and Mechatronic Engineering,
Building J07, The University of Sydney,
NSW 2006, Australia.

Summary

Critical technologies being developed to make micro Air Vehicle systems feasible include advanced aerodynamics, flight stability and control, lightweight propulsion and power generation, lightweight navigation, reliable communications and onboard processing systems, advanced structures and materials, and unique miniature sensor payload technologies. The challenge then remains to optimally integrate these technologies into systems to meet operational requirements. This paper presents some practical aspects of these challenges.

1 Introduction

Aerial surveillance or utility systems the size of insects or small birds appeal to the imagination of science fiction writers and engineers alike. So can it be done? What are some of the challenges to be faced when developing such flight platforms? How are some of these challenges being taken on? The idea is certainly not new. After all, the first recorded use of an aerial surveillance system can be found in Genesis 8 of the Bible: *"Noah sent out a raven and a dove to see if water had fully receded from the surface of the earth"*. Critical technologies are globally being developed to make micro Air Vehicle (150 mm or 6 inches maximum dimension) systems of the future feasible. The challenge then remains to optimally integrate these technologies into a system to meet operational requirements. This paper seeks more to outline some of these challenges in the local context, than to offer any specific solution in detail.

1.1 Unmanned Aerial Vehicles (UAVs)

Unmanned Aerial Vehicles (UAVs) are highly capable flight platforms operated without an onboard human pilot. They are robotic aircraft, which are often computerised and fully autonomous. Over the history of UAV operations, they have been shown to have unmatched qualities that often make them the only effective solution in specialised tasks where risks to flight crew are high, where beyond normal human endurance is required, or where human presence is not necessary. Thus they continue to offer new and cost-effective capabilities not previously attainable.

A casual browse on the Internet would reveal hundreds of UAV systems being developed or operated worldwide, e.g. The NASA Wallops Flight Facility's Unmanned Aerial Vehicles Web Site located at: http://uav.wff.nasa.gov/. UAVs have become subjects in the evening news reports since the Persian Gulf War in 1991, when soldiers were seen surrendering to *Pioneer* UAVs. The use of UAVs in more recent conflicts in the Balkans (*Predator* UAVs), and in Afghanistan (*Predator* and *Global Hawk* UAVs) have highlighted their usefulness and utility by the military. In the civilian sectors, NASA has been using UAVs for scientific research for many years, and an example of a successful commercial UAV programme is the Australian meteorological surveillance *Aerosonde* UAV.

Amongst many other reasons, motivation for aerospace research groups to be involved in UAV research include the increasing availability of enabling technologies spawned by the information revolution, where UAVs offer an aerospace instrument for this development. Besides, UAVs offer new opportunities to the modern aerospace engineer where bold innovations within a maturing manned aircraft sector are getting more constrained. Airframes, propulsion units, launch and recovery, ground control stations, navigation and guidance, autonomous flight controllers, fault tolerance, sensor payloads, data communication, storage, processing and dissemination are aspects of key UAV technologies, which must be integrated into operational systems suitable for the missions they are developed.

1.2 Micro Air Vehicles

Micro Air Vehicles (μAVs) are essentially small aerial robots with six degrees of freedom. They can be usefully deployed with a payload to remote/hazardous locations for the purpose of reconnaissance, targeting, tagging, bio-chemical sensing, policing, hostage situation awareness, and often does not need to beg one's imagination in what else they can be useful for. They also present a potential capability for indoor aerial reconnaissance and possibly other missions, which may require manoeuvring within tight and confined situations. The US Defence Advanced Research Projects Agency (DARPA) issued a requirement in 1992 for a surveillance platform that could fit within a sphere of 150 mm (6 in.), weigh less than 140 g (4 oz.), fly for up to 2 hours, have a range of 10 km (6 miles), operate in winds of up to 13 m/s (26 knots); have autonomous capabilities, carry daylight image sensor with a telemetry link, be stealthy, and cost no more than US$1000 in production quantities [1]. These stringent basic specifications have deliberately been chosen to push available technologies. The essential enabling technologies for μAVs include Low Reynolds Number (Re) aerodynamics, propulsion and energy storage, lightweight multi-functional airframe components, miniaturised sensors, navigation, flight stability and control, and secure communication.

Early studies by various research laboratories have demonstrated that the concept was feasible. Other than the obvious electronic miniaturisation issues, which are making rapid progress due to consumer electronic industry demands, there are major issues relating to the ability to successfully build and operate such a flight platform. Flight research has predominantly been on manned platforms over this first century of manned aviation. After the Wright Brothers' first flight, history of aviation has been noted with mankind's desire to fly faster, higher, and with bigger aircraft. Until fairly recently, small flight platforms were left to weekend hobby aero-modellers, whose claim to aeronautical expertise were often seen to be grossly exaggerated. However, when more of these modellers go on to complete Aerospace Engineering degrees, and start doing serious research into understanding the sciences of aeronautics applicable to these "toys", they actually became the source of much of the recent advances in micro and mini Air Vehicles.

2 Design Challenges

An operational μAV is often envisaged to be a small flight surveillance system which can be readily carried in a backpack, launched with minimal preparation, be operated with a minimal skill in any given situation which may include flight within confined spaces, be reasonably recoverable, and if not, be cheap enough to be disposable. This presents an enormous challenge, the totality of which is well beyond the scope of this paper, where challenges to airframe system design will mainly be discussed.

2.1 Low Reynolds Number Aerodynamics

Aerodynamic methodologies at low Reynolds numbers are not yet well defined, and yet are essential to the design and operation of μAVs, which typically operate in Re < 150,000. This is in the domain of understanding the aerodynamic and flight behaviour of birds, insects, and model aeroplanes. Indeed, much valuable research at such low Reynolds numbers are being done by either biologists [2] or aero-modelling engineers and researchers [3]. Furthermore, the compactness required of such flight platforms lead to very low aspect ratio wings (in the context of fixed wing configuration μAVs). Such low aspect ratio wings, which have unique characteristics, are not often used anywhere else. Computational tools are also notably inadequate in modelling flows around μAVs, particularly in the highly turbulent and gusty urban environment typically envisaged for these flight platforms.

2.2 Propulsion Systems

Propulsion and energy availability is another critical consideration. Having enough power to perform useful missions is much constrained by available size and weight, as well as energy density of available fuel, which also needs to be safe for routine operations. Propulsion units in this size range are simply not yet well understood. In nature, flight capable insects and birds spend a major part of their waking hours locating and consuming large quantities of food to replenish their energy source. To date, propulsive concepts that have been explored for use in μAVs include the use of high efficiency electric motors and high energy-density batteries, chemical muscles, and liquid fuelled turbine engines [4]. Ornithopters have made a recent "comeback" because the concept of flapping wings offer lifting capabilities as well as propulsion, which fits in well with the "multi-functional airframe component" driver. Whether ornithopters *per se*, are best suited to mAVs remains to be proven in practice, but the multi-functional concept remain a goal when optimising the design of such small flight platforms.

Power requirement to maximise endurance, for a typically propeller-driven μAV can be expressed as:

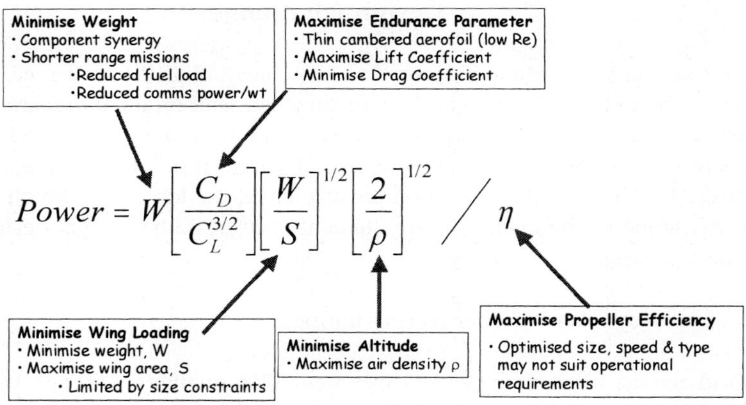

Noting the desirable attributes of each parameter of the above expression emphasises the complex nature of the problem when applied to μAVs. Similar expressions can be derived for other design requirements, which highlight the requirement for a multi-disciplinary optimisation (MDO) design approach when developing small flight platforms.

2.3 Stability and Manoeuvrability

The intended utility of micro Air Vehicles as portable remote sensing flight platforms makes it essential for them to be stable and yet manoeuvrable. They are required to handle turbulence typically encountered at low flight altitudes, and within urban surroundings, typically demanded of μAV missions currently envisaged. Furthermore, it is necessary for μAVs to be capable of manoeuvring within confined indoor environments. Operators of μAVs should not be expected to have specialised skills other than being able to prepare the flight platforms for launch, being able to direct them to intended locations, and being able to make good use of the acquired information. The requirement for simplicity-in-operation places demands on increasing levels of autonomy, which is not yet even common on much larger UAVs [4].

2.4 Configurations - Component Synergy and Integration

Since the DARPA-issued specifications for μAVs, several research teams have flown basic systems with modest capabilities. One such example is the 150mm electric motor powered AeroVironment *Black Widow*, which has an all-up-weight (AUW) of around 50g, having a 1 km operational radius, and over 20 minutes of endurance [5]. Numerous airframe concepts as well as propulsion technologies are being worked on as DARPA sponsored projects. Not surprisingly, much current activity has been inspired by nature, where it is noted that there is a large range of flight-capable insects, birds and other animals, whose capabilities are still far from being achieved by small machines. Due to the extreme size and volume constraints, a high level of component system synergy and integration is required. Budgeting for the various technology-driven aspects of such platforms, from aerodynamics to sensors, must be moderated by a need to optimally integrate these technologies from the outset, rather than develop each component separately. The bottom line in developing micro Air Vehicles is that it has to be done as a multi-functional design from the outset. One cannot (for example) develop the aerodynamics in an isolated fashion - having something that flies will not necessarily make a good μAV system.

Airframe related configurations that are currently being explored globally include:

- Electric-powered propeller-driven fixed wing platforms:
  - AeroVironment Black Widow (50g, 1km range, >20 minutes endurance); and
  - BAE SYSTEMS Sanders MicroSTAR (100g, 5km range, >30 minutes endurance).
- Internal Combustion engine powered, propeller-driven (which offer better energy density, but suffers from reliability of available small engines) platforms, which:
  - use Rotors and Ducted Fans;
  - are capable of vertical launch, recovery and flight;
  - use Micro Turbines, as in examples being developed at a laboratory in MIT, which:
    - are silicon MEMS based, fuelled with hydrogen, rotating at 1 million RPM, and providing 0.1 N thrust;

  - may be better used in generating electricity to power motors;
  - have poor efficiency now due to fabrication difficulties of microscopic components.
- Flapping Wing (ornithopter) platforms which include:
  - Entomopters from Georgia Tech Research Institute;
  - Micromechanical Flying Insect from Berkeley; and
  - μAVs with MEMS wings from CalTech and AeroVironment [6]

There are certainly many other configurations being explored. The exciting challenge for aerospace engineers is that opportunities exist for innovative configurations, pushing the need to "think outside the square" when developing μAV flight platforms. It has been noted that micro Air Vehicles are "…at the *Wright Flyer* stage in development…", and configuration variations are expected to, likewise, stabilise when understanding and technology matures.

Basic systems currently being flown have very basic and modest capabilities, with minimal flight and mission autonomy. Mission-path planning and navigation are issues which demand attention because the mode of operation (such as indoor environment) precludes the use of the typical Global Navigation Satellite Systems (GNSS). Similarly, urban and indoor environments pose significant problems to secure and reliable communication for sensor telemetry and platform control uplink. Airframe components would have to be multi-functional, such as aerodynamic surfaces having to double-up as electronic circuit boards, or sensors which may have be integrated as structural components. Adding autonomy presents yet another dimensional challenge. Given the physical scale constraints, there is no alternative than to use a global multi-disciplinary optimisation design approach when developing μAVs, with continuing sensitivity analysis of all design parameters, with an overriding requirement for high levels of component synergy and integration.

2.5 Mini Air Vehicles (mAV)

It is noteworthy that much of these technologies are currently being assessed on slightly larger mini Air Vehicle (mAV) systems (flight platforms with 200-1500 mm maximum dimension). This approach presents opportunities to develop analytical tools, acquire system design experience on small flight platforms, at least an order of magnitude smaller than what engineers are typically used to, and be able to make use of commercial off-the-shelf (COTS) equipment and structural material.

At present, even the highly resourced DARPA considers operational μAVs not yet feasible with contemporary miniaturisation technologies, and thus supports deployment of "larger" mAVs such as the *Dragon Eye*, which may create new markets and spur industry to develop lighter and smaller system components [4].

It is noteworthy that the US Marines are currently evaluating the *Dragon Eye* (1.14m wingspan, 1.8kg) mAV, to possibly replace much larger (manned helicopter-sized) *Fire Scout* VTOL UAVs. The motivation for researchers to consider the use of mAVs is that these slightly larger flight platforms provide the means to develop μAV related capabilities without the high expense of extreme miniaturisation.

Whilst the USA appears to lead in μAV technologies and development, there are certainly mAV-related research and development (R&D) activities globally, including several groups in Australia. At Sydney University, there has been an active Unmanned Aerial Vehicle (UAV) Research Group over the past decade. Over this time, the aim has been to build design and operational expertise on several small flight platforms, ranging from over 40 kilograms to less than 400 grams. Although the emphasis had been to integrate COTS resources into R&D flight platforms, the cumulated team expertise has placed the group well to take on some of the technological challenges in developing micro-sized flight platforms.

3 Related Research Activities at The University of Sydney

3.1 Background on UAV Research

UAV activities have been a sporadic part of the Department of Aeronautical Engineering at The University of Sydney since the department's inception in 1939. Although not much has been documented, archival photographs show various remotely piloted aircraft being developed and operated over its 62-year history. More recently, a UAV Research Group, which originated in 1988, has made good progress in incorporating UAV flight platform design, development, control, and operation to the department's core teaching and research activities. This include the instrumentation of COTS Remote-Controlled flight platforms, the design development and operation of flight research platforms, the design and development of innovative airframe concepts, the design and development of micro Air Vehicles and exploring commercial applications for UAVs. Projects are generally based on a low-budget approach to UAV flight research, which has so far provided flight platforms for wide variety of UAV-related research projects.

3.2 UAV *Ariel*

The UAV *Ariel* is a flight research platform, developed since 1989 as an exercise in UAV airframe design and development. As the development of the early derivative-UAVs progressed, it became apparent that a larger, modular and more capable UAV would have application not only to immediate research aims, but also to other applications. Numerous students, undergraduate and postgraduate, worked with academic staff on various aspects of the design, wind tunnel testing, flight simulation development, tooling design and construction, and finally the building of the flight prototype. The prototype *Ariel* first flew after approximately six years of developmental work. Flight operations were mainly conducted at a general aviation airport (Camden Airport), with a Civil Aviation and Safety Authority (CASA) issued Permit-to-Fly, and full cooperation from the relevant airport authority and airport flight controllers. This project provided the UAV Research Group with much valued experience in UAV airframe and system design and development, and UAV operations in controlled airspace. It also provided the subject for numerous research papers, undergraduate and postgraduate (Masters and PhD) thesis, inter-varsity projects [7], and research contracts.

Figure 1 UAV *Ariel* (left), and UAV *Brumby Mk I* (right) during test flights

3.3 UAV *Brumby*

The *Brumby* UAV is the natural follow-on of the UAV *Ariel*. Like its namesake, which is the Australian wild horse, the UAV *Brumby* is designed to be a rugged research 'workhorse'. After operational experience with the *Ariel*, the original *Brumby* was specifically developed to be a rapid prototype low cost research UAV, to provide a flight research platform in support of various research activities. The *Brumby Mk.I* was the first version built and flown and, as an indication of the success of rapid prototyping, the airframe was built and flown (albeit with remote pilot) in less than six weeks (including the fabrication of tooling and composite moulds). First flight was on 21 November 1997. It was demonstrated to be a stable flight platform well suited to research requiring the carriage of sensors on a flight platform. Two *Brumby Mk I* airframes have been built, with the University of Ohio, USA, currently operating one as a flight research platform.

The *Brumby Mk I* prototype research airframe concept has since been further developed, the latest *ANSER Brumby Mk III* being part of a BAE SYSTEMS led UAV project [8]. Amongst other airframe related activities, a mini Air Vehicle airframe patterned after a 40% scale model of the interim *Brumby Mk II*, named *mini-Brumby*, has been developed to explore mAV flight research applications. This was done to exploit wind tunnel data available [9] for the *Brumby Mk II*, obtained through several student research projects

3.4 *TWing/Mirli* VTOL UAV

The launch and recovery of a UAV remain critical phases of flight operations. At The University of Sydney, investigations into a novel Tail-sitter Vertical Take Off and Landing (VTOL) UAV design, using an MDO design approach were initiated in 1995 [10]. Tail-sitter UAVs promise greater operational flexibility than conventional UAVs while avoiding the performance penalties experienced by helicopter-like vehicles. This is done by allowing the vehicle to adopt a vertical attitude during takeoff and landing during which its weight is supported entirely by propeller thrust, thus in turn obviates the need for a runway. During the vertical flight phases, the aircraft is controlled by prop-wash over wing and fin mounted control surfaces. Once the vehicle reaches a certain altitude, it pitches over and performs a "stall-tumble" transition manoeuvre during which it trades altitude for forward speed. As the speed increases, the wings provide the lift for the vehicle. After performing a mission, the vehicle regains vertical flight via a controlled pull-up and descent. The *TWing* UAV project is currently supported by the Australian Research Council (ARC) and Sonacom Pty Ltd, to develop a flight-worthy technology demonstrator for investigating the feasibility and controllability of the launch and recovery phases [11].

Figure 2 *TWing/Mirli* VTOL UAV during tethered hover tests (left – *courtesy Dr Hugh Stone*), and *Bidule* mini Air Vehicle (right)

3.5 Mini Air Vehicle *Bidule*

The mini Air Vehicle *Bidule* is a research platform initiated as an exercise into developing micro-sized flight platforms. Although not a micro-sized vehicle based on DARPA definitions, it serves as a useful platform for investigating low Reynolds number aerodynamics, and particularly the unique configuration of having 90% of the wing immersed in prop-wash, giving the *Bidule* excellent low speed flight capabilities. Longitudinal and lateral controls are achieved utilising a pair of elevons, and twin rudders allow the directional control. As a technology demonstrator, its purpose was to initially validate the basic design through remote-controlled indoor free-flight tests before considering a possible reduction to a real "micro" size. Wind tunnel testing of a propelled model of the Bidule prototype showed quantitatively that the basic design benefits from the prop-wash effect in terms of increased lift. In the context of size reduction to the present vehicle, the results suggest keeping the idea of a wing-body immersed in a propeller slipstream, provided that the destabilising effects due to the power system and the prop-wash can be kept to an acceptable level [12]. From initial wind tunnel investigations, Reynolds number effect for the *Bidule* mAV configuration were found to be low, whilst the prop-wash effect was found to be highly beneficial, which supported observations of the flight characteristics of the prototype. A panel method numerical code is being adapted to this investigation, supported by ongoing wind tunnel experimentation.

3.6 UAV Flight Test Facility

Having ready access to a flight test facility is crucial to the success of UAV programmes. Sites that have been utilised by the USyd UAV Research Group for flight trials have included various hobby model aeroplane fields and the Camden Airport (general aviation). More recently, a 20,000-acre university farm has been made available, and work is underway to set up a permanent UAV Flight Test Facility there. Located in Marulan, NSW, the site has a 300m runway, a large shed suitable for use as workshop and storage, and accommodation for the flight test teams. The Australian CASA has issued an exemption to permit flight trials of (currently) *Brumby* UAVs over a designated area at the farm, with strict operational guidelines and procedures. Working closely with CASA from the early stages of projected flight trials of UAV projects has brought very positive outcomes with respect to flight operations.

3.7 What's next at The University of Sydney?

Current Aims of UAV Research Group – "Leading to the Future" include:

- Development of new analysis, design, manufacturing, and verification tools for small UAVs;

- Exploitation of available COTS technology when developing mAVs and μAVs;
- Exploration of new mAV and μAV airframe systems, and challenge conventional thinking in flight system design;
- Demonstration and implementation of autonomous flight control capabilities and mission control systems in small UAVs;
- Further demonstration of the *TWing* VTOL UAV concept, and developing it further into an operational system;
- The miniaturisation of the *Bidule* concept into a μAV system;
- The development and testing of miniaturised automatic flight control systems using the *mini-Brumby* mAV;
- The development and integration of "smart" actuators and airframe components for new μAV systems, based on lessons from nature;
- Development of Multi-Disciplinary Optimisation (MDO) Design Tools specifically for μAVs;
- Development of derivatives to existing small UAV designs;
- The cultivation of collaborative relationships with companies, research organisations, and universities; and
- The continuing design and development of mini- to micro-sized UAV airframes to suit a multitude of potential applications in civil and defence market domains.

4 Conclusions

Taking on the challenge to develop small flight platforms suitable for useful surveillance missions can provide an exciting multi-disciplinary research domain. This challenge has been taken on by the UAV Research Group at The University of Sydney through a gradual process, maximising the utilisation and integration of available off-the-shelf technologies. The experience base is now maturing to implementing more formal optimisation techniques to the design and development of small flight platforms as mini Air Vehicles and micro Air Vehicles.

References

[1] G. Goebel. "Unmanned Aerial Vehicles" v1.1.0, http://www.vectorsite.net/avuav.html.

[2] C.P. Ellington. "The Novel Aerodynamics of Insect Flight: Applications to Micro-Air Vehicles". In The Journal of Experimental Biology 202, 1999, pp.3439-3448.

[3] M.S. Selig, J.F. Donovan, D.B. Fraser. "Airfoils at Low Speeds". In Soartech 8 (published by H.A. Stokely, 1989).

[4] Office of the Secretary of Defense. "Unmanned Aerial Vehicles Roadmap". U.S. Department of Defence, April 2001.

[5] J.M Grasmeyer, M.T. Keennon. "Development of the Black Widow Micro Air Vehicle", In American Institute of Aeronautics and Astronautics Paper AIAA-2001-0127.

[6] T. Pornsin-Sisirak, S.W. Lee, H. Nassef, J. Grasmeyer, Y.C. Tai, C.M. Ho, and M. Keennon. "MEMS Wing Technology for A Battery-Powered Ornithopter". In proceedings of Thirteenth IEEE International Conference on Micro Electrical Mechanical Systems (MEMS '00), Miyazaki, Japan, Jan 23-27 (2000).

[7] M. Crump, C. Bil. "Autonomous Shipboard Launching of Unmanned Air Vehicles (UAVs) in Extreme Conditions". In Conference Proceeding of Information, Decision and Control 2002 (IDC2002), 11-13 February 2002, Adelaide, Australia, pp.95-100.

[8] K.C. Wong, H.J.H. Peters, P. Catarzi. "Adapting to Limitations of a Wind Tunnel Test Facility in the Aerodynamic Testing of a new UAV", In CDROM proceedings of the 9th Australian International Aerospace Congress (AIAC 2001), 5-8 March, 2001, Canberra, Australia.

[9] B. Yelland et. al., "UAV Technology Developmental: A Node within a System". In Flight International's UAV Australia conference proceedings, 8-9 February 2001, Melbourne, Australia.

[10] R.H. Stone. "Configuration Design of a Canard Configured Tail-Sitter Unmanned Vehicle Using Multidisciplinary Optimization". PhD Thesis, University of Sydney, 1999.

[11] R.H. Stone, G. Clarke. "The T-Wing: A VTOL UAV for Defence and Civilian Applications". In Flight International's UAV Australia conference proceedings, 8-9 February 2001, Melbourne, Australia.

[12] T. Spoerry, K.C. Wong. "Design and Development of a Micro Air Vehicle (mAV) Concept: Project Bidule". In CDROM proceedings of 9th Australian International Aerospace Congress (AIAC 2001), 5-8 March, 2001, Canberra, Australia.

Future Tools for Design and Manufacture of Innovative Products in the Aeronautics Industry

Moderated by R. SANDLAND, CSIRO (Chairman)
Prepared by M. SEFRIOUI, Dassault Aviation (Round Table *Rapporteur*)

This Round Table featured academic and industrial experts, as well as representatives of governmental institutions involved in aeronautical applications of flow, structure and waves. The presentations here are in the order used for the Round Table – from governmental laboratories and industrial users through to academia.

Ron Sandland (Chair of the Round Table Discussion)

Dr Sandland holds the degrees of BSc (Hons), PhD and is an Associate of the Institute of Actuaries. He is an Honorary Life Member of the Statistical Society of Australia and a Fellow of the Academy of Technological Sciences and Engineering. He is currently Deputy Chief Executive of CSIRO, where his role embraces overall strategy for CSIRO in the areas of Information and Communications Technology, as well as for research in services and manufacturing.

I think that this French - Australian workshop is a wonderful idea, and comes at a time when there is a widely accepted need for stronger interaction between modern mathematics and technological applications. The role mathematics should play in helping to solve industrial problems has already been recognised for some time. Now, the aim in industry is to do everything better, faster and cheaper. It is an ambitious goal, but the best way to hold a competitive advantage.

In industry, the need to reduce time in product design and development makes it necessary to have mathematics in the tool bag. One of the most challenging fields of applications is the aeronautical industry, especially since this application helps to define the future of manufacturing.

Our expert speakers will each make a short presentation to identify important considerations as they see them. After that, we will have a general discussion in which we will jointly identify important issues and themes.

Ian Mair

Dr Mair holds the degrees AM, BEng (Civil), MEngSc, PhD and is currently Chief Executive Officer in the CRC for Advanced Composite Structures. He is a Fellow of the Academy of Technological Sciences and Engineering, a Professorial Fellow in Civil Engineering at the University of Melbourne, and Chairman of Engineering Education Australia Pty Ltd. He was made a Member of the Order of Australia in 2000 for services to engineering. His areas of interest are research and development, materials science and engineering, technological industries and interface of technology and society.

Worldwide there is a lot of ongoing research on composites, and since this conference is oriented towards aeronautics, recent comments from Lockheed Martin present a good summary of industry's needs. The paramount needs of industry are: ongoing search for high performance materials that can be produced at low cost; today's manufacturing processes are costly and labour intensive; aerospace is searching for bigger, faster and lower cost composites without loss of properties. Whatever the answer to these challenges may be, the one certainty is that no single organisation can handle the challenge on its own.

There is a set of available computational design tools that help, but there is no comprehensive package that can single-handedly deal with all aspects of a design. Some capabilities of these tools cover, for example, topology optimisation and parametric design optimisation for structures which can be applied in the conceptual design stage.

This is important since 80% of the cost of a product is defined at the conceptual design stages. With appropriate software tools the time taken in design can be substantially reduced enabling many more concepts to be examined. For example on a recent spoiler design study, the normal nine man-month effort for the design was reduced to two man-weeks. However, such a dramatic improvement does not change the fact that there is still a fair amount of "art" involved in the design and software utilisation. Improved software tools are needed to streamline the design and manufacturing processes in order to cut down on cost, for instance through more "automation" in the design process.

One way to save on costs is by applying the concept of re-usability. It simply consists in re-using an experience that you are already familiar with. The speed-up can be quite impressive, as this approach exploits knowhow that has already been accumulated. Another approach is to consider a component-based tool. For instance, CATIA 5 is made of modular components that interact and help the engineering process. It is called knowledge-based engineering.

On design of composites, I'd like to emphasise that, in contrast to metals, the composite properties are established during the manufacturing process and may vary throughout a component. An example is resin infusion of dry fabrics, where you can actually change the fabric density, permeability and other attributes of the material as it is shaped. And that is not something people are used to in design. The design and manufacturing of such composites has markedly improved thanks to new software tools such as PAMFORM.

With respect to simulation and prediction of how the resin will flow and cure, I note that some software tools have started to appear. Also we now have software that is starting to predict what the mechanical properties may be after the resin has cured. An example is group interactive modelling, where the aim is to find how a resin is going to be changed by different additives. Perhaps the ultimate wish is to break free of physical testing and only run simulations, but that is still a goal for the future!

Q: What about materials standards as a way to reduce costs?
With standards several issues arise. Materials standards have a place but standardisation can discourage development of new materials since certification might be really hard to obtain. Standard tests cannot cover all stress conditions that may be experienced and use of the results has to be appropriately conservative. Effort to overcome this limitation in design involves development of material failure models, however, these are far from perfect. In the UK for instance, some work is being done on how well the various available models can predict onset of failure in a specimen with a wide range of results. There is of course also the question of the confidence in the repeatability of the manufacturing process itself, which adds variability and needs to be allowed for in the design.

Q: Aircraft design is at the edge of knowledge boundaries since you cannot afford to have a problem appear in flight. For that field, I would say that elimination of testing is a remote dream.
I agree with that observation, and there will always be a need for full component testing. However, it might be possible to work at decreasing the actual amount of testing required whilst not completely eliminating it. For instance, in the automotive industry, physical crash testing is performed on full vehicles, but the direction is to use fewer crash tests (1 instead of 3 for example) through increased confidence in the predictive capability of the design software.

Jean-Jacques Gagnepain

Dr Gagnepain holds the degree of *Doctorat Es Sciences Physiques.* In November 2001, he was appointed Scientific Adviser of the General Secretary for Defence. Previously he was Director of the Engineering Sciences Department of CNRS (1992 - 2001) and responsible for CNRS Industrial Affairs (1997 - 2001). He is President of the National Institute for Research on Transport and Security.

I wish to describe two major challenges in modelling, namely multi-disciplinary and multi-scale aspects.

Multi-disciplinary aspects

Science is generally trying to reduce complex problems by isolating objects from the surrounding world in order to gain a better understanding of the phenomena involved. Simplify a problem to solve it more easily. Most of the models are built in one single discipline.

The organisation of research, in CNRS but in other research organisations as well, is not well adapted to multi-disciplinary fields because of the walls between disciplines. For example, when two phenomena are coupled, sometimes one set of equations can be solved independently of the other, and the result is a source to feed the second one. In that case, independent codes can be merged (in principle), and we can have one model per discipline. But if coupling is two-way, the two sets of equations cannot be separated and the problem must be globally treated as a multi-disciplinary modelling. That means the model must be developed from the beginning to take both phenomena into account.

Why is it hard to merge? Research is organised like teaching. And since teaching is organised by disciplines, research is also organised by disciplines. As long as organisation of teaching does not change, research may not.

The needs of multi-disciplinarity come from *outside* the academic world, mostly from industry. In industry, the needs continually change and evolve. For instance, before WW2, car companies focused on mechanics and aircraft companies on aeronautics. Now, each is involved in a very large number of disciplines. The problems these industries face can no longer be solved by one discipline but need multi-design optimisation, and we need to adapt. The needs of industry involve increasingly complex systems, and are therefore less and less limited to elementary objects. Such complex systems are not isolated but coupled to its environment, which is in itself another system.

Multi-scaling aspects

The second challenge is in the level of description of the objects and the phenomena. The classical models do not work any more at very small scales, because inhomogeneities of the bulks, discontinuities at interfaces, random distributed characteristics, non-stationarity, *etc.* These multi-scale effects can be found for example in turbulent flows, or in materials like bulk pieces. Bulk pieces are homogeneous without too many surface effects, but at a smaller scale, problems arise with the appearance of inhomogeneities and defects: the classical model is no longer valid at that scale. That leads to another problem, the coupling of microscopic characteristics of a material and its macroscopic properties of usage (composite materials for instance).

Modelisation

In fact, these considerations lead to one main objective: modelling the reality, the real things and not some kind of artefact. And to do so, one has to choose between two opposed approaches.

- It is common to work on the modelling of simple objects in order to test the tools. Such was, for instance, the approach chosen by P. Le Quéré in his lecture, where he showed the coupling of two temperatures by a simple model in order to better test the model.
- Consider now a whole system with all its complexity, with the risk of not being able to validate the tools. This is similar to the problem addressed in the presentation of M. Mallet, where he underlined the difficulty of modelling almost a whole aeroplane. A tentative answer might come from the USA, where the *Advanced Strategic Computer Initiative (ASCI)* has started developing models of complete systems that takes into account the whole complexity.

However, the solution to this dilemma probably lies in a compromise between these two extremes.

Chris Guy

Dr Guy holds the degrees of BTech and PhD and is Acting Director of the Aeronautical and Maritime Research Laboratory, Defence Science and Technology Organisation (DSTO).

He previously held the positions of Chief of DSTO's Airframes and Engines Division, Chief of DSTO's Air Operations Division, Senior Defence Scientific Adviser and Scientific Adviser-Army. He is a Director of the CRC for Advanced Composite Structures.

I will briefly outline a small part of the work that we undertake in DSTO. The examples I present will give you an idea of some applications in which computational simulations are important.

Flight Mechanics

When a store is released from an aircraft, safety is important. To ensure safe release, we use a mixture of wind tunnel testing and CFD simulation to calculate the store trajectory when released, before undertaking any flight-testing of that store/aircraft combination. We also use computational techniques to ensure that flutter does not occur for a particular store/airspeed combination. The costs involved are quite significant - for example it cost $18 million to update our transonic wind tunnel. However, wind tunnels might become obsolete in 20 years time, and all work might become computational.

Another example concerns submarine hydrodynamic noise. The problem arises because the submarine hull sheds vortices. This can give rise to non-uniform inflow to the propeller, which in turn makes the propeller vibrate, thus causing noise. We addressed this problem using a combination of CFD and wind/water tunnel experimentation to investigate modifications to the hull shape. We subsequently reduced the hydrodynamic noise significantly.

Computational Electromagnetics (CEM)

It is necessary to ensure that problems do not occur with aircraft electrical/electronic systems when an aircraft encounters electromagnetic radiation. An example might be an undemanded elevator input when an aircraft flies through a radar beam, or that radar signal triggering the accidental release of a store. In order to avoid such events occurring, a comprehensive series of ground testing is undertaken. This often involves CEM techniques in conjunction with exposure of the aircraft to electromagnetic radiation.

Structural Design

Our interest in structural design revolves around the study of cracks and metal fatigue in aircraft. We have carried out experimental fatigue testing for most of our aircraft including our F/A-18, F-111 and P-3C fleets. The problem with such an approach is that it is expensive. While we currently use a number of mathematical modelling techniques for aircraft fatigue life estimation, the ultimate goal might be to replace full-scale fatigue testing of aircraft with its virtual counterpart. However, we are still a long way from this goal. It is interesting to note that the aircraft loads model that we use as part of our F-111 sole operator program is much superior to those used by the designers of the aircraft.

Operational Analysis and Simulation

We assist with the selection and acquisition of new aircraft and ships, especially in the areas of determining the required capability, contributing to specifications and helping with the evaluation of tenders. A number of sophisticated mathematical modelling and operational analysis techniques are used as part of this process, some using intelligent agents to represent the behaviour of aircrew in particular operational scenarios. In addition we use manned flight simulators to investigate both tactics and new equipment that might be brought into service.

Jacques Périaux

Professor Périaux completed his PhD in 1979 under the supervision of Professor Roland Glowinski at the Université de Paris. His Habilitation in Applied Mathematics, in a jury chaired by Professor J L Lions, had the specialty of numerical analysis. From 1970 to 1995, he worked for Dassault Aviation as Leader of the Numerical Analysis Group in the Theoretical Aerodynamic Department, before being appointed High Scientific Adviser in 1996- in *Division Etudes Avancee*. Since 1998, he has been Chairman of the Scientific Pole between Dassault Aviation and University Pierre et Marie Curie. His research interests are in aeronautics and numerical and evolutionary methods for the design of aircraft.

New Tools Needed for the Next Decade

Dassault Aviation is an Air Systems integrator with technical skills vital to the application of optimisation tools to whole systems and all their functions. In the past fifty years, Dassault Aviation has delivered more than 7,000 civil and military aircraft to 73 countries, logging some 15 million hours in flight to date. This vast experience has allowed Dassault Aviation to build up considerable expertise in the design, development, production, sales and support of all types of aircraft. The key values of such expertise are flexibility, efficiency, teamwork, customer-centredness, and a passion for aviation.

There is a strategic interest at Dassault Aviation to ensure competitiveness in the design and manufacture of new products via innovative methods and tools. This endeavour targets new solutions for challenging multidisciplinary design problems in aeronautics, which require the use of seamless methods at the interfaces of several disciplines (see Figure 1) such as applied mathematics, modelisation of physics and computer science. Examples abound to confirm that the main challenges are at interfaces: multiphase problems, composite materials, any system with its environment, and the human/machine interface. These new problems require new tools in numerical analysis: methods for multi-scale problems, new decompositions, new optimisation/control algorithms (because of multi-criteria related to the multi-system approach), the coupling of numerical analysis algorithms and CAO, and so on.

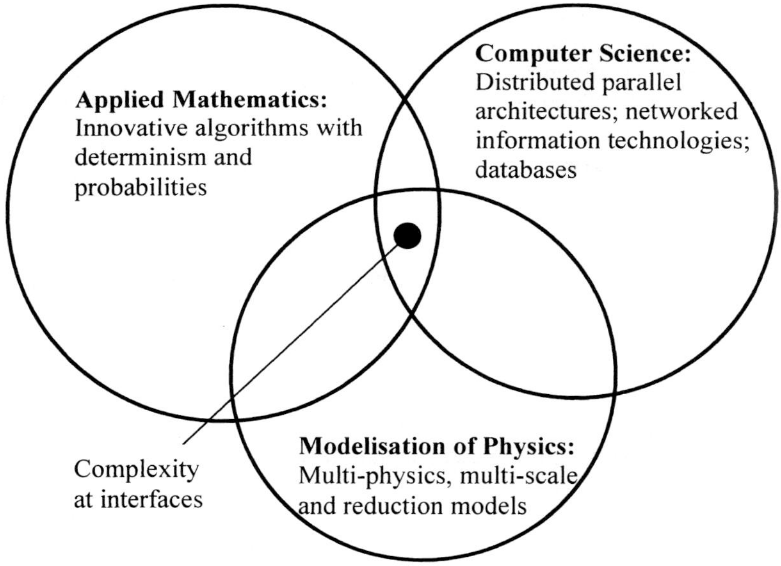

Figure 1 The need for seamless methods at the interface of disciplines.

The industrial trends for manned and unmanned vehicles, which will require new tools are:

- cost-effective design and manufacturing;
- for military aircraft: stealth and aerodynamic shapes, active aerodynamics, conformal antennas and integrated sensors;
- for civil aircraft: long range supersonic cruise, noise, environment comfort and safety.

In order to master the new challenges in aeronautics, numerical simulations play a key design role. The design loop is based on a virtual product (for cost reduction), optimisation and trade-off (for evolution of specifications), and demonstration before flight (for performance prediction and risk reduction). The new numerical simulation tools include generalised automated shape optimisation (hybridised deterministic/evolutionary optimisers), nonlinear 3-D and multi-physics models (such as aero-acoustics and aero-elasticity), decentralised control, hierarchical multidisciplinary design optimisation, affordable high performance grid computing, and validation of multi-physics software.

A sensible way to approach these challenges is through networks of research centres, SMEs and private companies. These networks also have critical relevance in aspects of information and communication for the education and training of specialists. The expected outcomes from such networks are: fostering emergence of research/industrial teams, providing opportunities for cross-fertilisation between different industrial areas (for instance aeronautics, automotive, ships, ...), structuring dynamic and comprehensive databases as first rank references, and integrating research capacities (6th European framework programme).

Q: The concept of integration of design and manufacture seems excellent, but it requires lots of time and energy: how to master it?
The key issues at Dassault Aviation about integration are centralised information and quality of communication, R&T cooperation, and networked cooperation between the Scientific Pôle Dassault Aviation/UPMC and universities and research centres. To give some examples:

Aircraft Design with "plateau".
We conducted recently a unique information-centred working process (in fact a model for effectiveness, flexibility and adaptability) with a real and virtual plateau for the design of the new FALCON 7X. The integrated tasks of the preliminary design phase of the program were performed by partners as a concurrent engineering team in an open space with a very fast and iterative work process and database sharing.

R&T Programmes of the European Commission.
Dassault Aviation's participation in the European programmes makes possible the development of new technologies that are financially accessible only to groups. The programmes serve to create testing or production resources beyond the reach of a single company. These cooperative ventures promote comparisons of various technologies that can help optimise solutions at reasonable cost.

Cooperation with Universities and Research Centres.
The links established between Dassault Aviation engineers and university/laboratory researchers worldwide allow for an optimised use of the most advanced technologies.

Pôle Scientifique Dassault Aviation /UPMC.
This open structure, shared with the University of Paris 6, provides an opportunity to welcome high-level scientists from the world and work with them within the industrial organisation itself. It makes it possible to prepare in practical terms for integration of the most advanced scientific techniques, with guidance from project managers responsible for the problems raised by critical technological challenges.

To meet the new requirements of the aeronautic sector, further integrated effort needs to be invested to develop and assess new mathematical and computational methods and resulting software systems. These new project management tools provide decision support for joint system optimisation and mastery of overall architectures, while optimising costs and lead times.

Sébastien Candel

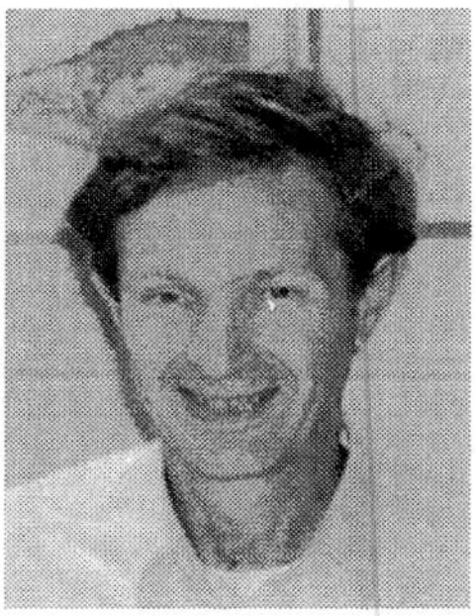

Since 1978, Professor Candel has been Professor of Fluid Mechanics and Aerospace Engineering at Ecole Centrale Paris. His current research interests are aeroacoustics, combustion and propulsion. He is a corresponding member of the French Academy of Sciences and member of the French Academy of Technology. Professor Candel is also Vice-President of the Combustion Institute since 1996, and Chairman of the Supersonic Aircraft Research Network since 1999.

Combustion is a combination of several disciplines - kinetics, thermodynamics, fluid dynamics and transport phenomena. As an example of the complexity, a simple propane air flame involves 40 different species and more than 100 reactions. Further complexities are encountered in practical systems: the flow is turbulent, the pressure is high (reaching 40 bars in jet engine combustors and 100 bars in rocket motors), the power density is quite large (50 GW per cubic metre in rocket chambers) and reactant is often injected as a liquid.

Future combustion codes

There is no unique software tool at present because there are different possible modelling levels (see Table 1 below). Because the physical processes are so complex and involve coupled phenomena, one has to proceed in a systematic fashion using a combination of experimental and numerical tools. The objective is first to:

- Develop a basic understanding using analytical tools and physical intuition based on pre-existing knowledge. For example, in coaxial injection: where is the flame, and is it stabilised?
- Design detailed experiments to understand the processes and provide guidelines for modelling against which codes can be validated; it is impossible to do without experiments in this domain
- Develop suitable basic building blocks for simulation codes and in particular with the sub-grid modelling required by large eddy simulations. (My view is that the future modelling tools will belong to the LES framework.)
- Validate the models in well-controlled experiments and subscale configurations using the modern laser diagnostic tools. Check the software against experimental test cases.
- Systematically envisage further developments to deal with the major challenges in the field of combustion. One of these challenges is related to

combustion dynamics. The objective is to develop predictive tools for combustion instabilities.

Table 1 Modelling levels in numerical combustion.

	Chemical kinetics	Transport	Flow
Detailed modelling (DNM)	Complex chemistry	Multispecies transport	Laminar flow described by Navier-Stokes equations
Physical modelling	Turbulent combustion model, simplified kinetics	Turbulent model for heat and species transport	Turbulent flow described by average Navier-Stokes equations
Large eddy simulation	Flame description on coarse grid	Sub-grid model for turbulent transport	Turbulent transitional flow described by filtered Navier-Stokes equations
Direct simulation (DNS)	Simplified or moderately complex kinetics	Simplified or multispecies transport	Low Re number turbulent flow described by Navier-Stokes equations

Q: Maybe combustion needs lots of experiments because it is still at the beginning, not like CFD?

Combustion is in fact an old science, but decisive progress has been made in the last period. The need for experiments and also of direct simulations or detailed numerical modelling stems from the fact that combustion involves a maze of physical and chemical phenomena requiring detailed understanding. You can generally compute the solution, but you cannot be sure that it is right because the modelling is only an approximate description of what goes on in the flame. Of course, there are cases where the calculations have become sufficient and need no further experiments. However, in the complicated cases involving turbulence, sprays, complex kinetics, supercritical conditions *etc.*, experiments and direct simulations provide essential indications.

Olivier Pironneau

Professor Pironneau holds the degrees of DEA, MSc, PhD, Thèse d'Etat and is Professor in the University Pierre et Marie Curie, in which he is Director of the Laboratory for Numerical Analysis. Professor Pironneau is a *Lauréat du prix Marcel Dassault de l'Académie des Sciences*, a corresponding Member of the Academy of Sciences and a Member of the University Institute of France. He holds the *Décoration de l'Ordre National du Mérite.*

Software development is not going to stop. In Japan for instance, they are developing the Whole Earth Simulator. This is highly multi-disciplinary since it involves physics, applied mathematics and computer science. It is very hard to find people simultaneously capable in all these disciplines. Another important element of software development, beyond the scientific aspects, is to take into account marketing and maintenance.

To produce new software, there is no essential need for mathematics. However here are three examples to show how new mathematical work might help you beat competitors:

- new numerical methods - wavelets, multi-pole expansions
- developments in numerical analysis - *a posteriori* error estimates
- judicious simplification based on mathematics - model reduction

The main question in this context is: who should write the software?

I am quite confident that in-house development, the way we see it now, at least that is from ground up, is going to stop because it is too expensive even for big companies. In CFD, for instance, really cheap development can be obtained if you work with companies or universities in China or India. And universities have for a long time produced high quality programs, such as: C++, f2c, Linux, Maple, *etc.* The list grows every day. Other codes have been developed at universities first before being marketed and subsequently sold to companies. The typical case is that of a professor who has developed a code and then goes out and takes it to market.

But are universities well equipped to write software? This is similar to the question "are universities well equipped to write books?" We know they are generally not systematically written on order, but they keep popping up.

CAD on the other hand faces other problems seen from the university point of view. CAD was actually born in university research centres, one of them being the LIMSI in France. But its success meant it migrated to private companies very fast, and now it is quite expensive and requires a large learning curve. In France, we are quite lucky because we have CATIA, but the developers do not have the time to interact properly with universities, and universities are consequently almost totally out of the loop.

So my question remains: who should write the good software?

Q: What about hardware? We are moving towards services, databases and such. But it is getting hard to get good enough machines, and the hardware is less and less suitable to numerical applications, mainly because of limitations on the bus.
The reason for that probably relates to computer games. Thanks to full 3D computer games, companies are building faster and more powerful PCs. And FFT on DMC chips on such computers might be faster than a supercomputer. The way I see it, software development is driven by leading edge scientific research, in the same way that development in the car industry is driven by Formula 1 racing.

Judith Raper

Professor Raper holds the degrees BE (Chem), PhD and is Dean of Engineering at the University of Sydney. She is the Former Director of the Centre for Particle and Catalyst Characterisation at the University of NSW, and a member of the ARC panel for Engineering and Environmental Sciences. She was NSW Professional Engineer of the Year in 1998. Her research interests include particle characterisation, air and water treatment processes, engineering education and utilisation of waste fly ash.

I want to describe how we are educating tomorrow's engineers so they are equipped with tools for innovation. The curriculum includes emphasis on

- *Hands-on design.* In additional to theoretical studies, students also have to complete a number of hands-on activities. Year 1: students work on design and construction of a light aeroplane (Jabiru, see Figure 2). Year 2 features an engineering design competition; Year 3 a task involving design/build/test; and Year 4 a research project, for example UAV design.
- *Teamwork, experiential learning.* The goals here are to provide for and motivate experiential learning in aircraft construction and design; to increase student exposure and involvement in practical aspects of aircraft design, development, certification and operation; and to provide motivation for the more theoretical courses.
- *Communication skills.*
- *Understanding of emerging technologies.* These include engineering computing applications, IT skills, flight simulators, virtual wind tunnels, analysis packages (computational fluid dynamics, finite element analysis, computer-aided design)
- *Sound theoretical background.*
- *An interdisciplinary approach.* Our streams of study are aeronautical, mechanical and mechatronics. These courses combine units in structural analysis, materials and design, electronics and communications systems, flight mechanics and control theory, thermodynamics and propulsion systems. For advanced students, we seek to provide challenges, to introduce them to engineering skills immediately, to provide opportunities for students from different disciplines to work together, and to show that engineering can be fun. In interdisciplinary projects for advanced students, they work in teams of 4-6, and are expected to learn about teamwork, project management, working to time and budget, presentation, as well as the theoretical and practical aspects of engineering.

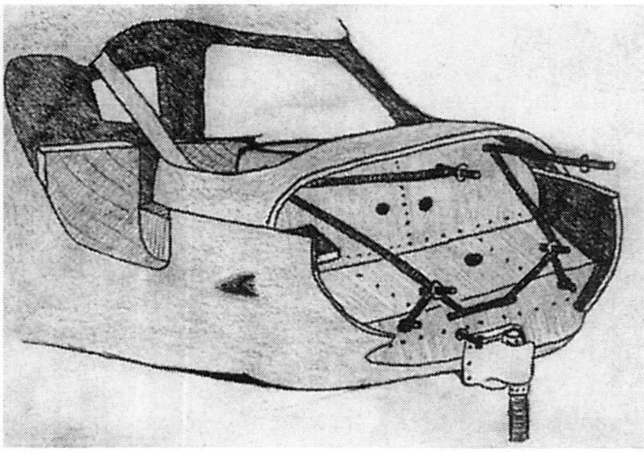

Figure 2 Sketch from a student logbook for the Year 1 project.

Messages heard at the Round Table Discussions

From the vigorous discussion that followed the round table session, we distilled the following important themes and issues:

- Multi-physics software tools are necessary to design aeroplanes that must be "*more affordable, safer, quieter and cleaner*" [Vision Report 2020 in Aeronautics]
- There is an inherent complexity at the interfaces of applied mathematics, computer science and multi-scale modelling of disciplines. According to sentiments of the late Professor J L Lions: "*Challenges are at now interfaces everywhere, and there is a need in the near future of seamless mathematical and numerical methods.*"
- To facilitate multi-physics simulations, there is an obvious need for communication between software packages.
- There is an important role of rapid prototyping to reduce the need for laboratory tests.
- Modern manufacturing practice involves a merging of design, prototype and testing.
- Multi-physics systems involve multi-disciplinary and multi-scale sub-systems, which cannot themselves be reduced any further.
- It is important that aeronautical systems undergo virtual testing of scenarios.
- Multi-physics codes must undergo verification and validation before confident application in manufacturing.
- It is an important fact that there is still incomplete modelisation for complex disciplines such as combustion.
- Human - computer interaction plays an important role in development of software packages for systemic applications.
- In the light of the above points on affordable and innovative software products, who will develop these products for the future? If SMEs are to play an increasing role, supportive government policy is required to link these SMEs to universities and industry.
- We must be attentive to the educational needs and requirements of students who will build these future systems.
- Finally, these software products need to be developed in a safe and certified way.

Notes on Numerical Fluid Mechanics and Multidisciplinary Design

Available Volumes

Volume 84: J. Ballmann: Flow Modulation and Fluid-Structure Interaction at Airplane Wings - Research Results of the Collaborative Research Center SFB 401 at RWTH Aachen, University of Technology, Aachen, Germany. ISBN 3-540-40209-8

Volume 83: L. Davidson, D. Cokljat, J. Fröhlich, M. A. Leschziner, C. Mellen, W. Rodi (eds.): LESFOIL: Large Eddy Simulation of Flow Around a High Lift Airfoil - Results of the Project LESFOIL supported by the European Union 1998 - 2001. ISBN 3-540-00533-1

Volume 82: E. H. Hirschel (ed.): Numerical Flow Simulation III - CNRS-DFG Collaborative Research Programme, Results 2000-2002. ISBN 3-540-44130-1

Volume 81: W. Haase, V. Selmin, B. Winzell (eds.): Progress in Computational Flow-Structure Interaction - Results of the Project UNSI, supported by the European Union 1998-2000. ISBN 3-540-43902-1

Volume 80: E. Stanewsky, J. Délery, J. Fulker, P. de Matteis (eds.): Drag Reduction by Shock and Boundary Layer Control - Results of the Project EUROSHOCK II, supported by the European Union 1996-1999. ISBN 3-540-43317-1

Volume 79: B. Schulte-Werning, R. Grégoire, A. Malfatti, G. Matschke (eds.): TRANSAERO - A European Initiative on Transient Aerodynamics for Railway System Optimisation. ISBN 3-540-43316-3

Volume 78: M. Hafez, K. Morinishi, J. Periaux (eds.): Computational Fluid Dynamics for the 21st Century. Proceedings of a Symposium Honoring Prof. Satofuka on the Occasion of his 60th Birthday, Kyoto, Japan, 15-17 July 2000. ISBN 3-540-42053-3

Volume 77: S. Wagner, U. Rist, H.-J. Heinemann, R. Hilbig (eds.): New Results in Numerical and Experimental Fluid Mechanics III. Contributions to the 12th STAB/DGLR Symposium Stuttgart, Germany 2000. ISBN 3-540-42696-5

Volume 76: P. Thiede (ed.): Aerodynamic Drag Reduction Technologies. Proceedings of the CEAS/DragNet European Drag Reduction Conference, 19-21 June 2000, Potsdam, Germany. ISBN 3-540-41911-x

Volume 75: E.H. Hirschel (ed.): Numerical Flow Simulation II. CNRS-DFG Collaborative Research Programme, Results 1998-2000. ISBN 3-540-41608-0

Volume 74: D. Henry, A. Bergeon (eds.): Continuation Methods in Fluid Dynamics. Contributions to the ERCOFTAC/EUROMECH Colloquium 383, Aussois, France, 6.-9. September 1998. ISBN 3-540-41556-4

Volume 73: M. Griebel, S. Margenov, P. Yalamov (eds.): Large-Scale Scientific Computations of Engineering and Environmental Problems II. Proceedings of the Second Workshop on „Large-Scale Scientific Computations", Sozopol, Bulgaria, June 2-6, 1999. ISBN 3-540-41551-3

Volume 72: W. Nitsche, H.-J. Heinemann, R. Hilbig (eds.): New Results in Numerical and Experimental Fluid Mechanics II. Contributions to the 11th AG STAB/DGLR Symposium Berlin, Germany 1998. ISBN 3-540-41552-1

Volume 70: W. Hackbusch, G. Wittum (eds.): Numerical Treatment of Multi-Scale Problems. Proceedings of the 13th GAMM-Seminar, Kiel, January 24-26, 1997. ISBN 3-540-41562-9

Volume 69: D. P. Hills, M. R. Morris, M. J. Marchant, P. Guillen (eds.): Computational Mesh Adaptation. ECARP - European Computational Aerodynamics Research Project. ISBN 3-540-41548-3

Volume 67: C. H. Sieverding, G. Cicatelli, J. M. Desse, M. Meinke, P. Zunino: Experimental and Numerical Investigation of Time Varying Wakes behind Turbine Blades. Results of the Industrial and Material Technology Aeronautics Research Project AER2-CT92-0048, 1992-1996. ISBN 3-540-41546-7

Volume 66: E. H. Hirschel (ed.): Numerical Flow Simulation I. CNRS-DFG Collaborative Research Programme. Results 1996-1998. ISBN 3-540-41540-8

Volume 64: R. Friedrich, P. Bontoux (eds.): Computation and Visualization of Three-Dimensional Vortical and Turbulent Flows. Proceedings of the Fifth CNRS-DFG Workshop on Numerical Flow Simulation, München, Germany, December 6 and 7, 1996. ISBN 3-540-41557-2

Volume 63: M. Fiebig, N. K. Mitra (eds.): Vortices and Heat Transfer. Results of a DFG-Supported Research Group. ISBN 3-540-41558-0

Volume 62: M. Griebel, O. Iliev, S. D. Margenov, P. S. Vassilevski (eds.): Large-Scale Scientific Computations of Engineering and Environmental Problems. Proceedings of the First Workshop on „Large-Scale Scientific Computations" Varna, Bulgaria, June 7-11, 1997. ISBN 3-540-41560-2

Volume 57: H. Deconinck, B. Koren (eds.): Euler and Navier-Stokes Solvers Using Multi-Dimensional Upwind Schemes and Multigrid Acceleration. Results of the BRITE/EURAM Projects AERO-CT89-0003 and AER2-CT92-00040, 1989-1995. ISBN 3-540-41535-1

Druck:	Strauss Offsetdruck, Mörlenbach
Verarbeitung:	Schäffer, Grünstadt